NMR Basic Principles and Progress
Grundlagen und Fortschritte 15

Editors: P. Diehl E. Fluck R. Kosfeld

Dynamic NMR Spectroscopy

With Contributions by
Alois Steigel
and Hans Wolfgang Spiess

Springer-Verlag
Berlin Heidelberg New York 1978

ISBN 3-540-08784-2 Springer-Verlag Berlin Heidelberg New York
ISBN 0-387-08784-2 Springer-Verlag New York Heidelberg Berlin

Library of Congress Cataloging in Publication Data. Steigel, Alois, 1943—. Dynamic NMR spectroscopy. (NMR, basic principles and progress; v. 15). Bibliography: p. Includes index. CONTENTS: Steigel, A. Mechanistic studies of rearrangements and exchange reactions by dynamic NMR spectroscopy.–Spiess, H. W. Rotation of molecules and nuclear spin relaxation. 1. Nuclear magnetic resonance spectroscopy. I. Spiess, Hans Wolfgang, 1942—. II. Title. III. Series. QC490.N2. vol. 15. [QC762]. 538'.3s. [538'.3]. 78-15777

Printing and bookbinding: Brühlsche Universitätsdruckerei, Lahn-Gießen.
2152/3140-543210

Table of Contents

List of Editors

Mechanistic Studies of Rearrangements and Exchange Reactions by Dynamic NMR Spectroscopy

Alois Steigel

Institut für Organische Chemie der Universität Düsseldorf, D-4000 Düsseldorf

Contents

1. Introduction

Since the first successful NMR experiments in 1946 it was well appreciated that dynamic processes play an important role in the NMR spectroscopy of bulk matter [*1*]. Early theories on the dependence of the relaxation parameters T_1 and T_2 on the motions of nuclear spins were successful in explaining the dipolar broadening of the NMR signal in solids and the motional narrowing in liquids [*2*]. With the discovery of chemical shifts and spin-spin couplings another type of dynamical process affecting the NMR line shape became apparent, the chemical exchange. As a consequence, dynamical NMR studies split into two groups differing not only in the dynamical topics but also in the method of investigation: physical studies of the motion of spins in liquids and solids by measurement of the relaxation times of single resonances and, on the other hand, chemical studies based on band shape analysis of NMR spectra recorded under steady state conditions. The two fields of research lost some of their basic differences with the development of the Fourier transform NMR method [*3*], which allows the measurement of relaxation times of different resonances at the same time, *i.e.* the study of differential motional behavior of different parts of molecules, thus providing a new tool in conformational analyses. For example, information can be obtained by this method on the relative importance of overall motions and internal motions [*4*]. On the other hand, recent theories on band shapes now also allow the extraction of detailed relaxation information from the spectra of coupled spin systems [*5*]. These new developments make the distinction between relaxation and band shape studies arbitrary and therefore it is justified to classify all the topics mentioned above by the same term "Dynamic NMR Spectroscopy" (DNMR). Reviews on several aspects of DNMR have been given in a book edited recently by Jackman and Cotton [*6*]. In this report the characteristics of one branch of DNMR will be described, the use of band shape analyses to obtain mechanistic information on rearrangements and chemical exchange processes.

In contrast to other spectroscopic methods, which in kinetics can only be used to study irreversible reactions, NMR spectroscopy, in addition to its use in conventional kinetics, allows for kinetic studies of systems in chemical equilibrium. Reversible rate processes with activation energies between 20 and 100 kJ/mole can be studied by theoretical analysis of exchange-modified NMR band shapes. In this band shape analysis, spectra calculated for different values of the pseudo-first-order rate constants k = rate/[A], *i.e.* different inverse lifetimes of the species A present in the dynamic equilibrium, are compared with the experimental spectra taken at different temperatures. For single intramolecular processes, agreement between the calculated and the experimental spectra yields directly the rate constants of the exchange process at the given temperatures. In the case of intermolecular exchange reactions, the concentrations of the reaction components need to be considered to calculate the specific rate constants from the determined pseudo-first-order rate constants. In the latter case therefore, the NMR band shapes also depend on the concentration of the reaction partners. Activation parameters of the exchange process may be obtained from the temperature dependence of the rate constants by use of the Arrhenius or Eyring equations.

Earlier studies of intermolecular and intramolecular exchange processes, reviewed for example by Loewenstein and Connor [7] and by Binsch [*8*], respectively, have been limited mostly to band shape analyses of simple spin systems (*e.g.* $A \rightleftharpoons B$), to band shape analyses based on approximative equations, or to the determination of the rate constant at the coalescence temperature [*9*]. Recent progress in the theory of exchange-modified band shapes [*10*] led to the development of versatile computer programs. The program DNMR 3 for instance, written by Binsch and Kleier [*11*], allows for complete band shape analysis of non-first-order spin systems of up to six spins.

In systems for which several exchange possibilities can be foreseen, information on the reaction mechanism may be obtained by the band shape analysis. The agreement of band shapes, calculated for a certain exchange mode, with the experimental spectra, can prove the postulated mechanism, and at the same time rate constants are determined as mentioned above. In the present report the basic principles of the mechanistic analysis by NMR band shape calculations are illustrated by using representative examples. Special emphasis is laid on the derivation of kinetic exchange matrices corresponding to the mechanistic alternatives. Permutational analysis of the rearrangements in phosphoranes will be performed to show the problems encountered in DNMR studies of polytopal rearrangements. The relation of classical and quantum mechanical calculations of exchange-modified band shapes is illustrated by discussing AX and AB, as well as A_2X_2 and A_2B_2 spin systems.

2. General Comments on Band Shape Analyses

2.1. Experimental Requirements

The preconditions for accurate band shape analyses have been extensively discussed in the literature (see for instance [*8, 12, 13*]). Progress in evaluating the various sources of errors can be historically followed by the numerous attempts to obtain reliable activation energies for the hindered rotation around the carbonyl-nitrogen bond in N,N-dimethylamides, which has been of interest from the time of the first chemical applications of DNMR [*9, 14*] twenty years ago, until now [*15*]. The reason for the immense effort in this area is to be found in the small chemical shift differences of the methyl resonances which necessitate a careful consideration of all factors of relevance to the band shape analysis, such as the recording of the spectra, evaluation of the temperature dependence of chemical shifts, couplings and line widths without exchange, and the measurement of the temperature inside the sample. As was demonstrated in recent studies (see for instance [*16, 17*], complexity in NMR spectra helps in reducing the experimental uncertainties by allowing several checks during the simulation of the manifold changes in band shapes.

In mechanistic DNMR studies conclusions can often only be reached on the basis of subtle differences of the corresponding band shapes. In these cases, therefore, all factors affecting the band shapes must be carefully taken into consideration. The influence of medium effects in mechanistic studies should also be emphasized: before postulating a mechanism one should ensure that the collapse of the resonance lines is not caused by impurities of by a catalytic process. Especially in the case of inorganic fluorine compounds carefully purified samples have to be used to prove for instance the nonoccurrence of intramolecular mechanisms.

2.2. Mathematical Procedure

The methods of calculating the band shapes of complex DNMR spectra have been developed by extending the classical [*18, 19*] and quantum mechanical [*20, 21*] theories of exchange of nuclear spins. Both extended theories are based on matrix methods, the classical one having been derived from exchange modified Bloch equations in 1958 [*22*]. In this progress article the theories will not be derived, but instead, the resulting equations will be discussed and compared to facilitate their application by using the corresponding computer programs. At first sight, the classical Eqs. (1) [*23*] and (2) [*24*] and the quantum mechanically based Eqs. (3) [*5, 10*] and (4) [*25*] look very similar. In both groups of equations the same mathematical operations are used to calculate the complex magnetization G: multiplication of a row vector (P, $\mathit{1}$, I_b^-, and I_e^-, respectively) with the inverse of a complex square matrix and a column vector ($\mathit{1}$, P, σ, and $I_e^{-\prime}$, respectively); C is a scaling factor.

$$G = -\mathrm{i}CPA^{-1}\mathit{1} \tag{1}$$

$$G = -\mathrm{i}C\mathit{1}B^{-1}P \tag{2}$$

$$G = -\mathrm{i}CI_b^-C^{-1}\sigma \tag{3}$$

$$G = -\mathrm{i}CI_e^-D^{-1}I_e^{-\prime} \tag{4}$$

The band shape $S(\omega)$, *i.e.* the functional dependence of the absorption intensity on frequency, is simply given as the imaginary part of G. Although the meaning of the vectors and matrices will be discussed in detail in the next section and in Section 9.1, some common features of the equations are described here.

In the classical Eqs. (1) and (2) the matrices A and B are transpose to one another, thus explaining the reversed order of the population vector P and the unit vector $\mathit{1}$ in the two equations. This should be called to mind when setting up the kinetic exchange matrices (cf. next section). The order of the matrices A and B is equal to the number of resonance lines involved in exchange.

Except for simple cases, the quantum mechanically based Eqs. (3) and (4) have to be used to calculate the exchange effects on coupled spin systems. The similarity of the two equations is only a formal one, since different representations are used, the representation of basis functions in Eq. (3) and the representation of eigenfunctions in Eq. (4).

In the representation of basis functions the static spin Hamiltonian is not diagonalized, as in the representation of eigenfunctions, but instead, the DNMR spectra are calculated in one step by taking into account the exchange relations between the spin basis functions. For this reason matrix C, in contrast to matrix D, contains off-diagonal imaginary elements, which correspond to the coupling constants. Another characteristic feature is that, even if factorization is performed by using the transition condition $\Delta F_z = -1$, the spin lowering vector I_b^- and vector σ contain zero elements in addition to elements of magnitude one. For nonmutual exchange processes the elements of vector σ have to be multiplied by the appropriate populations of the species involved in exchange. The zero elements in the two vectors correspond to the so-called combination transitions in which more than one individual spin is changed, while the total spin F_z only changes by one. In practice the approach of using the representation of basis functions is easily carried out since versatile computer programs, such as DNMR 3 [*11*] have been written.

On the other hand, the representation of eigenfunctions [Eq. (4)] has the advantage of describing the DNMR problem in a more illustrative way. In this method the eigensystem of the static spin Hamiltonian and the allowed transitions between the spin eigenfunctions are first calculated before setting up the band shape equation. The calculated allowed transitions also comprise combination transitions, mentioned above. Since they are of negligible intensity even for non-first-order cases, they need not to be included in the band shape Eq. (4). As a consequence, matrix D can be constructed in such a way that its order is equal to the number of multiplet lines observed in the static NMR spectrum (slow exchange region). Similarly the elements of the spin lowering vector I_e^- correspond to the observed resonances in being equal to the square root of the intensity of the lines. The primed vector $I_e^{-\prime}$ is identical to I_e^- for mutual exchange problems, but includes the populations of different species for nonmutual intramolecular and intermolecular reactions [cf. relation between I_b^- and σ in Eq. (3)].

The feature of the latter approach allows for a generalization of the term site, originally coined for DNMR studies of uncoupled systems to define the resonance lines involved in exchange. Thus the multiplet lines of first-order as well as of non-first-order cases can also be characterized as sites. A limited extension of the term site to certain coupled systems (cf. Section 8.1) was suggested in 1973 [*26*]. In the next section the site exchange of multiplet lines will be illustrated for a first-order system by giving a more detailed description of the character of the band shape equations.

The actual calculation of the band shapes of a DNMR spectrum by one of the Eqs. (1)–(4) requires solving the equation for each frequency point in the frequency range desired. In the first multi-site exchange computer programs this had been done by inverting the complex matrix every time for each of the 100 to 1 000 points. Newer programs, such as DNMR 3 [*11*], PZDMF [*27*], EXCHSYS [*28*], and CLSFIT [*29*], use a more efficient procedure which is based on the method of Gordon and McGinnes [*30*] of calculating the eigensystem; by diagonalization of the frequency-independent part of the complex matrix and inversion of the transformation matrix, the calculation of the band shape points is reduced to simple multiplication operations.

3. Formulation of the Exchange Problem

In this section an illustrative example, a first-order spin system, is used to describe in further detail the character of Eqs. (1), (2), and (4). The more complicated Eq. (3) will be discussed when dealing explicitly with non-first-order spin systems (Section 9.1.).

As stated in the previous section, the order of the matrix in the classical approach [Eqs. (1) and (2)] as well as in the representation of eigenfunctions [Eq. (4)] is identical to the number of sites, *i.e.* the number of resonance lines, which are affected by the exchange process. Thus in general, the order of the matrix depends on the number of nonequivalent nuclei observed, on spin-spin couplings, and on the number of compounds or stereoisomers present in the dynamic equilibrium.

In the case of an A_2X_2 spin system, in which the nuclei A and X are supposed to be interchanged by a dynamical process, the resonance lines of the two triplets correspond to six sites (Fig. 1), *i.e.* the order of the matrix to be constructed is six. This situation can be found in octahedral cis-complexes of type ML_2Z_4, or in compounds with trigonal bipyramidal geometry of type MLZ_4, where the observed nuclei Z are split into two groups of magnetic equivalent nuclei, A_2 and X_2. Examples which come close to this first-order problem are given, for instance, by the ^{19}F–NMR studies of TiF_4L_2 *1* (*e.g.* L = N,N-diethylformamide [*31*], or L = dimethylether [*32*]) and of SF_4 *2* [*33, 34*]. In the latter case the free electron pair of the central sulfur atom is responsible for the C_{2v}-geometry of the molecule [*35*]. This system will be discussed later in further detail, since it was demonstrated that mechanistic conclusions are easier to derive when the ^{19}F-NMR spectrum of SF_4 has non-first-order character, *i.e.* when recording the DNMR spectra by use of a smaller magnetic field strength [*25*].

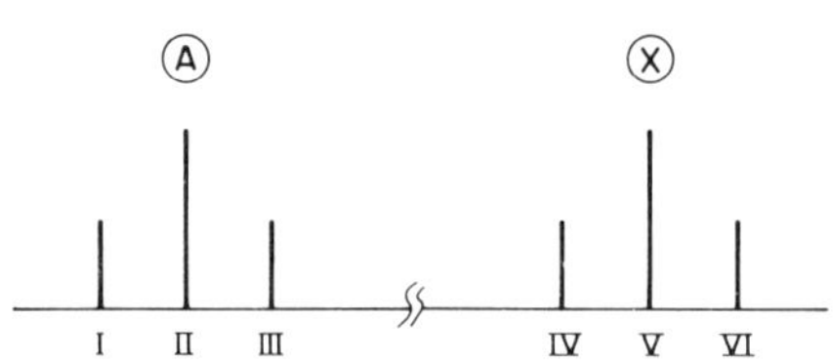

Fig. 1. Schematic A_2X_2 spectrum

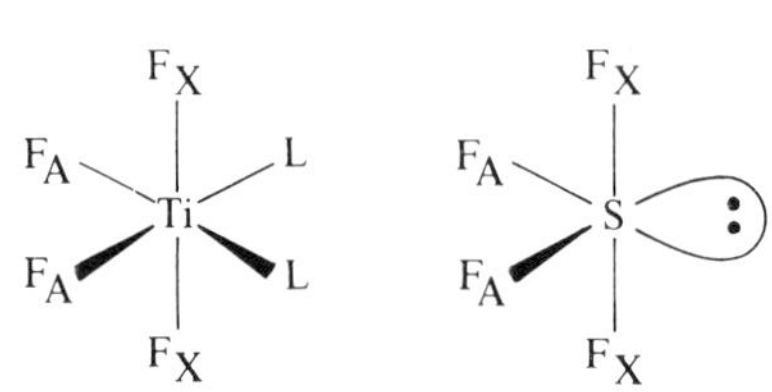

1 *2*

The matrices in Eqs. (1), (2), and (4) can be obtained by summation of two matrices. For the A_2X_2 case the first matrix is of the diagonal form I, the elements containing the line widths at half height W_i in the absence of exchange and the corresponding resonance frequencies $\omega_i = 2\pi\nu_i$. Instead of line widths, some computer programs require as input the effective transverse relaxation time T_2, which is related to W by $T_2 = 1/(\pi W)$. In several classical DNMR studies of coupled systems (cf. next section) the parameters W_i have been chosen to simulate the resonances broadened or even split by small couplings. When matrix I is introduced in Eqs. (1), (2), or (4) and the vectors P and $I_e^- = I_e^{-\prime}$ are taken as (1, 2, 1, 1, 2, 1) and $(1, \sqrt{2}, 1, 1, \sqrt{2}, 1)$, respectively, the static A_2X_2 spectrum can be calculated.

$$\begin{pmatrix} -\pi W_{\mathrm{I}} + i(\omega_{\mathrm{I}}-\omega) & & & & & \\ & -\pi W_{\mathrm{II}} + i(\omega_{\mathrm{II}}-\omega) & & & & \\ & & -\pi W_{\mathrm{III}} + i(\omega_{\mathrm{III}}-\omega) & & & \\ & & & -\pi W_{\mathrm{IV}} + i(\omega_{\mathrm{IV}}-\omega) & & \\ & & & & -\pi W_{\mathrm{V}} + i(\omega_{\mathrm{V}}-\omega) & \\ & & & & & -\pi W_{\mathrm{VI}} - i(\omega_{\mathrm{VI}}-\omega) \end{pmatrix}$$

I

In order to calculate the desired DNMR spectrum observed at a certain temperature, a so-called kinetic exchange matrix, which characterizes the transfer of magnetization between the sites, has to be added to matrix I. The elements of this matrix correspond to the respective exchange probabilities multiplied by a pseudo-first-order rate constant k, which is the inverse of the lifetime of the species involved in exchange at the temperature considered [7]. Before giving examples for the kinetic exchange matrix an important fact concerning the treatment of coupled exchange systems is emphasized. Since compounds such as *1* and *2* do not represent ideal first-order A_2X_2 cases, the numbers ascribed here to the populations and exchange probabilities are not exactly those corresponding to the actual systems. In fact, in order to calculate accurate band shapes also for the fast-exchange region, it is important to account for deviations in these numbers caused by non-first-order effects. The reason for this will be discussed when dealing explicitly with coupled systems (Section 9.1.). Here we will use the ideal A_2X_2 case to describe the characteristics of kinetic exchange matrices and their relation to different types of dynamic processes.

A priori, band shape studies do not give information on the physical pathway of a reaction or rearrangement, but instead, DNMR is suited to characterize the exchange of nuclei observed. Thus the approach of describing the dynamical process is based on the evaluation of the permutations of nuclei possible in the system studied [*36*]. While this is usually straightforward for organic systems, polytopal rearrangements of inorganic complexes are more difficult to assess. Here only two types of exchange are considered for the systems *1* and *2*, the first one being a process interchanging one nucleus of the X_2 group and one nucleus of the A_2 group ("one-pair exchange"), *i.e.* the permutation (23) shown or permutations (13), (14), and (24), and the second also interchanging the remaining two fluorine nuclei ("two-pair exchange"), *i.e.* the permutation (23)(14) shown or the permutation (13)(24).

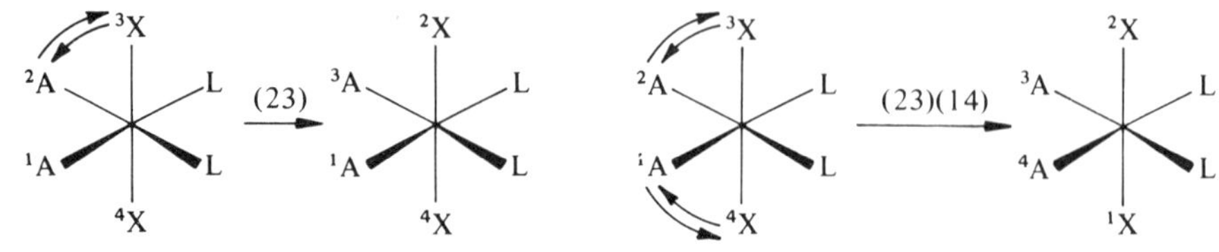

Other types of permutations including ligands L can also result in net interchange of the nuclei A and X. To obtain the maximal information on the rearrangement mechanism of these or other coordination compounds, a complete permutational analysis and classification of the permutations according to differentiability by NMR is helpful [*36*]. For instance, the interchange of one fluorine of the X_2 group and one ligand L has the same effect on the NMR spectrum as the permutation (23) depicted above. The complete permutational approach to polytopal rearrangements will be discussed in Section 6.

$$
\begin{array}{c} \\ \mathrm{I} \\ \mathrm{II} \\ \mathrm{III} \\ \mathrm{IV} \\ \mathrm{V} \\ \mathrm{VI} \end{array}
\begin{array}{c}
\begin{array}{cccccc} \mathrm{I} & \mathrm{II} & \mathrm{III} & \mathrm{IV} & \mathrm{V} & \mathrm{VI} \end{array} \\
\begin{pmatrix}
-k & & & k & & \\
 & -k & & & k & \\
 & & -k & & & k \\
k & & & -k & & \\
 & k & & & -k & \\
 & & k & & & -k
\end{pmatrix}
\end{array}
$$

II

$$
\begin{array}{c} \\ \mathrm{I} \\ \mathrm{II} \\ \mathrm{III} \\ \mathrm{IV} \\ \mathrm{V} \\ \mathrm{VI} \end{array}
\begin{array}{c}
\begin{array}{cccccc} \mathrm{I} & \mathrm{II} & \mathrm{III} & \mathrm{IV} & \mathrm{V} & \mathrm{VI} \end{array} \\
\begin{pmatrix}
-3k/4 & k/4 & & k/4 & k/4 & \\
k/8 & -3k/4 & k/8 & k/8 & k/4 & k/8 \\
 & k/4 & -3k/4 & & k/4 & k/4 \\
k/4 & k/4 & & -3k/4 & k/4 & \\
k/8 & k/4 & k/8 & k/8 & -3k/4 & k/8 \\
 & k/4 & k/4 & & k/4 & -3k/4
\end{pmatrix}
\end{array}
$$

III

The two NMR-distinguishable exchange modes taken in consideration, *i.e.* the two-pair and the one-pair fluorine exchange, are characterized–as derived in Section 8.2.– by the corresponding kinetic exchange matrices II and III, respectively. The two-pair exchange results in very simple site exchange relations. Recalling Fig. 1, matrix II implies complete transfer of magnetization of a triplet resonance line to the corresponding line of the other triplet, thus, for instance, of site I to site IV (first matrix row) and reversely (fourth matrix row). Matrix III, characterizing the one-pair exchange, is of more complex nature. Since it is unsymmetrical, it must be emphasized that for a classical type DNMR calculation, Eq. (1) and not (2) has to be used (cf. previous section). Matrix III shows that the magnetization of site I, for instance, is retained by a probability of 1/4, and transferred by the same probability to each of the sites II, IV, and V (first row of the matrix). For site II, as well as for site V, transfer of magnetization is predicted as occurring to all A_2X_2 sites.

The difference between the kinetic exchange matrices II and III suggests that the two exchange types result in different DNMR patterns. In the next section two classical DNMR studies will illustrate that unlike kinetic exchange matrices are a necessary but not sufficient precondition for a distinctly different DNMR behavior.

4. Classical Multi-Site Exchange Analysis

The classical multi-site exchange studies performed in the last decade were based on matrix formulations of the exchange-modified Bloch equations, *i.e.* on Eqs. (1) and (2) given in Section 2.2. Spin-spin splittings were taken into account only by adjusting the linewidth parameters W_i to the expansion of the multiplets, which means that the splitted resonances had been treated as single sites. Therefore in these studies, the number of sites is given by the number of nuclei or groups of nuclei in chemically different environments. Although this is a rough method, valuable information on the dynamics of several systems has nevertheless been obtained. An early, fascinating example is the study of bullvalene by Saunders in 1963 [*37*], who was able to show that single Cope rearrangements can explain the observed DNMR spectra.

Another study by Saunders [*38*], the degenerate methyl shift in the heptamethylbenzonium ion *3*, will be used to show how exchange matrices can be derived in the classical approach and to show the corresponding calculated band shapes.

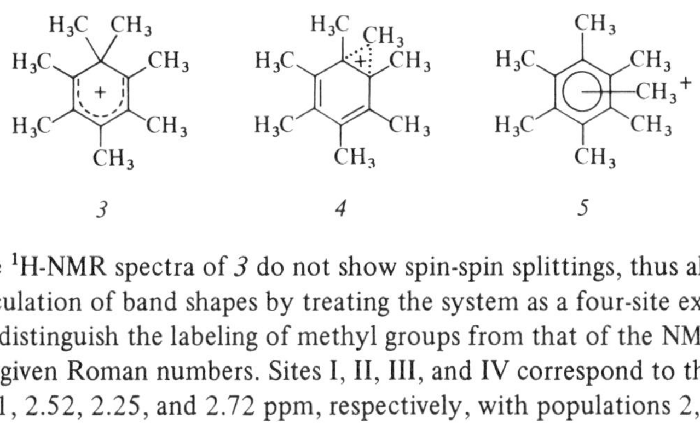

3 *4* *5*

The ^{1}H-NMR spectra of *3* do not show spin-spin splittings, thus allowing an exact calculation of band shapes by treating the system as a four-site exchange problem. To distinguish the labeling of methyl groups from that of the NMR sites, the latter are given Roman numbers. Sites I, II, III, and IV correspond to the resonances at 1.41, 2.52, 2.25, and 2.72 ppm, respectively, with populations 2, 2, 2, and 1. As seen from the labeling of the methyl groups the 1,2 methyl shift of methyl group 1, which could proceed via intermediate *4*, can be described by the permutation (1)(24)(36)(57).

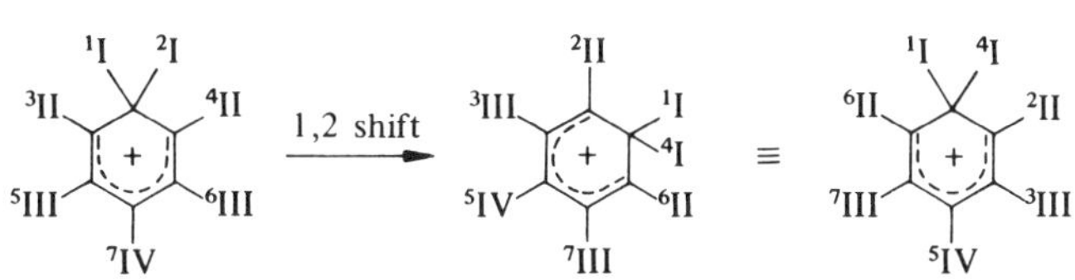

Similarly the 1,2-shift in the other direction, *i.e.* anticlockwise, is characterized by the permutation (1)(23)(45)(67). The corresponding permutations of the sites are obtained from the methyl permutations by substitution of the numbers of the methyls by the sites these methyl groups initially occupied, yielding (I)(I II)(II III) (III IV) for both 1,2-shifts. From this permutation the exchange probabilities of the sites are easily derived, leading to the kinetic exchange matrix IV. For instance, site I is transferred with half probability (cf. matrix element 1,1) to site II (element 1,2), and site IV, being the resonance of only one methyl group (para environment), is transferred with probability one to site III (cf. fourth matrix row), if the rearrangement process is a 1,2-methyl shift.

$$\underset{\text{IV}}{\begin{pmatrix} -k/2 & k/2 & & \\ k/2 & -k & k/2 & \\ & k/2 & -k & k/2 \\ & & k & -k \end{pmatrix}} \qquad \underset{\text{V}}{\begin{pmatrix} -k/2 & k/5 & k/5 & k/10 \\ k/5 & -4k/5 & 2k/5 & k/5 \\ k/5 & 2k/5 & -4k/5 & k/5 \\ k/5 & 2k/5 & 2k/5 & -k \end{pmatrix}}$$

For a random shift mechanism, which could proceed either by an intramolecular rearrangement via an intermediate of type *5* or by an intermolecular exchange process, in addition to the two permutations given above, one also must take into account the two 1,3-shift possibilites (cw and ccw) and the 1,4-methyl shift. The corresponding permutations of the methyl groups are (1)(26)(4)(5)(37), (1)(25)(3)(6)(47), and (1)(27)(35)(46), respectively, while those of the sites are (I)(I III)(II)(III)(II IV) for the 1,3-shifts and (I)(I IV)(II III)(II III) for the 1,4-shift. The elements of the kinetic exchange matrix V for the random shift mechanism are then derived by summing the site exchange probabilities for the five possible permutations, corrected by a statistical factor. For instance, the contribution of the 1,2-shift to this mechanism is simply obtained by multiplying the elements of matrix IV by 2/5.

Since the kinetic exchange matrices IV and V are markedly different, the corresponding spectra calculated with Eq. (1) should differ too. They are shown together with the experimental DNMR spectra in Fig. 2.

At first sight the differences between the calculated band shapes are hard to detect. However, from the exchange matrices we know that, in contrast to the 1,2-shift mechanism, the random shift interchanges sites I and III (resonances at 1.41 and 2.25 ppm, respectively) and also sites II and IV (2.52 and 2.72 ppm, respectively). Considering these characteristic site exchanges between the two low field lines and the two high field lines in the random shift mechanism, the differences between the calculated band shapes for the two shift possibilites become evident and can be used to derive the mechanism actually occurring. Thus by comparison with the experimental spectra, Saunders concluded that the rearrangement proceeds by the 1,2-methyl shift mechanism. The difficulty in distinguishing the calculated band shapes in this case is due to the fact that the random mechanism statistically allows for all possible shifts including the 1,2-shifts from which it has to be discriminated. It is, for instance, much easier to distinguish the 1,2-shift from a 1,3-shift. In the latter, exchange relations only exist between sites I and III and between sites II and IV (permutation

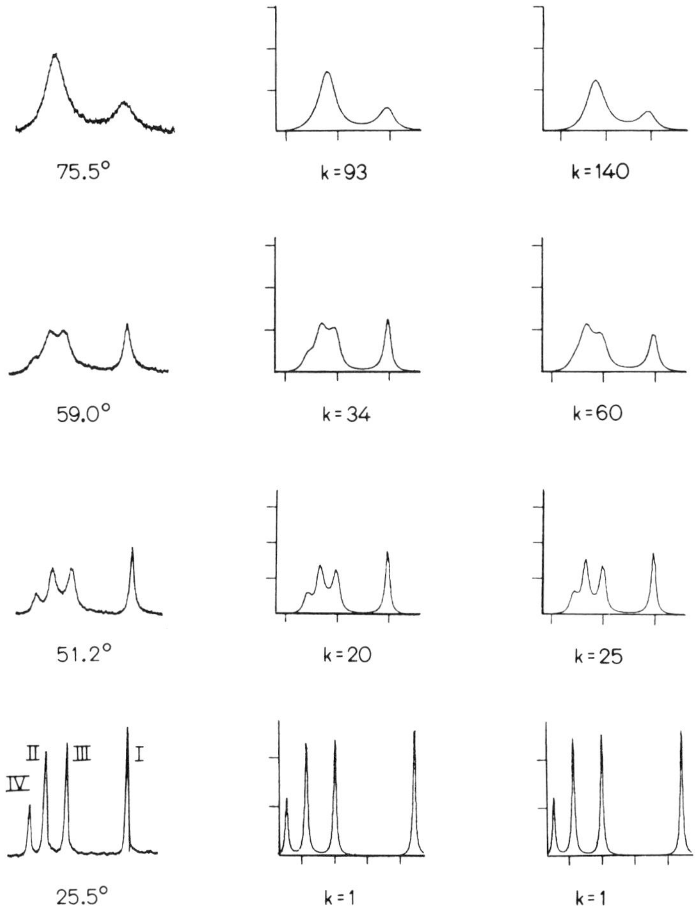

Fig. 2. Comparison of experimental and calculated ^{1}H-DNMR spectra of heptamethylbenzonium ion *3* (M. Saunders [38])
Left column: Experimental spectra; *Middle column:* Spectra calculated for a 1,2 methyl shift; *Right column:* Spectra calculated for a random shift

(I)(I III)(II)(III)(II IV), *i.e.* the middle NMR lines (sites II and III) would not exchange, in contrast to the 1,2-shift.

In a similar exchange problem, the intramolecular shift of the σ-bound $CuPEt_3$-group in C_5H_5-$CuPEt_3$ *6* studied by Whitesides and Fleming in 1967 [*39*], greater differences in calculated band shapes have been obtained. For this reason, the approximate treatment as a three-site exchange by neglection of spin-spin couplings between the protons is justified.

Proceeding in the same way as above, the kinetic exchange matrices are derived. The permutations corresponding to the 1,2- and 1,3-shifts are seen to be for the protons (13)(25)(4) and (15)(24)(3), and for the sites (I II)(II III)(III) and (I III)(II III)(II), leading to the exchange matrices VI and VII, respectively. Matrix VIII, describing

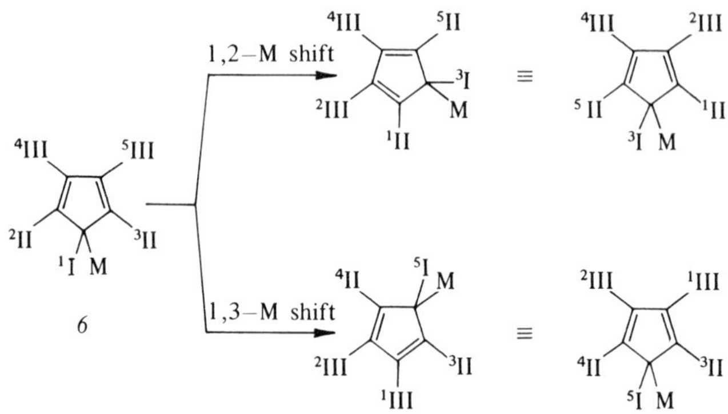

6

the exchange for the random shift mechanism, is obtained by adding the matrices VI and VII and dividing the elements of the resulting matrix by two.

$$\underset{\text{VI}}{\begin{pmatrix} -k & k & \\ k/2 & -k & k/2 \\ & k/2 & -k/2 \end{pmatrix}} \quad \underset{\text{VII}}{\begin{pmatrix} -k & & k \\ & -k/2 & k/2 \\ k/2 & k/2 & -k \end{pmatrix}} \quad \underset{\text{VIII}}{\begin{pmatrix} -k & k/2 & k/2 \\ k/4 & -3k/4 & k/2 \\ k/4 & k/2 & -3k/4 \end{pmatrix}}$$

The three exchange matrices show characteristic differences for the olefinic sites II and III, which is most clearly seen by comparing the corresponding diagonal elements. Thus for the 1,2-shift, the exchange probability of site II is twice that of site III, while the reverse is true for the 1,3-shift. For the random exchange, the exchange propability is the same for both sites, *i.e.* 3/4. This mechanism can be discarded at once, since in the experimental spectra (Fig. 3), taken between –60 ° and –40 °C, the up-field olefinic site (6.57 ppm) is collapsing more rapidly, implying a selective shift mechanism. The olefinic sites II and III had been tentatively assigned by Whitesides [*39*] to the resonances at 6.95 and 6.57 ppm, respectively. This implied that matrix VII (1,3-shift mechanism) would describe the rearrangement process. However, analyzing the fine structure of the resonances, Cotton [*40*] came to the conclusion that the assignment has to be reversed, *i.e.* the more rapidly collapsing up-field olefinic resonance at 6.57 ppm is in reality site II. Therefore, in the calculation of the band shapes performed by Whitesides (Fig. 3), the resonance frequencies specified in the diagonal matrix (cf. Section 3, matrix I) did not correctly correspond to the sites. Since the kinetic exchange matrix VII can be transformed to matrix VI by interchanging the rows and the columns corresponding to sites II and III, the theoretical band shapes in the middle column of Fig. 3, originally calculated by using matrix VII, should be thought of as having been calculated by using matrix VI (1,2-shift mechanism) and the correct diagonal matrix.

Classical-type calculations have been performed in many mechanistic studies. Most experimental fields of interest, such as conformational changes and carbonium ion rearrangements in organic chemistry, and stereochemical nonrigidity in inorganic and organo-metallic compounds, have been covered in the book edited by Jackman and Cotton [*6*].

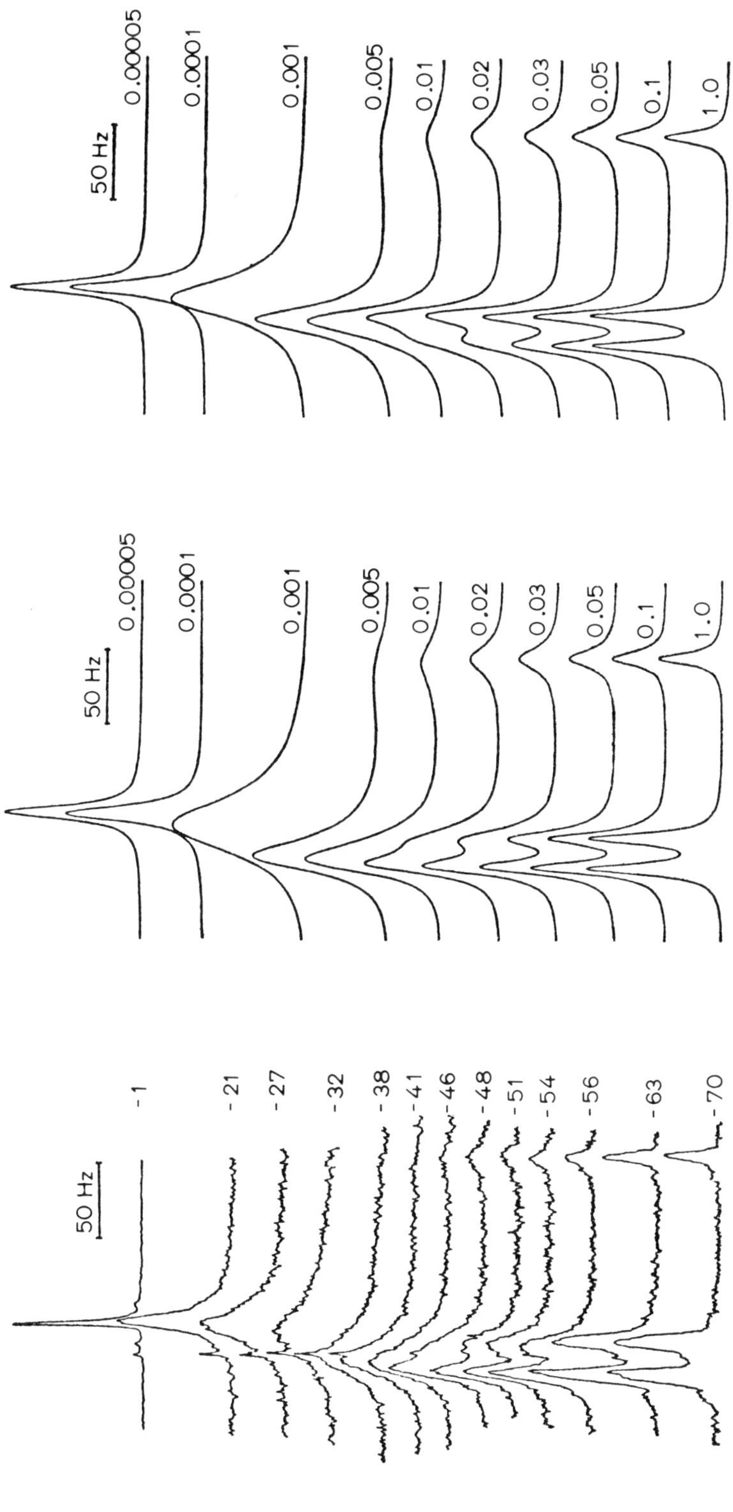

Fig. 3. Comparison of experimental and calculated ^{1}H-DNMR spectra of σ-cyclopentadienyl(triphenylphosphine)copper(I) *6* (Whitesides *et al.* [39]).
Left column: Experimental spectra; *Middle column:* Theoretical spectra corresponding to the 1,2-shift mechanism;
Right column: Theoretical spectra corresponding to the 1,3-shift mechanism

To improve further the informative power of the classical analyses based on the exchange modified Bloch equations, new techniques have been developed which allow a closer look at certain complex exchange processes. These methods will be discussed in the following sections.

5. Use of Prochiral CX_2Y-Groups as Mechanistic Probes

The phenomenon of diastereotopism [*41, 42*] of the groups X in compounds of the type R-CX_2Y, where R represents a chiral residue, has been used in recent years in numerous DNMR studies. In addition to the kinetic investigation of the enantiomerization in residue R, it can be used as a mechanistic probe when a complex exchange behavior in R is anticipated. Thus, by comparing the DNMR pattern of the exchanging nuclei in R with that for the diastereotopic groups X, processes which involve enantiomerization in R can be characterized and distinguished from processes in which the chirality of R is not affected.

An early example of such an approach had been given in 1964 by Anet *et al.* [*43*], in which the dynamical behavior of cyclooctatetraene was elucidated.

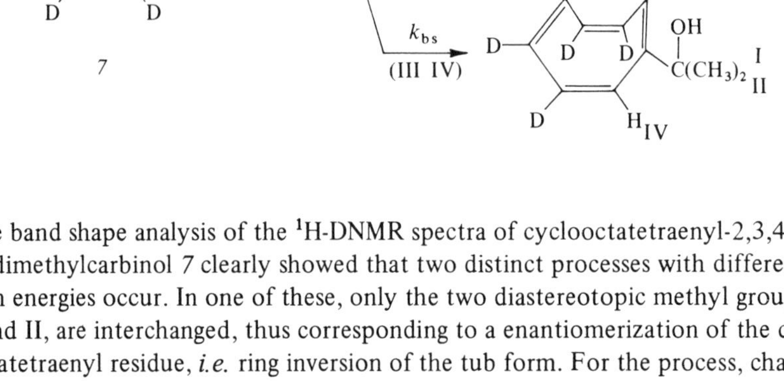

7

The band shape analysis of the ^{1}H-DNMR spectra of cyclooctatetraenyl-2,3,4,5,6,7-d_6-dimethylcarbinol 7 clearly showed that two distinct processes with different activation energies occur. In one of these, only the two diastereotopic methyl groups, sites I and II, are interchanged, thus corresponding to a enantiomerization of the cyclooctatetraenyl residue, *i.e.* ring inversion of the tub form. For the process, characterized by the olefinic site permutation (III IV), a bond shift is responsible, the rate constant k_{bs} being smaller than k_{ri}. In the band shape analysis, the two processes can either be treated simultaneously as a 4-site problem by using the kinetic exchange matrix IX, or separately by splitting matrix IX in two matrices, each of order two.

$$\begin{pmatrix} -k_{ri} & k_{ri} & & \\ k_{ri} & -k_{ri} & & \\ & & -k_{bs} & k_{bs} \\ & & k_{bs} & -k_{bs} \end{pmatrix} \quad \begin{pmatrix} -k_r & k_r/2 & k_r/2 \\ k_r/2 & -k_r & k_r/2 \\ k_r/2 & k_r/2 & -k_r \end{pmatrix} \quad \begin{pmatrix} -k_i & k_i \\ k_i & -k_i \end{pmatrix}$$

IX X XI

In other CX_2Y-probes commonly used in DNMR, the diastereotopic groups X are coupled to Y, such as in $CH(CH_3)_2$, or coupled to each other giving rise to an AB-system, such as in CH_2-C_6H_5 or CH_2-CD_3. The calculation of the corresponding 4-site exchange is easily performed by computer programs such as QUABEX [*8*] and needs no further comment here (for the calculational procedure of exchanging coupled spin systems see Sections 8 and 9). Therefore in the following examples, as well as in the polyhedral rearrangements discussed in the next two sections, we do not need to specify the exchange relations of the four sites of the diastereotopic groups, but instead characterize this exchange as if only two sites, corresponding to the two nuclei or groups of nuclei, were present.

As in the cyclooctatetraenyl case discussed above, two distinct dynamic processes are observed for N-tert-butyl-N-(ethyl-2,2,2-d_3)N-chloroamine *8*, which was studied by Bushweller and co-workers in 1975 [*44*]. The faster process corresponds to a rotation of the methyl groups around the C-N bond, leaving the AB-pattern of the diastereotopic protons unchanged. The exchange matrix X is easily derived by adding the exchange probabilities due to the two modes of rotation, *i.e.* permutations (123) and (321), or in terms of the sites, (III II I) and (I II III).

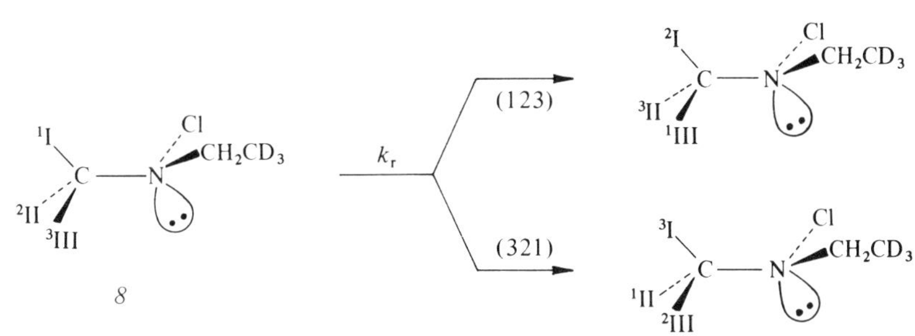

8

The exchange of the diastereotopic protons is caused by another slower process, as seen by comparing the rates of the AB exchange (labeled k_i in matrix XI) with those of the 3-site exchange of the methyl groups (Fig. 4).

Simple nitrogen inversion could account for this slower exchange process, but eclipsed conformations would result. Indeed, Bushweller *et al.* could show for the similar compounds $(CH_3)_3C$-$N(CH_3)$-CH_2-C_6H_5 [*45*] and $(CH_3)_3C$-$N(CD_3)$-CH_2-CD_3 [*46*] that the nitrogen inversion is accompanied by rotation of the methyls of the tertiary butyl group, probably by 60° instead of 120° as in the simple rotation depicted above.

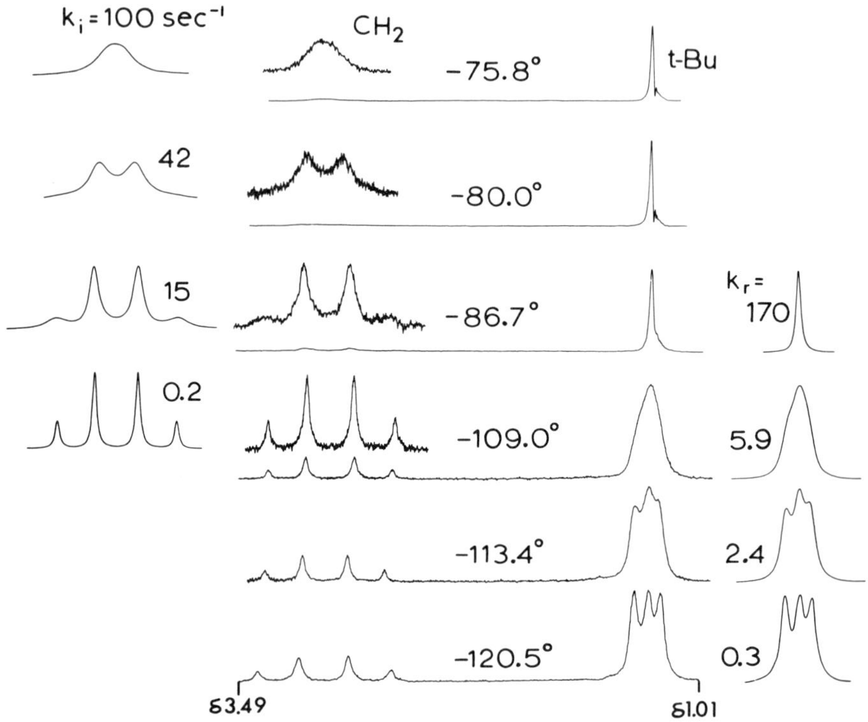

Fig. 4. Comparison of experimental and calculated ¹H-DNMR spectra of N-tert-butyl-N-(ethyl-2,2,2-d_3)-N-chloroamine *8* (Bushweller *et al.* [44]).
Left column: Theoretical spectra calculated as a function of the rate of nitrogen inversion (k_i);
Middle column: Experimental spectra with irradiation at the ²H resonance frequency;
Right column: Theoretical spectra as a function of the rate of rotation (k_r)

In contrast to the two examples discussed above, the study of bis(2,6-xylyl)-1-(3-isopropyl-2,4,6-trimethylphenyl) borane *9* by Mislow and co-workers [*47*] revealed equal rates for the exchange of the diastereotopic isopropyl methyls and the exchange of the xylyl methyls. This can be seen from Fig. 5, in which the experimental spectra are compared with the theoretical spectra calculated by using the kinetic exchange matrix XII.

Although the aromatic methyl region consists of seven methyl resonances, two of which accidentally have the same chemical shift, the corresponding part of matrix XII is only of order four, since the three methyl groups of the isopropyl-substituted ring are not involved in exchange. The latter, were, however, incorporated in the diagonal matrix (see Section 3.) to facilitate the comparison with the experimental spectra. Similar to the derivation of matrix X in the previous example, the 4 x 4 submatrix of matrix XII is constructed from the two exchange matrices XIII and XIV, which correspond to the site permutations (I II III IV) and (IV III II I), respectively. As mentioned above, the exchange of the diastereotopic methyl groups of the isopropyl residue may be characterized by permutation (V VI).

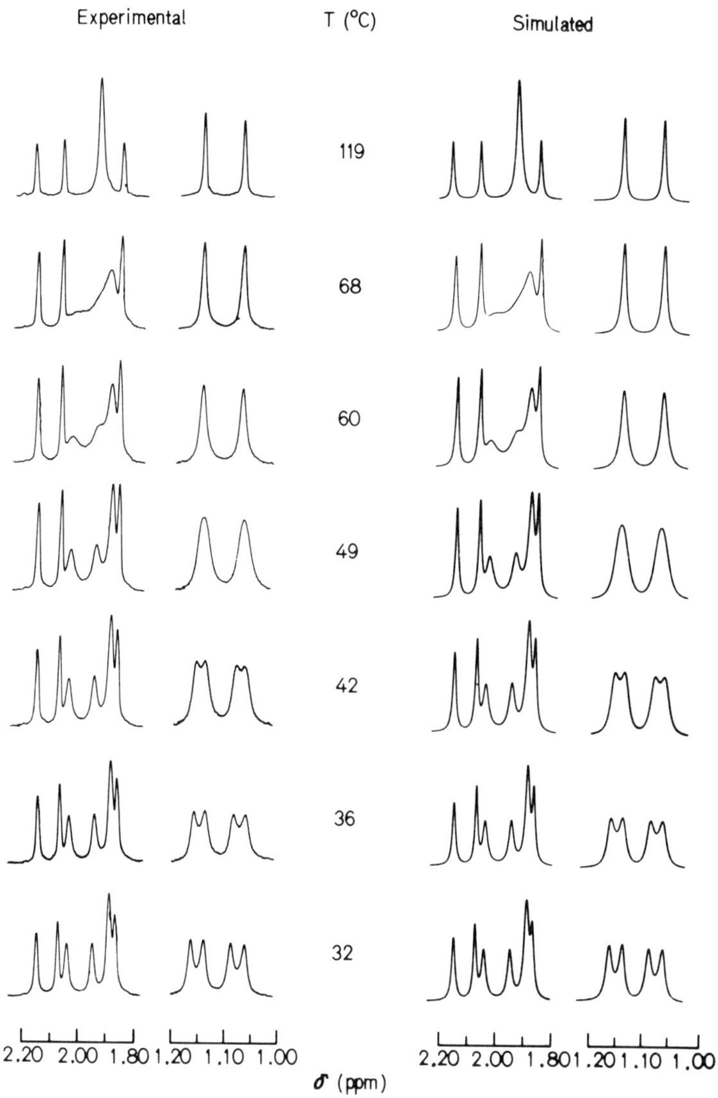

Fig. 5. Comparison of experimental and simulated ^{1}H-DNMR spectra of bis(2,6-xylyl)-1-(3-isopropyl-2,4,6-trimethylphenyl)borane *9* (Mislow *et al.* [47])

$$\begin{pmatrix} -k & k/2 & & k/2 & & \\ k/2 & -k & k/2 & & & \\ & k/2 & -k & k/2 & & \\ k/2 & & k/2 & -k & & \\ & & & & -k & k \\ & & & & k & -k \end{pmatrix} \quad \begin{pmatrix} -k & & & k \\ k & -k & & \\ & k & -k & \\ & & k & -k \end{pmatrix} \quad \begin{pmatrix} -k & k & & \\ & -k & k & \\ & & -k & k \\ k & & & -k \end{pmatrix}$$

XII XIII XIV

These site permutations were interpreted by Mislow and co-workers as arising from two-ring flips, a flip describing the rotation by 90 ° of an aryl group around the B-Ar bond through a plane perpendicular to the reference plane, which is defined by the three carbon atoms attached to boron. Thus, while the flips of rings 1 and 2 lead to the site permutation (IV III II I)(V VI), another two-ring flip involving rings 1 and 3 occurs with the same probability (enantiomeric rearrangements) leading to the site permutation (I II III IV)(V VI).

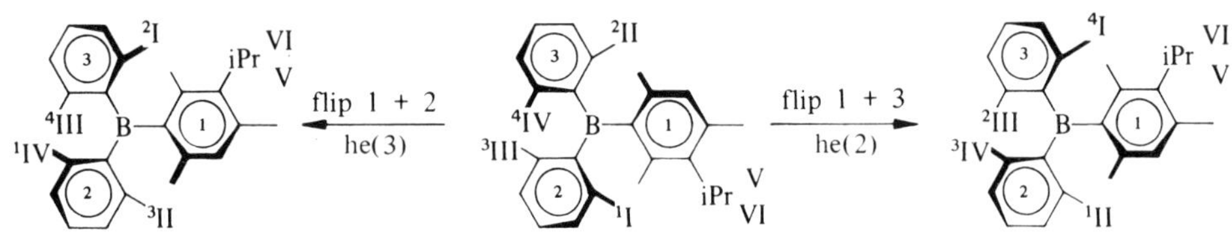

9

To ensure that other motions of the three rings do not lead to the same site permutations, a complete analysis of all 16 possible ways of conformational changes was necessary [*47*]. Here we present a newer analysis by Mislow [*48*], which is based on group-theoretical arguments. In this approach all rearrangements are characterized by appropriately combining the edge interchange operation e with the reversal of helicity h. While the first operation corresponds to a rotation of an aryl ring by 180 °, the operation h implies a three-ring flip, *i.e.* all rings are rotated by 90 ° through planes perpendicular to the reference plane. The two rearrangements depicted above thus correspond to the operator products he(3) and he(2). The classification of the possible rearrangements and the resulting site permutations are listed in Table 1.

It is seen that the 16 rearrangements all lead to different site exchanges. The only NMR nondifferentiable mechanisms are the enantiomeric nonflip rearrangements e(1) e(2) and e(1) e(3) and the enantiomeric flip rearrangements h e(3) and h e(2).

Table 1.

Nonflip rearrangements		Flip rearrangements		
Type	Site exchange	Type	Flipping rings	Site exchange
I		h e(1) e(2) e(3)		(V VI)
e(1)	(I II) (III IV)	h e(2) e(3)	1	(I II) (III IV) (V VI)
e(2)	(I III)	h e(1) e(3)	2	(I III) (V VI)
e(3)	(II IV)	h e(1) e(2)	3	(II IV) (V VI)
e(1) e(2)	(IV III II I)	h e(3)	1, 2	(IV III II I) (V VI)
e(1) e(3)	(I II III IV)	h e(2)	1, 3	(I II III IV) (V VI)
e(2) e(3)	(I III) (II IV)	h e(1)	2, 3	(I III) (II IV) (V VI)
e(1) e(2) e(3)	(I IV) (II III)	h	1, 2, 3	(I IV) (II III) (V VI)

The latter two are seen to be the only ones which account for the two experimental findings, *i.e.* equal exchange rates of the four xylyl methyls and the two isopropyl methyls and secondly the collapse of the four xylyl methyl resonances to only one line. While the one-flip rearrangements h e(1) e(3) and h e(1) e(2) are not consistent with the criteria of equal rates, the flip rearrangements h e(2) e(3), h e(1), and h fulfil this criteria, but the four xylyl methyl resonances would collapse to two lines in the fast exchange region.

In the examples discussed so far, the procedure has been to define the mechanistic alternatives, then setting up the permutations of the groups involved in exchange, and finally deriving the corresponding permutations of the sites. To study the mechanism of polytopal rearrangements [*49*] of five- or higher-coordinate inorganic complexes, a different kind of approach is necessary, since a priori reasonable mechanistic alternatives are unknown. In the next two sections the feature of complete permutational analysis is illustrated by discussing the rearrangements of phosphoranes.

6. Permutational Approach to Polytopal Rearrangements

The study of inorganic coordination compounds has been a major topic in NMR spectroscopy since the discovery of chemical shifts in 1950. While polyhedral geometries, particularly of inorganic fluorine compounds, could be derived, observed exchange phenomena were difficult to interpret. For instance in the case of SF_4, although the exchange of the fluorine nuclei was established already in 1958 [*33, 34*] by low temperature ^{19}F-NMR spectra, the mechanism of this rearrangement was not elucidated until recently (cf. Section 9.2.). Similar difficulties account for the fact that in PF_5 all fluorine atoms remained magnetically equivalent even at low temperatures [*50*]. For this compound, Berry in 1960 [*51*] proposed a degenerate rearrangement process of the trigonal bipyramidal structure via a square pyramid leading to an exchange of two of the three equatorial fluorines with the two axial fluorines. This Berry mechanism indeed seems to explain the easy rearrangement in many five-coordinate compounds. Nevertheless, the approach of guessing the most reasonable rearrangement mechanisms and simulating the experimental DNMR spectra with the corresponding kinetic exchange matrices necessarily involves some arbitrariness. In this approach the possibility always exists that for five-, six, or higher-coordinate compounds the true mechanism will be overlooked. Therefore, to allow a proper treatment of exchange processes in coordination compounds, a systematic approach is needed.

The use of mathematical methods was promoted by Muetterties [*49*], who gave a topological representation of polytopal rearrangements, the term polytopal implying that the rearrangements proceed via intermediates or transition states (polytopal isomers) whose spatial arrangements can be described in terms of idealized polyhedra, the vertices of which correspond to the ligand positions.

It will not be attempted here to evaluate the general validity of this assumption. For strained systems, however, it may be difficult to reach symmetrical intermediates or transition states. In this report we will use the term polytopal rearrangement as a collective name for the rearrangements of coordination compounds, in which the ligands are permuted among the vertices or skeletal positions of the coordination polyhedron. Instead of a topological approach to polytopal rearrangements, it is more advantageous from the NMR point of view to investigate systematically the effects of all possible permutations of the ligands on the NMR sites. Once the pattern of site exchange is established, possible mechanisms fitting the observed exchange behavior can be discussed. In recent years group theoretical arguments have been used to facilitate this kind of approach. While Klemperer [*36*] developed a procedure by which NMR-differentiable permutational isomerization reactions of the specific coordination compound of interest can be derived, another type of approach [*52–54*] allows a common treatment of all cases having the same polyhedral geometry by deriving distinguishable types of rearrangements, for which the term "modes of rearrangement" was coined. The latter approach will be described here for the case of trigonal bipyramids, and the DNMR studies on spirocyclic phosphoranes by Whitesides and co-workers [*55, 56*] are used as specific examples. Similar features are encountered in the polytopal rearrangements of tris-chelated octahedral complexes which, in recent years, have been studied very actively both theoretically by permutational analysis [*57, 58*] and experimentally by DNMR [*59, 60*].

In a set of five elements, there are 5! = 120 permutational possibilities to interchange them. When the five elements (ligands) occupy the skeletal positions of a trigonal bipyramid, the 120 permutations can be classified as depicted in Table 2. The numbering of the ligands is seen in the reference arrangement *10*.

Classification of the permutations by rotational equivalence yields 20 cosets, each consisting of six rotationally equivalent permutations. The further decomposition of the cosets into six double cosets, M_0–M_5, which are called modes of rearrangement [*53, 54*], is achieved by taking advantage of symmetry equivalence. For instance, mode M_1, characterized as exchange of two axial-equatorial ligand pairs (see underlined letters in the permutation denotation) accompanied by enantiomerization, comprises three cosets, each consisting of six rotationally equivalent permutations, *i.e.* two of type aeae and four of type ae x aee. Therefore, these two types of permutation, corresponding to the Berry mechanism and to the turnstile rotation [*61*], respectively, cannot be distinguished by physical methods. One has to resort to theoretical methods, such as MO calculations [*62*], to differentiate between them. Although the term mode of rearrangement has been coined to indicate that these classes can be distinguished by physical methods, the differentiability by NMR varies greatly with the type of ligand pattern present.

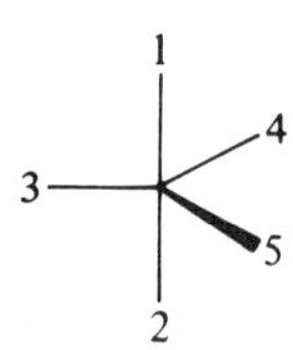

10

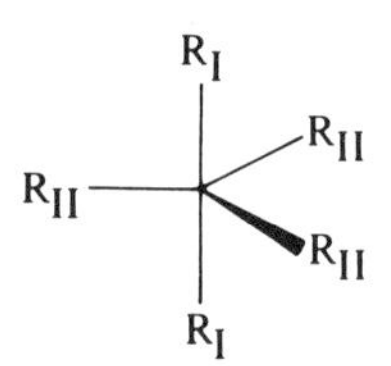

11

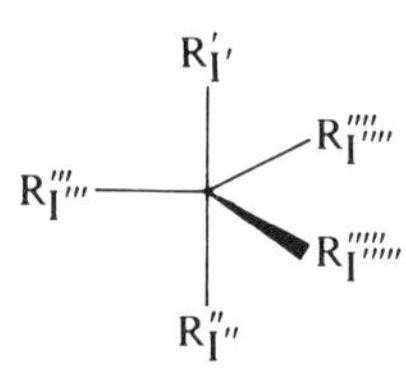
12

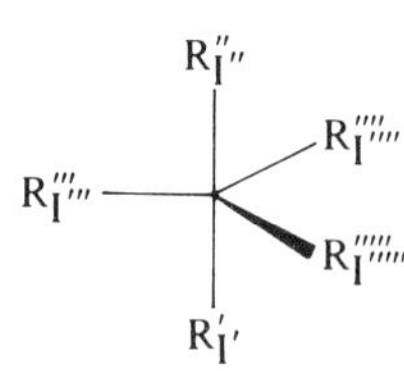
13

Table 2.

Modes of rearrangement not involving enantiomerization

Type	Ligand permutation			
	I	eee	aa × ee	
M_0		(345)	(12)(34)	
		(543)	(12)(35)	
			(12)(45)	
	ae	aeee	aaee	aae × ee
	(13)	(1453)	(2143)	(213)(45)
		(1543)	(2153)	
	(14)	(1354)	(2134)	(214)(35)
		(1534)	(2154)	
	(15)	(1345)	(2135)	(215)(34)
		(1435)	(2145)	
M_2	(23)	(2453)	(1243)	(123)(45)
		(2543)	(1253)	
	(24)	(2354)	(1234)	(124)(35)
		(2534)	(1254)	
	(25)	(2345)	(1235)	(125)(34)
		(2435)	(1245)	
	ae × ae		aeaee	
	(13)(24)		(13254)	
	(14)(23)		(24153)	
			(14253)	
			(23154)	
	(13)(25)		(13245)	
	(15)(23)		(25143)	
M_5			(15243)	
			(23145)	
	(14)(25)		(14235)	
	(15)(24)		(25134)	
			(15234)	
			(24135)	

Modes of rearrangement involving enantiomerization

Type	Ligand permutation			
	aa	aa × eee		ee
M_3	(12)	(12)(345)		(34)
		(12)(543)		(35)
				(45)
	ae × ee	aee	aaeee	aae
	(13)(45)	(143)	(21453)	(213)
		(153)	(21543)	
	(14)(35)	(134)	(21354)	(214)
		(154)	(21534)	
	(15)(34)	(135)	(21345)	(215)
		(145)	(21435)	
M_4	(23)(45)	(243)	(12453)	(123)
		(253)	(12543)	
	(24)(35)	(234)	(12354)	(124)
		(254)	(12534)	
	(25)(34)	(235)	(12345)	(125)
		(245)	(12435)	
	aeae	ae × aee		
	(1324)	(13)(254)		
	(1423)	(24)(153)		
		(14)(253)		
		(23)(154)		
	(1325)	(13)(245)		
	(1523)	(25)(143)		
M_1		(15)(243)		
		(23)(145)		
	(1425)	(14)(235)		
	(1524)	(25)(134)		
		(15)(234)		
		(24)(135)		

This is easily seen by comparing the two extreme cases, the coordination type *11* with five identical ligands and type *12*, in which all ligands are of different nature. In the case of identical ligands *11*, there are only two different NMR sites, I and II, provided of course that spin-spin couplings are negligible (DNMR studies on coupled ML_5-systems have been reported by Jesson, Meakin, and co-workers [*63, 64*]). As a consequence, only three types of rearrangements can be distinguished: the identity rearrangement (modes M_0 and M_3), the one-pair exchange (modes M_2 and M_4) and the two-pair exchange (modes M_5 and M_1). There is no way for the symmetrical ML_5 compounds to differentiate by DNMR between the enantiomeric modes, whose permutations differ by the permutational operation *aa* or *ee*. On the other hand, in the case of systems with five different ligands *12*, in principle all 20 cosets can be differentiated by NMR. This is so, because each of the 20 cosets of permutations leads to a different stereoisomer. Thus the five sites I′, I″, I‴, I⁗, and I′′′′′ of the reference stereoisomer *12* are exchanging with a different set of five sites of the new stereoisomer, reached by the respective ligand permutation. Enantiomerizations such as the rearrangement of *12* to *13* apparently cannot be seen by NMR as evidenced by inspection of the corresponding sites. However, if one of the ligands contains a prochiral group (cf. Section 5.), the enantiomerization process can be detected. By this method the enantiomeric cosets become NMR-differentiable.

From the description given for compounds of type *12* it is clear that the differentiation between all 20 cosets requires all theoretically possible NMR sites, including those of the prochiral group, to be seen in the NMR spectrum (slow exchange region). Besides the problem of resolving all resonances and assigning them, the greatest limitation to the study of systems with several different ligands lies in the fact that certain stereoisomers will dominate, thus not only reducing the number of observable stereoisomers, but also enhancing the chance of rigidity of the preferred stereoisomers. This does not imply that unsymetrically substituted coordination compounds cannot be studies at all by DNMR. In fact, several NMR studies of conclusive mechanistic power have been reported for systems of the type PR_4R' *14*, where the ligand R′ occupies an equatorial position of the trigonal bipyramid. Due to this stereochemical preference, all the permutations leading to an axial R′ can be discarded for a characterization of the overall exchange of the ligands. Thus, using the same labeling as in *10*. *i.e.* R′ being ligand 5, any of the permutations of the cosets containing the (15) and (25) permutations (mode M_2), the (13)(25), (15)(23) and (14)(25), (15)(24) permutations (mode M_5) and of the corresponding enantiomeric cosets in modes M_4 and M_1 must be immediately followed by another permutation to yield the same equatorial stereoisomer *14*, which means that the overall permutation, observable by DNMR, can be characterized exclusively by the remaining cosets. One example of equatorial preference, $PF_4(NMe_2)$ will be discussed in further detail in Section 8.1.

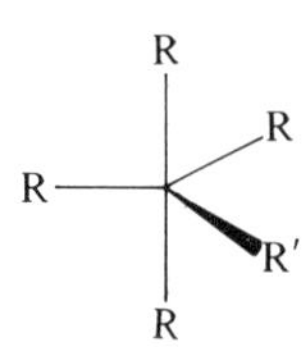

14

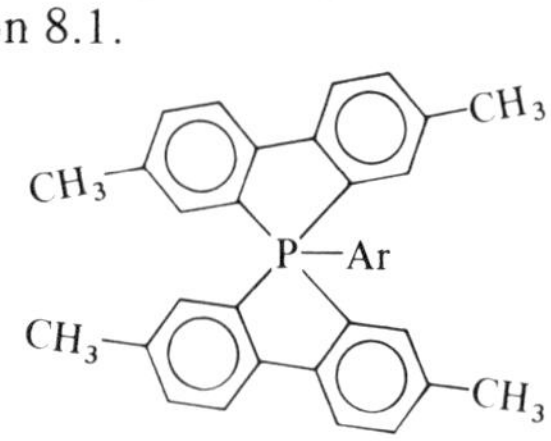

Ar = 2,4,6–(iPr)$_3$Ph

15

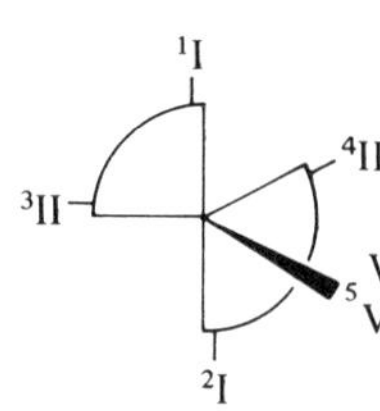

16

Here we consider another type of PR_4R',the spirocyclic phosphorane *15*, studied by Whitesides and co-workers [*56*]. In this compound the triisopropyl-phenyl group occupies the equatorial position, since in order to minimize angle strain the 2,2′-biphenylene groups are preferentially bridging one axial and one equatorial position of the trigonal bipyramid instead of two equatorial positions. The labeling of the methyl groups and of the DNMR sites is given in the schematic structure *16*. Herein the diastereotopic isopropyl methyls of the aryl residue 5 are labeled as V and VI to allow an easier comparison with the case discussed in the next section. For convenience the two site exchanges, I and II, and V and VI, were simulated separately as shown in Fig. 6.

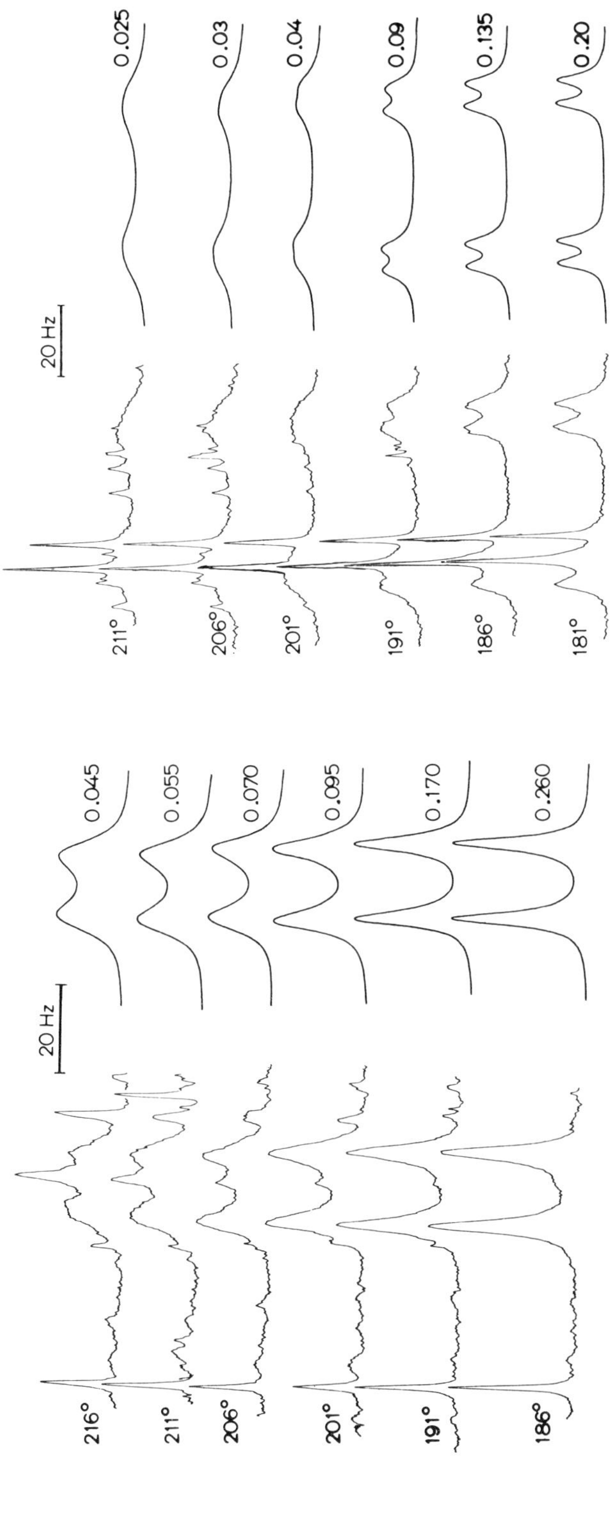

Fig. 6. Comparison of experimental and calculated ¹H-DNMR spectra of the spirocyclic phosphorane *15* (Whitesides *et al.* [56]).
Left side: Experimental (T, °C) and calculated (τ, sec) spectra for the biphenylene methyl resonances;
Right side: Experimental (T, °C) and calculated (τ, sec) spectra for the isopropyl methyl resonances

It is evident that the exchange of the biphenylene methyls and of the isopropyl methyls, which probe the reversal of chirality of the phosphorane (cf. Section 5.), occur with different rates at the same temperature, *i.e.* with $1/\tau_b$ and $1/\tau_i$, respectively. Due to thermal decomposition at higher temperatures, only a tentative establishment of a one to two ratio of the corresponding rates was possible. This suggests that the two exchange matrices XV and XVI can be joined together to give the total kinetic exchange matrix XVII, corresponding to the site permutation (I II)(I)(II) (V VI).

$$\underset{\text{XV}}{\begin{pmatrix} -k_b & k_b \\ k_b & -k_b \end{pmatrix}} \qquad \underset{\text{XVI}}{\begin{pmatrix} -k_i & k_i \\ k_i & -k_i \end{pmatrix}} \qquad \underset{\text{XVII}}{\begin{pmatrix} -k/2 & k/2 & & \\ k/2 & -k/2 & & \\ & & -k & k \\ & & k & -k \end{pmatrix}}$$

The approach of complete permutational analysis described earlier in this section is well suited to discuss the implications of this result. In Table 3 only those cosets are given which conform to the stereochemical requirements of the spirocyclic phosphorane. Thus, besides the omission of the cosets which lead to an axial aryl group (cf. discussion above), the cosets containing the (14) and (23) permutations and their enantiomeric counterparts are also discarded, since they would require a bridging of two axial positions by a biphenylene group.

Table 3.

Modes of rearrangement not involving enantiomerization

Type and site exchange	Ligand permutation			
	I	eee	aa × ee	
M_0		(345)	(12)(34)	
		(543)	(12)(35)	
			(12)(45)	
	ae	aeee	aaee	aae × ee
	(13)	(1453)	(2143)	(213)(45)
M_2		(1543)	(2153)	
(I II)	(24)	(2354)	(1234)	(124)(35)
		(2534)	(1254)	
		ae × ae	aeaee	
M_5		(13)(24)	(13254)	
(I II) (I II)		(14)(23)	(24153)	
			(14253)	
			(23154)	

Modes of rearrangement involving enantiomerization

Type and site exchange	Ligand permutation			
	aa	aa × eee		ee
M_3	(12)	(12)(345)		(34)
(V VI)		(12)(543)		(35)
				(45)
	ae × ee	aee	aaeee	aae
	(13)(45)	(143)	(21453)	(213)
M_4		(153)	(21543)	
(I II) (V VI)	(24)(35)	(234)	(12354)	(124)
		(254)	(12534)	
		aeae	ae × aee	
M_1		(1324)	(13)(254)	
(I II) (I II) (V VI)		(1423)	(24)(153)	
			(14)(253)	
			(23)(154)	

From Table 3 one sees that the observed site permutation (I II)(I)(II)(V VI) is accounted for only by mode M_4, *i.e.* as a net result of the actual mechanism, chirality reversal is accompanied by interchange of one a,e methyl pair belonging to the same biphenylene group. Therefore one can discard a Berry mechanism (cf. the two-pair exchange of mode M_1) formulated either via a square pyramid with the aryl group at the top of the pyramid or as a pseudorotation with the aryl group as pivot. The actual mechanism must not necessarily correspond directly to one of the ligand permutations of mode M_4. Besides these possibilities, the net exchange observed may also originate from a two-step rearrangement via a short-lived intermediate or transition state. Whitesides and co-workers [*56*] suggested a mechanism of the latter type, namely a Berry pseudorotation with one of the biphenylene skeletal positions as pivot, leading to an intermediate trigonal bipyramid with the aryl group axial, followed by a second Berry rearrangement to reform the preferred stereoisomer.

7. Split Modes of Rearrangement

In the previous section it was shown for five-coordinated compounds with several different ligands that the modes of rearrangement in principle are not the limit of differentiability. Ultimately all the cosets could be differentiated, since in the case of the ligands all being different, each coset leads to a different stereoisomer. Modes of rearrangement are also not sufficient to define the differentiability for tris-chelated octahedral complexes of the type $M(AB)_3$, as was shown by Musher in 1972 [*57*]. These complexes occur as mixtures of four stereoisomers, one of which is depicted schematically as *17*. To account for this further differentiability, Musher proposed the term split modes of rearrangement.

In this section another type of splitting is described, which does not only involve the modes of rearrangement but also the cosets of the ligand permutations themselves. The compound *18*, for which such a splitting occurs, is similar to the spirocyclic phosphorane *15*, the only difference being in the aryl group.

17

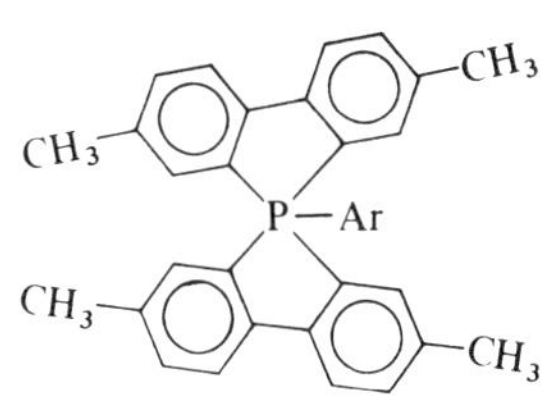

Ar = 2–iPr–Ph

18

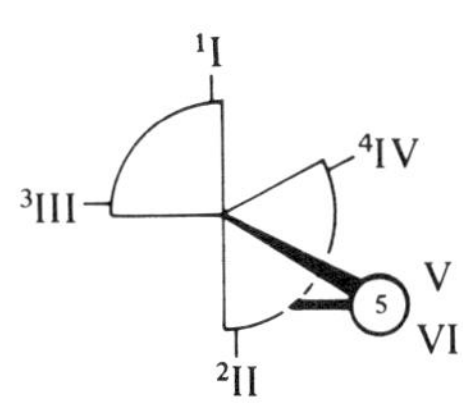

19

This phosphorane was studied by Whitesides and co-workers in 1967 [*55*] and 1974 [*56*] using in the latter paper the approach of Klemperer [*36*] to enumerate the NMR-differentiable permutational isomerization reactions. Here we will take advantage of the permutational analysis described in the previous section to characterize the dynamical behavior of the compound. Due to steric restriction, the unsymmetrical aryl group has two possibilities of orientation towards the spirocyclic part of the phosphorane. The left-handed orientational alternative is shown in the schematic structure *19*. In contrast to the spirocyclic phosphorane *15*, all four biphenylene methyl groups give rise to different resonances, *i.e.* sites I to IV. In Table 4 the classification of the ligand permutations by split modes of rearrangement is given together with the corresponding site exchanges. The comparison with Table 3 reveals that each coset is now split into two halves leading to 16 NMR-distinguishable types of permutations, which were also derived by Whitesides *et al.* [*56*] using the approach of Klemperer. Of course, the splitting must be a consequence of the additional stereoisomerism caused by the unsymmetrical aryl group. In fact, the permutations in one half of a splitted coset can be transformed to the permutations of the other half by performing the permutational operation (12)(34), which corresponds to a rotation of the aryl group by 180 °, *i.e.* change of the left-handed isomer to the right-handed one, and the reverse.

Table 4.

Split modes of rearrangement not involving enantiomerization			Split modes of rearrangement involving enantiomerization		
Type and site exchange	Ligand permutation		Type and site exchange	Ligand permutation	
	I	eee		aa	aa × eee
M'_0		(345)	M'_3	(12)	(12)(345)
		(543)	(V VI)		(12)(543)
	aa × ee			ee	
M''_0	(12)(34)		M''_3	(34)	
(I II) (III IV)	(12)(35)		(I II) (III IV) (V VI)	(35)	
	(12)(45)			(45)	
	ae	aeee		aaeee	aae
M'_2	(13)	(1453)	M'_4	(21453)	(213)
(I III)		(1543)	(I III) (V VI)	(21543)	
M''_2	(24)	(2354)	M''_4	(12354)	(124)
(II IV)		(2534)	(II IV) (V VI)	(12534)	
	aaee	aae × ee		ae × ee	aee
M'''_2	(2143)	(213)(45)	M'''_4	(24) (35)	(234)
(IV III II I)	(2153)		(IV III II I) (V VI)		(254)
M''''_2	(1234)	(124)(35)	M''''_4	(13)(45)	(143)
(I II III IV)	(1254)		(I II III IV) (V VI)		(153)
	ae × ae	aeaee		aeae	ae × aee
M'_5	(13)(24)	(13254)	M'_1	(1324)	(13)(254)
(I III) (II IV)		(24153)	(I III) (II IV) (V VI)		(24)(153)
M''_5	(14)(23)	(14253)	M''_1	(1423)	(14)(253)
(I IV) (II III)		(23154)	(I IV) (II III) (V VI)		(23)(154)

As was concluded by Whitesides and co-workers [*56*], only two permutational isomerization reactions, labeled here as M_4''' and M_4'''' (Table 4), would be compatible with the observed DNMR spectra (see Fig. 7). Since they could not conceive a physically reasonable mechanism leading to these permutations, they suggested a Berry-type rearrangement pathway via a square pyramidal intermediate with the aryl group at the top of the pyramid, which would allow rapid rotation of the aryl group around the carbon-phosphorous bond, before reforming the trigonal bipyramid. Such a mechanism gives rise to four ligand permutations corresponding to the equally probable occurrence or nonocurrence of helicity reversal and/or aryl rotation, *i.e.* the permutations I, (12)(34), (1324), and (1423). As seen from Table 4, these permutations belong to the modes M_0', M_0'', M_1', and M_1'' and cause the site permutations (I)(II)(III)(IV)(V)(VI), (I II)(III IV)(V)(VI), (I IV)(II III)(V VI), and (I III)(III V)(V VI), respectively. Setting up the net site exchange probabilities, the kinetic exchange matrix XVIII is obtained.

$$\begin{pmatrix} -3k/4 & k/4 & k/4 & k/4 & & \\ k/4 & -3k/4 & k/4 & k/4 & & \\ k/4 & k/4 & -3k/4 & k/4 & & \\ k/4 & k/4 & k/4 & -3k/4 & & \\ & & & & -k/2 & k/2 \\ & & & & k/2 & -k/2 \end{pmatrix} \qquad \begin{pmatrix} -k & k/2 & & k/2 & & \\ k/2 & -k & k/2 & & & \\ & k/2 & -k & k/2 & & \\ k/2 & & k/2 & -k & & \\ & & & & -k & k \\ & & & & k & -k \end{pmatrix}$$

XVIII XIX

For comparison, the kinetic exchange matrix XIX, which corresponds to the two split modes M_4''' and M_4'''' as mentioned above, is also given. The matrix XVIII has been used by Whitesides *et al.* in the band shape calculation of both the biphenylene methyl and the isopropyl methyl region. Actually the simulation of the experimental spectra was performed independently for both regions (Fig. 7) by using the corresponding submatrices, in order to test the assumption that the rates of the site exchange of the two regions are really related by the ratio 3/4 to 1/2.

The derived τ-values (Fig. 7), which are the lifetimes multiplied by 4/3 and 2, respectively, are indeed very close for a given temperature. However, from the Arrhenius plots of the two sets [*56*], it appears that the isopropyl methyl exchange is faster than assumed. Thus it might well be that band shape simulations using matrix XIX would lead to an even better agreement. Indeed, a physically reasonable mechanism could be found for the modes M_4'''' and M_4''''' in the mechanism postulated by Whitesides *et al.* for the previously discussed phosphorane *15*, *i.e.* a Berry pseudorotation leading to an intermediate trigonal bipyramid with an axial aryl group, followed by a second pseudorotation (cf. previous section). The sterical analysis of all ligang permutations which conform to this latter mechanism suggests that only in those cases where the resulting ligand permutations belong to modes M_0'', M_4', and M_4'', is the unsymmetrical aryl group rotated simultaneously by 180 °, while for the remaining permutations, belonging to modes M_0', M_4''', and M_4''''', the aryl rotation is sterically hindered. Since the aryl rotation transforms the modes M_0'', M_4', and M_4'' into modes M_0', M_4''', and M_4''''' (see above), this mechanism would indeed be properly

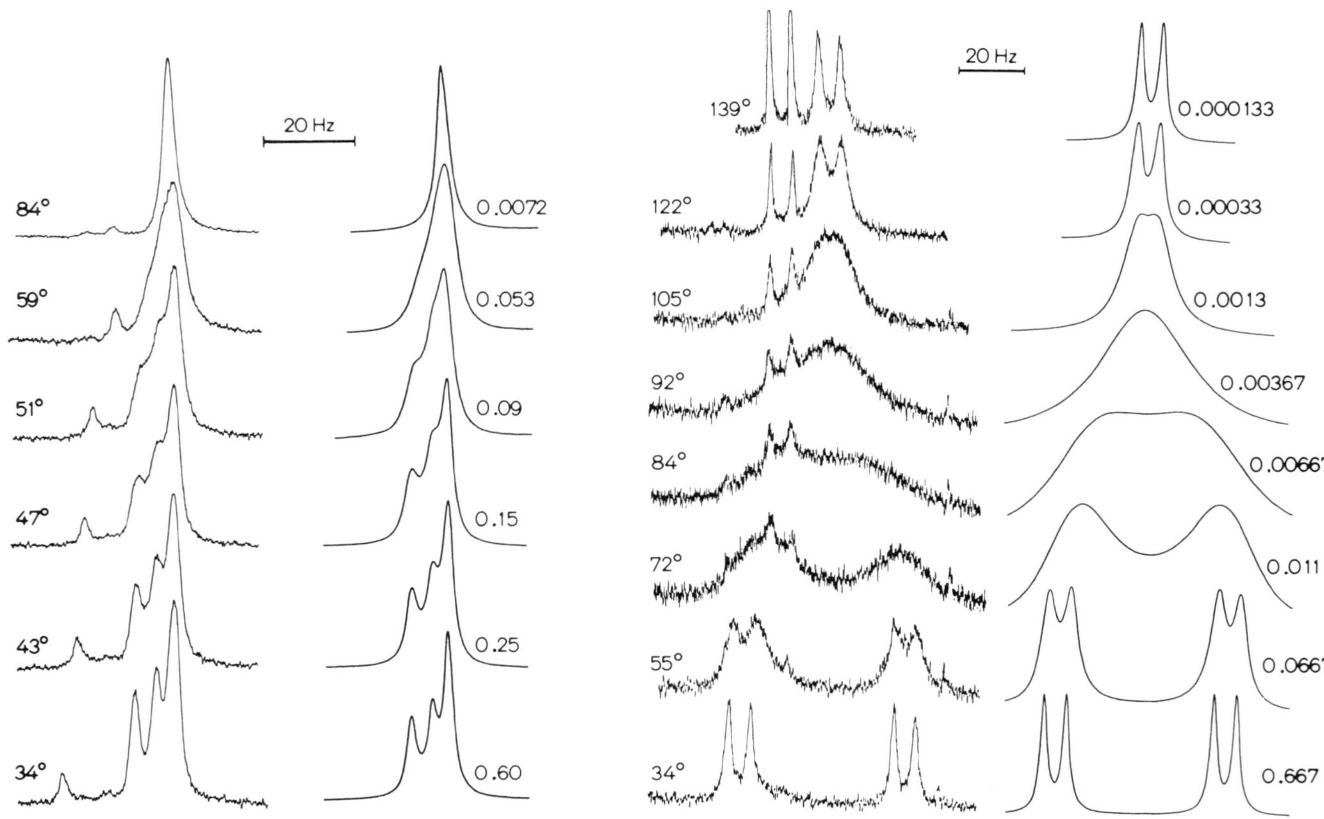

Fig. 7. Comparison of experimental and calculated ^{1}H-DNMR spectra of the spirocyclic phosphorane *18* (Whitesides *et al.* [56]).
Left side: Experimental (T, °C) and calculated (τ, sec) spectra for the biphenylene methyl resonances;
Right side: Experimental (T, °C) and calculated (τ, sec) spectra for the isopropyl methyl resonances

represented by matrix XIX. Unfortunately, the use of this exchange matrix in band shape simulations requires the assignment of the biphenylene methyl resonances, and therefore it seems questionable if a clear distinction between the exchange relations given in matrices XVIII and XIX will be possible.

8. Site Exchange in First-Order Spin Systems

8.1. Sites Determined by Spin-Spin Couplings

In recent years many examples have been described in which spin-spin splittings were used as a tool in mechanistic analysis of exchanging systems. In the case of nuclei of spin 1/2, scalar spin-spin coupling is caused by the spin states α and β of neighboring nuclei leading to a splitting of the resonance of the nucleus under consideration. First-order spin systems, for which the chemical shift differences are greater than the coupling constants by at least an order of magnitude, do not necessarily require a quantum mechanical treatment, and classical band shape equations, such as Eq. (1) (cf. Section 2.2.), may be used for simulation of the DNMR spectra. In this section DNMR studies are discussed in which the exchanging nuclei are not observed directly, but instead their splitting effect on the resonance of the nucleus observed is used to obtain information on the exchange process.

In one group of this type of DNMR problems the observed nucleus is of a different kind than the exchanging nuclei. For instance, the dynamical behavior of the fluorines in the phosphorane $PF_4(NMe_2)$ was studied by ^{31}P-NMR by Whitesides and co-workers in 1969 [*65*] and 1974 [*66*]. This study will be described here to show the approach of deriving mechanistic conclusions from the temperature dependence of the NMR spectra. The phosphorane, which has the trigonal bipyramidal geometry *20* with the amine group equatorial, gives rise to a nine-site ^{31}P-NMR spectrum in the slow exchange region, as schematically depicted in Fig. 8.

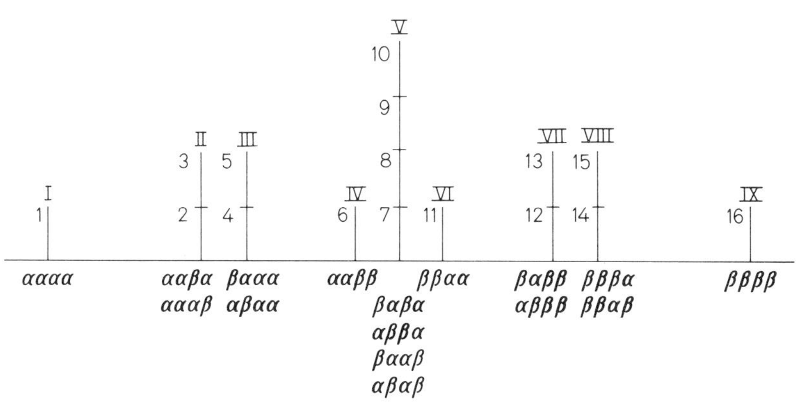

Fig. 8. Assignment of the ^{31}P transitions of $(CH_3)_2NPF_4$ *20* to fluorine spin combinations

The triplet of triplet is caused by coupling of the phosphorous with the two equatorial fluorines ($J = 915$ Hz) and the two axial fluorines ($J = 778$ Hz). As shown in Fig. 8, the nine ^{31}P-NMR sites can be characterized by the 16 different spin combinations of the fluorine atoms. In each spin combination the first two spins correspond to the axial fluorines and the last two spins to the quatorial fluorines. Although an accurate characterization of the resonances requires the use of symmetrized spin functions, such as $1/\sqrt{2}\,\alpha\alpha(\beta\alpha + \alpha\beta)$ and $1/\sqrt{2}\,\alpha\alpha(\beta\alpha - \alpha\beta)$ instead of $\alpha\alpha\beta\alpha$ and $\alpha\alpha\alpha\beta$, the exchange relations of the resonances of first-order spectra can equally well be derived by using the simplified spin combination description. The analysis of ligand permutations for the polytopal rearrangement of compounds of type *20* has already been described in Section 6. It is evident that for such an achiral compound, one can at best differentiate by DNMR between two intramolecular rearrangement types, the first of which includes the modes M_2 and M_4 and the second M_5 and M_1. These two rearrangement types correspond to the one-pair fluorine exchange, *i.e.* permutations (13), (14), (23), and (24), and to the two-pair fluorine exchange, *i.e.* permutations (13)(24), and (14)(23).

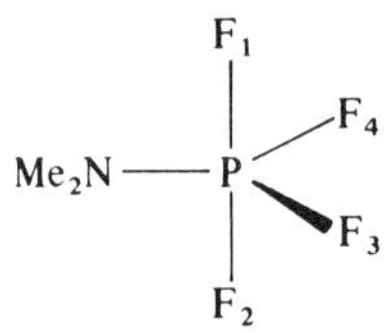

20

$$\begin{array}{c} \\ \text{IV} \\ \text{V} \\ \text{VI} \end{array}\begin{array}{c} \begin{array}{ccc} \text{IV} & \text{V} & \text{VI} \end{array} \\ \begin{pmatrix} -k & k & 0 \\ k/4 & -k/2 & k/4 \\ 0 & k & -k \end{pmatrix} \end{array} \qquad \begin{array}{c} \\ \text{IV} \\ \text{V} \\ \text{VI} \end{array}\begin{array}{c} \begin{array}{ccc} \text{IV} & \text{V} & \text{VI} \end{array} \\ \begin{pmatrix} -k & 0 & k \\ 0 & 0 & 0 \\ k & 0 & -k \end{pmatrix} \end{array}$$

XX XXI

Applying these permutations on the fluorine spin combinations characterizing the ^{31}P-sites (Fig. 8), it is easy to derive the site exchange probabilities for the two exchange types. For the three resonances in the center of the spectrum, *i.e.* sites IV, V, and VI, which are all charaterized by fluorine spin combinations of the total spin $F_z = 0$, the exchange relations for the one-pair and two-pair exchange are given by the corresponding kinetic exchange matrices XX and XXI. It is seen that there are distinct differences in the site exchange pattern. Thus site IV (spin combination $\alpha\alpha\beta\beta$) is transferred completely to site V (spin combinations $\beta\alpha\beta\alpha$, $\alpha\beta\beta\alpha$, $\beta\alpha\alpha\beta$, and $\alpha\beta\alpha\beta$) by the one-pair exchange, but to site VI (spin combination $\beta\beta\alpha\alpha$) by the two-pair exchange. Even more characteristic is the different behavior of the central site V, which is not affected at all by the two-pair fluorine rearrangement. Whitesides and co-workers [*66*] also included sites I, II, and III for the simulation of the experimental DNMR spectra. Figure 9 definitely proves the occurrence of the two-pair exchange, most distinctly seen in the lack of exchange of the resonance with the greatest intensity, *i.e.* site V.

Although this is in accordance with the Berry mechanism with the amine group as the pivot, the other permutations of the allowed coset of modes M_5 and M_1 (cf. Tables 2 and 3 in Section 6.) cannot be excluded by this DNMR study.

There is another important aspect in DNMR studies of this type, concerning the distinction between intra- and intermolecular rearrangement mechanisms. In the dis-

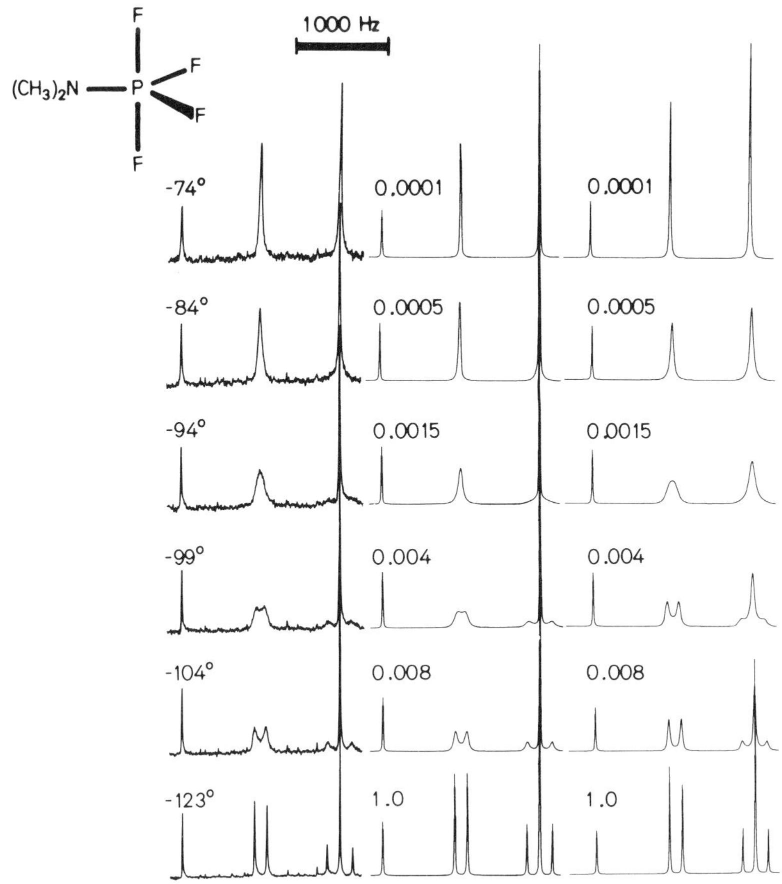

Fig. 9. Comparison of experimental and calculated ^{31}P-DNMR spectra of $(CH_3)_2NPF_4$ *20* (Whitesides *et al.* [66]).
Left column: Experimental spectra; *Middle column:* Spectra calculated for the two-pair exchange; *Right column:* Spectra calculated for the one-pair exchange

cussion above, we made use of factorization of the full 9-site problem according to the total spin F_z of the spin combinations. This was possible only because the rearrangement is of intramolecular nature, *i.e.* the total spin of any spin combination is not changed by the rearrangement. If the exchange process were an intermolecular one, this condition would not be satisfied and a total collapse of the nine sites to one line would occur in the fast exchange region, *i.e.* the ^{19}F-^{31}P coupling would not be preserved. Criteria of this type have been used as evidence for intermolecularity of the rearrangement of phosphoranes *21* and *22* [*67*]. Later investigations [*68, 69*] however showed that this behavior apparently is due to the reaction of the phosphoranes with the glass of the NMR tube: in teflon cells fluorine exchange is not accompanied by a collapse of the ^{19}F-^{31}P coupling, thus proving that the fluorophosphoranes *21* and *22* like $PF_4(NMe_2)$ are undergoing an intramolecular polytopal rearrangement.

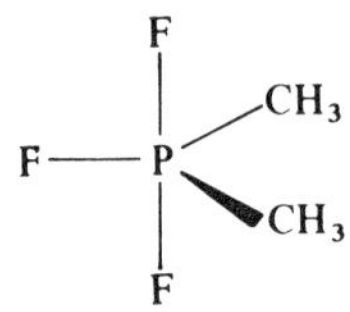

21

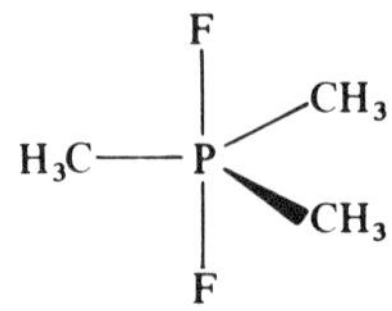

22

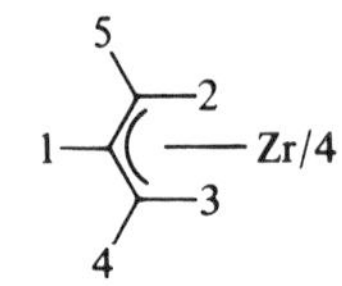

23

One final example of the class of DNMR problems, in which the sites are solely determined by spin-spin couplings, will be mentioned to show that this type of study is not restricted to fluorophosphoranes. The interchange of the syn (H_2 and H_3) and the anti allyl protons (H_4 and H_5) in tetraallylzirconium *23* has been studied by observing the temperature effect on the spectral pattern of the nonexchanging central proton H_1 [*28*]. To a first approximation by assuming first-order character the same kinetic exchange matrices as for *20* can be used to discriminate between the one-pair and two-pair rearrangement. It has been shown [*28*] that in this case the one-pair exchange occurs, which has been interpreted by a σ,π-mechanism.

8.2. Sites Determined by Spin-Spin Couplings and Chemical Shifts

In this section first-order cases are discussed in which the observed nuclei are involved in the exchange process. This class can be further divided into two groups of DNMR problems. In the first, the exchanging nuclei observed are not coupled to each other, but instead to nuclei not involved in the exchange process. The second group comprises cases in which the mututally exchanging nuclei are coupled to each other. Simple examples for both types have already been encountered in the DNMR studies involving prochiral groups. Thus the interchange of the methyls of prochiral isopropyl groups (Sections 5, 6, and 7) is a representative of the first class, while the interchange of the protons in the CH_2-CD_3 residue (Section 5) belongs to the second type, although it has pronounced non-first-order character. The feature of these two classes will be illustrated by analyzing three cases in order to facilitate the comparison of classical and quantum mechanical DNMR calculations (cf. next two sections).

The first example to discuss is an AMX $\rightleftharpoons$ XMA exchange problem with $J_{AX} = 0$. This situation is related to the exchange of the diastereotopic methyls (*i.e.* actually A_3 and X_3 spin groups instead of single spins A and X) of prochiral $CH(CH_3)_2$ groups (cf. Figs. 5, 6, and 7). One possibility to study the interchange of nuclei A and X is to observe the DNMR pattern of the nonexchanging nucleus M. This indirect method has already been described in the previous section. The other possibility of interest here is to observe the exchanging nuclei themselves. In Fig. 10 the resonances of the AX region are charaterized by the corresponding transitions of the spin functions.

As is seen in Fig. 10, the α and β spins of nucleus M of course lead to the splitting of the A and X resonances. In the case of intramolecular exchange the total spin F_z of the spin functions is not altered during the exchange process (cf. previous section). Thus, by interchanging the A and X labels of the spin functions, it is evident that the only site exchanges occurring are between sites I and III (F_z: $1 \rightarrow 0$)

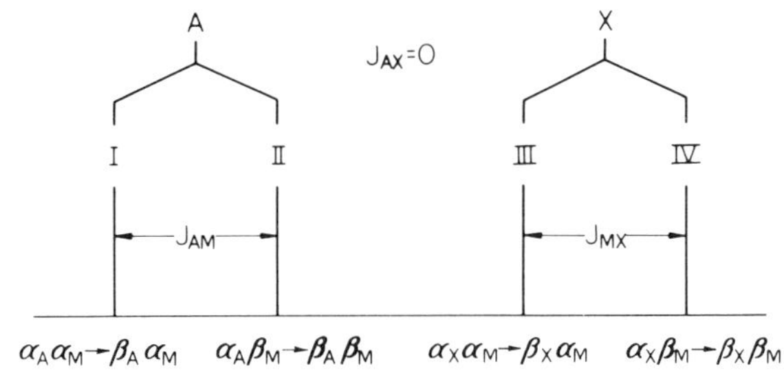

Fig. 10. Assignment of the transitions of the AX part of an AMX spectrum with $J_{AX} = 0$ to spin combinations

and between II and IV (F_z: $0 \rightarrow -1$). Theoretical band shapes, therefore, can be calculated either by using the kinetic exchange matrix XXII, or separately as two-site problems by using the trivial matrices XXIII and XXIV (cf. Section 5) and adding the two band shapes.

$$\begin{array}{c} \\ \text{I} \\ \text{II} \\ \text{III} \\ \text{IV} \end{array}\begin{array}{c} \begin{array}{cccc} \text{I} & \text{II} & \text{III} & \text{IV} \end{array} \\ \begin{pmatrix} -k & & k & \\ & -k & & k \\ k & & -k & \\ & k & & -k \end{pmatrix} \end{array} \qquad \begin{array}{c} \\ \text{I} \\ \text{III} \end{array}\begin{array}{c} \begin{array}{cc} \text{I} & \text{III} \end{array} \\ \begin{pmatrix} -k & k \\ k & -k \end{pmatrix} \end{array} \qquad \begin{array}{c} \\ \text{II} \\ \text{IV} \end{array}\begin{array}{c} \begin{array}{cc} \text{II} & \text{IV} \end{array} \\ \begin{pmatrix} -k & k \\ k & -k \end{pmatrix} \end{array}$$

$$\text{XXII} \qquad\qquad \text{XXIII} \qquad\qquad \text{XXIV}$$

As a simple example of the second type of problem (mutual exchange of coupled nuclei), we will discuss a first-order two-spin case, *i.e.* an AX ⇌ XA exchange problem. In contrast to the example above, the splitting of the resonances of nuclei A and X is now caused by the spins of A and X themselves (Fig. 11).

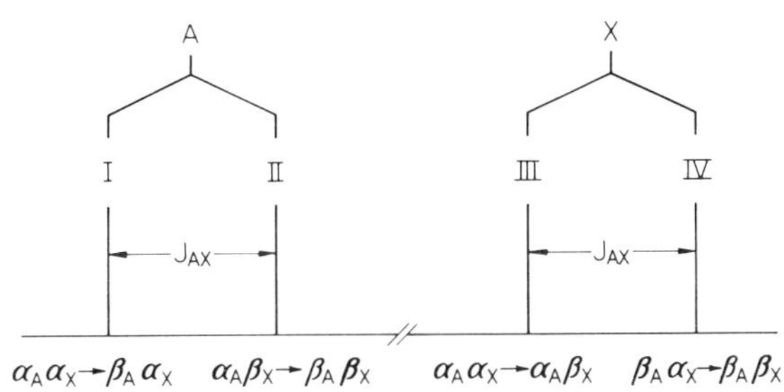

Fig. 11. Assignment of the transitions of an AX spectrum to spin combinations

Again the only site exchanges are between sites I and III and between II and IV, as seen by interchanging the A and X labels of the spins, or, describing it in another way, by interchanging the spins of each spin function, *e.g.* $\alpha\alpha \rightarrow \beta\alpha$ (site I) is transformed by the interchange into $\alpha\alpha \rightarrow \alpha\beta$ (site III). The latter simpler description of exchange will be used in the four-spin example discussed below.

A warning has to be given when applying the relations derived in the previous example to acutally occurring systems. Thus, even if the chemical shift difference of the two nuclei is very large compared to the coupling constant, the use of the kinetic exchange matrix XXII and of identical populations of the four sites will not allow a proper simulation of the fast exchange region. In fact, only the consideration of non-first-order character, small as it may be, can properly account for the fact that in the fast exchange region the spin-spin splitting is collapsing to yield finally a single line. This problem will be discussed in further detail in the next section.

The same limitation is encountered in the last example of this section, the exchange of nuclei A and X in an A_2X_2 system, which has been used in Section 3 to illustrate the character of the band shape equations. Nevertheless, the derivation of the exchange matrices II and III will be detailed, since it allows an interesting comparison with the exact quantum mechanical formulation. In Fig. 12. the sites of the A_2X_2 system are characterized by the various possible spin combinations, the first two spins of each combination corresponding to the A nuclei and the remaining two to the X nuclei.

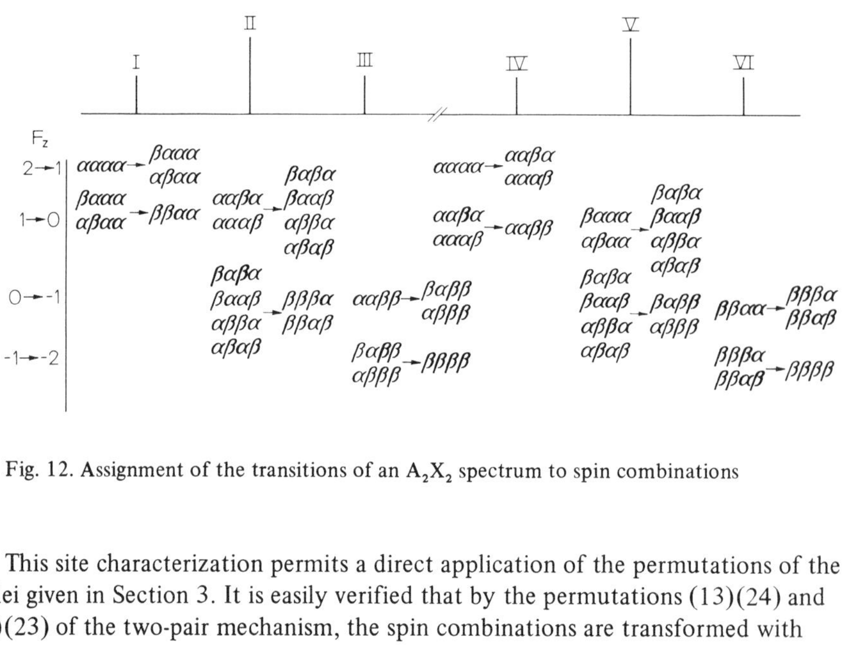

Fig. 12. Assignment of the transitions of an A_2X_2 spectrum to spin combinations

This site characterization permits a direct application of the permutations of the nuclei given in Section 3. It is easily verified that by the permutations (13)(24) and (14)(23) of the two-pair mechanism, the spin combinations are transformed with probability one to those which belong to the corresponding line of the other triplet. For instance, the transition F_z 1 → 0 of site I, *i.e.* $\beta\alpha\alpha\alpha$, $\alpha\beta\alpha\alpha \rightarrow \beta\beta\alpha\alpha$, is transformed by the two permutations into $\alpha\alpha\beta\alpha$, $\alpha\alpha\alpha\beta \rightarrow \alpha\alpha\beta\beta$ (site IV). On the other hand, the permutations (13), (14), (23), and (24) of the one-pair mechanism lead to very different exchange relations. In addition to transfers to the other triplet, the magnetization is also transferred to the sites of the same triplet. Taking again site I as an example, one sees that the F_z 2 → 1 transition is retained or transferred to site IV with the same probability corresponding to the site permutation (I)(I IV), while the magnetization of the 1 → 0 transition is transferred to site II and V, each transfer occurring with probability 1/2, which therefore corresponds to the site permutation (I II)(I V). The net result of these two permutations is that given in the first matrix row of matrix III (Section 3). In the same way the other site exchange relations are derived.

Although the exchange pattern is very different for the two mechanisms, the corresponding calculated spectra [*32*] only show slight differences, which can only be detected by carefully comparing the valley to height ratios of the triplet components for the same width of the total triplet. This comparison allows the conclusion that the one-pair mechanism leads to a more rapid collapse of the triplet structure, but the difference is too small to be of value in mechanistic studies of actual systems which could be treated as first-order problems for slow and moderate fast rates. It will be seen in Section 9.2. that in non-first-order systems (A_2B_2), in which the degeneracy of the individual transitions of the triplet lines is lifted, more distinct differences in DNMR band shapes are obtained.

9. Site Exchange in Non-First-Order Spin Systems

9.1. Mutual Exchange in a Two-Spin System

While the exchange-modified line shapes of first-order spectra may be calculated by the classical approach (see previous two sections), one must resort to quantum mechanical methods [*20, 21*] when dealing with non-first-order spectra, since strong couplings affect the simple multiplets of first-order spectra by changing the number, position, and intensity of the resonances. Before discussing the calculation of mechanistically interesting many-spin systems, we will use in this section the two-spin case $AB \rightleftharpoons BA$ to describe the feature of two quantum mechanically based methods, the representation of basis functions and the representation of eigenfunctions, the band shape equations of which were given in a formal way in Section 2.2. [Eqs. (3) and (4), respectively].

First we will calculate the static NMR spectrum of an AB system. The Hamilton operator of interest here for the calculation of the energy of a spin system in a magnetic field is given by

$$\mathcal{H} = -\sum_i \nu_i \mathfrak{J}_z(i) + \sum_{i<j}\sum J_{ij}\mathfrak{J}(i)\mathfrak{J}(j)$$

Besides the resonance frequencies ν and coupling constants J of the nuclei, this operator contains the nuclear spin operators $\mathfrak{J}_z$ and $\mathfrak{J}$. When applied on the spin product functions $\alpha\alpha$, $\beta\alpha$, $\alpha\beta$, and $\beta\beta$, the Hamilton matrix XXV of an AB system is obtained.

$$\begin{array}{c} \\ \alpha\alpha \\ \beta\alpha \\ \alpha\beta \\ \beta\beta \end{array} \begin{array}{c} \begin{array}{cccc} \alpha\alpha & \beta\alpha & \alpha\beta & \beta\beta \end{array} \\ \begin{pmatrix} -\nu_A/2-\nu_B/2+J/4 & & & \\ & \nu_A/2-\nu_B/2-J/4 & J/2 & \\ & J/2 & -\nu_A/2+\nu_B/2-J/4 & \\ & & & \nu_A/2+\nu_B/2+J/4 \end{pmatrix} \end{array}$$

XXV

To discuss a specific example, we assume the chemical shifts to be $\nu_A = 20$ Hz and $\nu_B = 40$ Hz and the coupling constant to be $J_{AB} = 10$ Hz (Fig. 13).

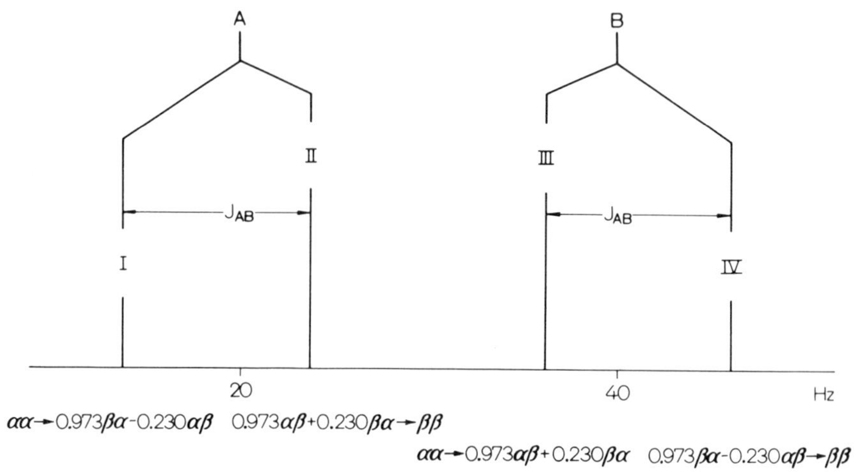

Fig. 13. Characterization of an AB spectrum with the parameters $\nu_A = 20$ Hz, $\nu_B = 40$ Hz, and $J_{AB} = 10$ Hz

For this case the Hamilton matrix XXV takes the form XXVI and the eigenvalues E_1–E_4 and eigenfunctions ψ_1–ψ_4 are easily obtained.

$$\begin{pmatrix} -27.5 & & & \\ & -12.5 & 5 & \\ & 5 & 7.5 & \\ & & & 32.5 \end{pmatrix} \qquad \begin{aligned} E_1 &= -27.5 \\ E_2 &= -13.680 \\ E_3 &= 8.680 \\ E_4 &= 32.5 \end{aligned} \qquad \begin{aligned} \psi_1 &= \alpha\alpha \\ \psi_2 &= 0.973\,\beta\alpha - 0.230\,\alpha\beta \\ \psi_3 &= 0.230\,\beta\alpha + 0.973\,\alpha\beta \\ \psi_4 &= \beta\beta \end{aligned}$$

XXVI

Thus the sites I–IV in Fig. 13 are characterized as

Site I	$\psi_1 \to \psi_2$	$E_2 - E_1 = 13.82$	$<\psi_2 \vert \mathfrak{J}^-(1) + \mathfrak{J}^-(2) \vert \psi_1>^2 = 0.5528$
Site II	$\psi_3 \to \psi_4$	$E_4 - E_3 = 23.82$	$<\psi_4 \vert \mathfrak{J}^-(1) + \mathfrak{J}^-(2) \vert \psi_3>^2 = 1.4472$
Site III	$\psi_1 \to \psi_3$	$E_3 - E_1 = 36.18$	$<\psi_3 \vert \mathfrak{J}^-(1) + \mathfrak{J}^-(2) \vert \psi_1>^2 = 1.4472$
Site IV	$\psi_2 \to \psi_4$	$E_4 - E_2 = 46.18$	$<\psi_4 \vert \mathfrak{J}^-(1) + \mathfrak{J}^-(2) \vert \psi_2>^2 = 0.5528$

It is evident that first-order character can be expected only if the chemical shift difference is very large compared to the coupling, causing the mixing of the $\alpha\beta$ and $\beta\alpha$ basis functions to become negligible. The frequencies would then be determined solely by the diagonal elements of the Hamilton matrix, *i.e.* $\nu_A \pm J/2$ and $\nu_B \pm J/2$, and the four transitions would have the same intensity.

As indicated in Section 2.2., there are two methods to calculate DNMR spectra of non-first-order spin systems. In the representation of basis functions, the band shape equation [Eq. (3), Section 2.2] does not specify the sites themselves and their exchange relations, but instead characterizes the primitive transitions between basis functions, the number of which is given by the possibilities of lowering the total spin F_z of the basis functions by one. For the AB case there are four possibilities, specified in the spin lowering matrix XXVII. In this case, of course, combination transitions (cf. Section 2.2.) cannot occur.

$$\begin{array}{c|cccc} & \alpha\alpha & \beta\alpha & \alpha\beta & \beta\beta \\ \hline \alpha\alpha & & & & \\ \beta\alpha & 1 & & & \\ \alpha\beta & 1 & & & \\ \beta\beta & & 1 & 1 & \end{array} \qquad \text{XXVII}$$

$$\begin{array}{c|cccc} & \alpha\alpha\to\beta\alpha & \alpha\beta\to\beta\beta & \alpha\alpha\to\alpha\beta & \beta\alpha\to\beta\beta \\ \hline \alpha\alpha\to\beta\alpha & \nu_A - J/2 & & J/2 & \\ \alpha\beta\to\beta\beta & & \nu_A + J/2 & & -J/2 \\ \alpha\alpha\to\alpha\beta & J/2 & & \nu_B - J/2 & \\ \beta\alpha\to\beta\beta & & -J/2 & & \nu_B + J/2 \end{array} \qquad \text{XXVIII}$$

As in the classical method (cf. Section 3.), matrix C of Eq. (3) (Section 2.2.) is constructed from a matrix which characterizes the static NMR spectrum, and an exchange matrix. The first matrix is the so-called Liouville matrix [*5*, *10*], specifying the primitive transitions. These transitions are connected by spin-spin couplings as seen in the Liouville matrix XXVIII for the AB case. The elements of this matrix are easily obtained from the corresponding elements of the Hamilton matrix XXV (cf. subroutine TRAMAT of the program DNMR 3 [*11*]). To derive, for instance, the first matrix row, the diagonal element 1,1 is given by $-H(1,1) + H(2,2)$ and the off-diagonal element 1,3 is equal to $H(2,3)$. The multiplication of the Liouville matrix by $2\pi i$ and subtraction of ωi and the natural line width $\pi W = 1/T_2$ from the diagonal elements yields the quantum mechanical counterpart of the classical diagonal matrix I (Section 3.), which allows the calculation of the static NMR spectrum. The exchange relations between the primitive transitions are derived in the same way as described in the previous section for first-order spin systems, *i.e.* by performing the corresponding permutations on the spins of the basis functions of a transition. Thus for the AB case, the kinetic exchange matrix is identical to matrix XXII, which in the previous section was derived for AX systems to allow classical type DNMR calculations for slow and moderately fast rates. For the specific AB example given above, the complete band shape equation in the representation of basis functions can now be specified as

$$G = -iC(1,1,1,1)\begin{pmatrix} -1-k+(30\pi-\omega)i & 0 & k+10\pi i & 0 \\ 0 & -1-k+(50\pi-\omega)i & 0 & k-10\pi i \\ k+10\pi i & 0 & -1-k+(70\pi-\omega)i & 0 \\ 0 & k-10\pi i & 0 & -1-k+(90\pi-\omega)i \end{pmatrix}^{-1}\begin{pmatrix}1\\1\\1\\1\end{pmatrix}$$

In this equation a natural line width πW of one is assigned to all transitions. The elements of the I_b^- vector, which are identical to those in the σ vector (mutual exchange; cf. Section 2.2.), were already specified in the corresponding spin lowering matrix XXVII. On the left side of Fig. 14 the calculated DNMR spectra using this band shape equation are depicted schematically by specifying the imaginary part of the spectral and the real part of the shape vector for the four exchange-modified transitions, which correspond to their frequency and intensity, respectively [*10, 11*]. The additional information inherent to these vectors—broadness and deviation from Lorentz shape—will not be considered here.

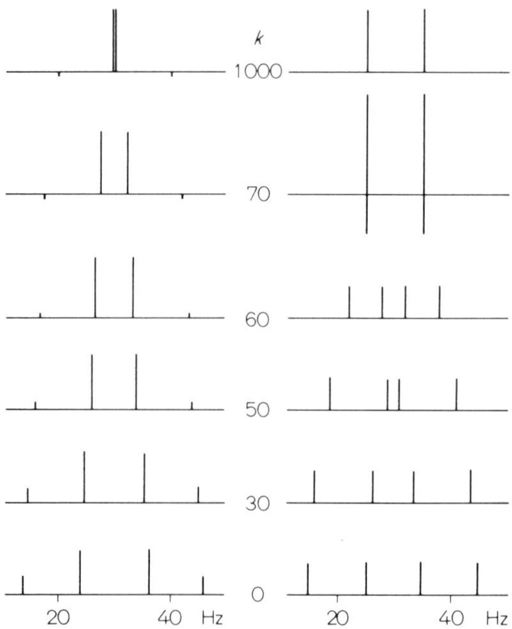

Fig. 14. Comparison of the schematic theoretical DNMR spectra of an AB system calculated by the exact band shape equation (*left side*) and by a band shape equation which neglects non-first-order character (*right side*)

It is seen that, as observed experimentally (cf. Fig. 4), the four sites collapse to a single line in the fast exchange region. This is not the case when the off-diagonal imaginary elements of the Liouville matrix (*i.e.* $10\pi i$) are omitted. This omission transforms the problem by force into a first-order case, *i.e.* into AX $\rightleftharpoons$ XA, and thus serves to illustrate the limitation encountered in classical approaches to this type of exchange problem (see previous section). As shown in the right side of Fig. 14, the spin-spin splitting now does not collapse in the fast exchange region; the crossover of the two inner sites occurs at k ca. 53.

Nevertheless there is a method [*25, 28*] which, in spite of the absence of off-diagonal imaginary matrix elements, properly accounts for non-first-order character. This method [cf. Eq. (4) in Section 2.2.] uses the representation of eigenfunctions instead of the representation of simple spin product functions and thus permits direct information on the exchange behavior of the sites. In this representation the equation characterizing the static AB spectrum is rapidly set up from the site frequencies and the square roots of the site intensities. To derive the kinetic exchange matrix, the basis function exchange matrix XXX must be transformed into the representation of eigenfunctions. For the AB example described above this transformation is effected by use of the coefficient matrix XXIX of the eigenfunctions and the corresponding inverse matrix XXXI.

$$\underset{\text{XXIX}}{\begin{pmatrix} 1 & & & \\ & 0.973 & -0.230 & \\ & 0.230 & 0.973 & \\ & & & 1 \end{pmatrix}} \quad \underset{\text{XXX}}{\begin{array}{c} \\ \alpha\alpha \\ \beta\alpha \\ \alpha\beta \\ \beta\beta \end{array}\begin{array}{c} \begin{array}{cccc} \alpha\alpha & \beta\alpha & \alpha\beta & \beta\beta \end{array} \\ \begin{pmatrix} 1 & & & \\ & & 1 & \\ & 1 & & \\ & & & 1 \end{pmatrix} \end{array}} \quad \underset{\text{XXXI}}{\begin{pmatrix} 1 & & & \\ & 0.973 & 0.230 & \\ & -0.230 & 0.973 & \\ & & & 1 \end{pmatrix}}$$

Thus by multiplication of matrix XXIX with XXX and with the inverse coefficient matrix XXXI, matrix XXXII is obtained, which characterizes the exchange between the eigenfunctions by the permutation of nuclei A and B. For instance ψ_2, *i.e.* $0.973\ \beta\alpha - 0.230\ \alpha\beta$, is transformed by the permutation into $-0.447\ \psi_2 + 0.894\ \psi_3$, which corresponds to $-0.230\ \beta\alpha + 0.973\ \alpha\beta$. The site exchange relations, given in the kinetic exchange matrix XXXIII, are derived by multiplying the respective numbers of the eigenfunction exchange matrix XXXII. We will, for instance, derive the site exchange behavior of site I, *i.e.* the transition $\psi_1 \rightarrow \psi_2$.

$$\underset{\text{XXXII}}{\begin{array}{c} \\ \psi_1 \\ \psi_2 \\ \psi_3 \\ \psi_4 \end{array}\begin{array}{c} \begin{array}{cccc} \psi_1 & \psi_2 & \psi_3 & \psi_4 \end{array} \\ \begin{pmatrix} 1 & & & \\ & -0.447 & 0.894 & \\ & 0.894 & 0.447 & \\ & & & 1 \end{pmatrix} \end{array}} \qquad \underset{\text{XXXIII}}{\begin{array}{c} \\ \text{I} \\ \text{II} \\ \text{III} \\ \text{IV} \end{array}\begin{array}{c} \begin{array}{cccc} \text{I} & \text{II} & \text{III} & \text{IV} \end{array} \\ \begin{pmatrix} -1.447k & 0 & 0.894k & 0 \\ 0 & -0.553k & 0 & 0.894k \\ 0.894k & 0 & -0.553k & 0 \\ 0 & 0.894k & 0 & -1.447k \end{pmatrix} \end{array}}$$

It is seen from matrix XXXII that site I only exchanges with site III ($\psi_1 \rightarrow \psi_3$). The corresponding coefficient 0.894 of element 1,3 in the kinetic exchange matrix XXXIII is obtained by multiplying 1 (transfer of $\psi_1 = \alpha\alpha$ to $\psi_1 = \alpha\alpha$ by permutation of the two spins) by 0.894 (transfer of $\psi_2 = 0.973\ \beta\alpha - 0.230\ \alpha\beta$ to $\psi_3 = 0.230\ \beta\alpha + 0.973\ \alpha\beta$). The corresponding diagonal coefficient -1.447 is calculated by summing -1 and the product of 1 and -0.447 (cf. elements 1,1 and 2,2 of matrix XXXII). Compared to the corresponding diagonal coefficient -1 in the kinetic exchange matrix of the representation of basis functions, this more negative value arises from the negative element 2,2 in the eigenfunction exchange matrix XXXII which is zero in the basis function exchange matrix XXX.

There is an interesting relation between the diagonal coefficients of the kinetic exchange matrix XXXIII and the intensities of the sites, which were given previously. It is seen that the coefficient for any site is identical to the intensity of the site to which the site considered is transferred by exchange, implying that the smaller outer lines of the AB spectrum are exchanging more rapidly than the inner lines, which indeed is confirmed by the DNMR spectra (see Figs. 4 and 14). This situation resembles the relationship between first-order rate constants and populations ($k_1p_1 = k_2p_2$) for systems with two isomers or compounds in dynamic equilibrium. Thus the diagonal elements of matrix XXXIII could be thought of as being first-order rate constants, *e.g.* $1.447\ k = k_{\text{I}}$. It must be emphazised however, that in contrast to classical DNMR formulations, the coefficients no longer represent exchange probabilities, not only because of the larger numbers than one, but also because the sum of coefficients in a matrix row is unequal to zero. Furthermore, as will be seen in the next section, even negative off-diagonal numbers are possible for more complicated non-first-order exchange systems.

With the kinetic exchange matrix XXXIII the complete band shape equation for our AB example in the representation of eigenfunctions can now be specified as

$$G = -\mathrm{i}CI_\mathrm{e}^- \begin{pmatrix} -1-1.447\,k + (27.64\,\pi - \omega)\mathrm{i} & 0 & 0.894\,k & 0 \\ 0 & -1-0.553\,k + (47.64\,\pi - \omega)\mathrm{i} & 0 & 0.894\,k \\ 0.894\,k & 0 & -1-0.553\,k + (72.36\,\pi - \omega)\mathrm{i} & 0 \\ 0 & 0.894\,k & 0 & -1-1.447\,k + (92.36\,\pi - \omega)\mathrm{i} \end{pmatrix}^{-1} I_\mathrm{e}^-$$

As mentioned earlier, the I_e^- vectors contain the square root of the intensities of the corresponding sites, *i.e.* the elements 0.744, 1.203, 1.203, and 0.744. These elements can also be obtained by transformation of the basis function spin lowering matrix XXVII into the representation of eigenfunctions, *i.e.* by the operation $CI_\mathrm{b}^- C^{-1}$, where C represents the coefficient matrix of the eigenfunctions XXIX. The DNMR spectra calculated with this equation by using the subroutines ALLMAT, NVRT, and CONVEC of DNMR 3 [*11*] are identical with those schematically depicted on the left side of Fig. 14, *i.e.* the numerical values of the corresponding shape and spectral vectors are identical.

From the mathematical procedure of the method using the representation of eigenfunctions it is clear that the less non-first-order character the two-spin system has, the more the band shape equation will approach the form of the approximate equation obtained from the band shape equation in the representation of basis functions (see above) by omitting the off-diagonal imaginary matrix elements. But even very small deviations from first-order character are necessary to account properly also for the fast exchange limit, *i.e.* for the collapse of the coupling (cf. Fig. 14).

9.2. Mechanistic Studies of Non-First-Order Spin Systems

In the previous section the two-spin case was used to illustrate the feature of two quantum mechanical methods based on the representation of basis functions and of eigenfunctions. Mechanistic implications, *i.e.* elucidation of permutational schemes, of course can only be derived for many-spin systems. In this section we will discuss the site exchange pattern of A_2B_2 systems for two exchange types of the A and B nuclei, in order to allow a comparison with the first-order limit (cf. Section 8.2.) and to illustrate the mechanistic study on SF_4 by Klemperer *et al.* [*25*].

Since in the A_2B_2 case the nuclei A and nuclei B are magnetically equivalent, it is sufficient to use only one coupling constant, J_{AB}, to set up the Hamilton matrix.

Factorizing this matrix with respect to the total spin F_z of the basis functions, the five submatrices XXXIV–XXXVIII are easily derived (cf. previous section).

In the basis representation our A_2B_2 problem is characterized by a 56 x 56 matrix, since the order of matrix C [Eq. (3) in Section 2.2.] is given by the number of transitions between all those pairs of basis functions in which the total spin is reduced by one (cf. previous section). With the help of F_z factorization this matrix can be split into four submatrices of the order 4 x 4, 24 x 24, 24 x 24, and 4 x 4, where the larger submatrices also contain combination transitions ($I_b^- = 0$), in which simultaneously the spin of three nuclei is changed (*e.g.* $\alpha\alpha\beta\alpha \rightarrow \beta\beta\alpha\alpha$). A further factorization by use of magnetic equivalence is not of interest here, since it does not allow a treatment of the one-pair exchange mechanism. The set-up of the static Liouville matrices and the corresponding kinetic exchange matrices is performed automatically by programs such as DNMR 3 [*11*] and the spectrum is calculated by summing the band shape contributions of each submatrix.

$$\begin{array}{cc} & \alpha\alpha\alpha\alpha \\ \alpha\alpha\alpha\alpha & (-\nu_A - \nu_B + J) \end{array}$$

XXXIV

$$\begin{array}{c} \\ \beta\alpha\alpha\alpha \\ \alpha\beta\alpha\alpha \\ \alpha\alpha\beta\alpha \\ \alpha\alpha\alpha\beta \end{array} \begin{array}{c} \begin{array}{cccc} \beta\alpha\alpha\alpha & \alpha\beta\alpha\alpha & \alpha\alpha\beta\alpha & \alpha\alpha\alpha\beta \end{array} \\ \begin{pmatrix} -\nu_B & 0 & J/2 & J/2 \\ 0 & -\nu_B & J/2 & J/2 \\ J/2 & J/2 & -\nu_A & 0 \\ J/2 & J/2 & 0 & -\nu_A \end{pmatrix} \end{array}$$

XXXV

$$\begin{array}{c} \\ \beta\beta\alpha\alpha \\ \beta\alpha\beta\alpha \\ \beta\alpha\alpha\beta \\ \alpha\beta\beta\alpha \\ \alpha\beta\alpha\beta \\ \alpha\alpha\beta\beta \end{array} \begin{array}{c} \begin{array}{cccccc} \beta\beta\alpha\alpha & \beta\alpha\beta\alpha & \beta\alpha\alpha\beta & \alpha\beta\beta\alpha & \alpha\beta\alpha\beta & \alpha\alpha\beta\beta \end{array} \\ \begin{pmatrix} \nu_A - \nu_B - J & J/2 & J/2 & J/2 & J/2 & 0 \\ J/2 & 0 & 0 & 0 & 0 & J/2 \\ J/2 & 0 & 0 & 0 & 0 & J/2 \\ J/2 & 0 & 0 & 0 & 0 & J/2 \\ J/2 & 0 & 0 & 0 & 0 & J/2 \\ 0 & J/2 & J/2 & J/2 & J/2 & -\nu_A + \nu_B - J \end{pmatrix} \end{array}$$

XXXVI

$$\begin{array}{c} \\ \beta\beta\beta\alpha \\ \beta\beta\alpha\beta \\ \beta\alpha\beta\beta \\ \alpha\beta\beta\beta \end{array} \begin{array}{c} \begin{array}{cccc} \beta\beta\beta\alpha & \beta\beta\alpha\beta & \beta\alpha\beta\beta & \alpha\beta\beta\beta \end{array} \\ \begin{pmatrix} \nu_A & 0 & J/2 & J/2 \\ 0 & \nu_A & J/2 & J/2 \\ J/2 & J/2 & \nu_B & 0 \\ J/2 & J/2 & 0 & \nu_B \end{pmatrix} \end{array}$$

XXXVII

$$\begin{array}{cc} & \beta\beta\beta\beta \\ \beta\beta\beta\beta & (\nu_A + \nu_B + J) \end{array}$$

XXXVIII

A discussion of the kinetic exchange matrices for the one-pair and two-pair exchange mechanisms in the basis representation is of no help here, firstly because of the large order of the matrices and secondly because they do not reveal the consequences of non-first-order character. Since the spin combinations, used in Section 8.2. to characterize the sites of A_2X_2 spectra, can be thought to represent basis product functions, the effect of the permutations of the two mechanisms on the primitive transitions between basis functions is the same as was described previously.

The analysis of the consequences of strong couplings on the exchange behavior of the lines is more illustrative in the representation of eigenfunctions. Using this method we will detail the exchange relations of the transitions of the A_2X_2 and A_2B_2 cases for the one-pair and two-pair mechanisms described in Section 3. For convenience we restrict the discussion to the transitions corresponding to the change of the total spin F_z from 2 to 1 and from 1 to 0, since the remaining transitions are given by symmetry.

In the hypothetical limit case A_2X_2, the off-diagonal elements ($J/2$) of the Hamilton submatrices XXXIV–XXXVIII can be discarded. Thus the eigenvalues E_1–E_{11} of the first three matrices are simply given by the diagonal elements and the corresponding eigenfunctions are the symmetrized wave functions ψ_1–ψ_{11} *[28]*.

$$E_1 = -\nu_A - \nu_B + J \qquad \psi_1 = \alpha\alpha\alpha\alpha$$

$$\begin{aligned}
E_2 &= -\nu_B & \psi_2 &= 1/2\,(\beta\alpha\alpha\alpha + \alpha\beta\alpha\alpha)\\
E_3 &= -\nu_B & \psi_3 &= 1/2\,(\beta\alpha\alpha\alpha - \alpha\beta\alpha\alpha)\\
E_4 &= -\nu_A & \psi_4 &= 1/2\,(\alpha\alpha\beta\alpha + \alpha\alpha\alpha\beta)\\
E_5 &= -\nu_A & \psi_5 &= 1/2\,(\alpha\alpha\beta\alpha - \alpha\alpha\alpha\beta)
\end{aligned}$$

$$\begin{aligned}
E_6 &= \nu_A - \nu_B - J & \psi_6 &= \beta\beta\alpha\alpha\\
E_7 &= 0 & \psi_7 &= 1/2\,(\beta\alpha\beta\alpha + \beta\alpha\alpha\beta + \alpha\beta\beta\alpha + \alpha\beta\alpha\beta)\\
E_8 &= 0 & \psi_8 &= 1/2\,(\beta\alpha\beta\alpha + \beta\alpha\alpha\beta - \alpha\beta\beta\alpha - \alpha\beta\alpha\beta)\\
E_9 &= 0 & \psi_9 &= 1/2\,(\beta\alpha\beta\alpha - \beta\alpha\alpha\beta + \alpha\beta\beta\alpha - \alpha\beta\alpha\beta)\\
E_{10} &= 0 & \psi_{10} &= 1/2\,(\beta\alpha\beta\alpha - \beta\alpha\alpha\beta - \alpha\beta\beta\alpha + \alpha\beta\alpha\beta)\\
E_{11} &= 0 & \psi_{11} &= \alpha\alpha\beta\beta
\end{aligned}$$

The allowed transitions between the eigenstates are derived by transforming the basis function spin lowering matrix I_b^- (cf. matrix XXVII of the previous section) into the representation of eigenfunctions, *i.e.* by calculating the eigenfunction spin lowering matrix $I_e^- = CI_b^- C^{-1}$, where C is the coefficient matrix of the eigenfunctions. As seen below, there are two transitions (I_1 and IV_1) corresponding to the change of the total spin F_z from 2 to 1, and six transitions (I_2, II_2, II_2^a, IV_2, V_2^a, and V_2) in which the eigenstates with $F_z = 1$ are transferred to those with $F_z = 0$. The labeling of the sites has been chosen to be in accordance with that given previously (cf. Figs. 1 and 12). The subscripts indicate the submatrix to which they belong and the label a stands for antisymmetry.

Site I_1	$\psi_1 \rightarrow \psi_2$	$E_2 - E_1 = \nu_A - J$	$I_e^- = 1.414$
Site IV_1	$\psi_1 \rightarrow \psi_4$	$E_4 - E_1 = \nu_B - J$	$I_e^- = 1.414$

Site I_2	$\psi_2 \rightarrow \psi_6$	$E_6 - E_2 = \nu_A - J$	$I_e^- = 1.414$
Site II_2	$\psi_4 \rightarrow \psi_7$	$E_7 - E_4 = \nu_A$	$I_e^- = 1.414$
Site II_2^a	$\psi_5 \rightarrow \psi_9$	$E_9 - E_5 = \nu_A$	$I_e^- = 1.414$
Site IV_2	$\psi_4 \rightarrow \psi_{11}$	$E_{11} - E_4 = \nu_B - J$	$I_e^- = 1.414$
Site V_2^a	$\psi_3 \rightarrow \psi_8$	$E_8 - E_3 = \nu_B$	$I_e^- = 1.414$
Site V_2	$\psi_2 \rightarrow \psi_7$	$E_7 - E_2 = \nu_B$	$I_e^- = 1.414$

It is seen that the eight transitions all have the same intensity, which is the square of the I_e^- value. When combined with the transitions of the two other submatrices not considered here, *i.e.* II_3, II_3^a, III_3, V_3^a, V_3, VI_3, III_4, and VI_4, the two triplets of the A_2X_2 spectrum are obtained.

For a representative A_2B_2 example we choose one of the cases given in the book of Wiberg and Nist with the parameters $\nu_A = 97$ Hz, $\nu_B = 103$ Hz and $J_{AB} = 2$ Hz [*70*]. The eigenvalues and eigenfunctions of the Hamilton submatrices XXXIV–XXXVI for this case are the following

$E_1 = -198$	$\psi_1 = \alpha\alpha\alpha\alpha$
$E_2 = -103.61$	$\psi_2 = 0.677\ (\beta\alpha\alpha\alpha + \alpha\beta\alpha\alpha) - 0.205\ (\alpha\alpha\beta\alpha + \alpha\alpha\alpha\beta)$
$E_3 = -103$	$\psi_3 = 1/2\ (\beta\alpha\alpha\alpha - \alpha\beta\alpha\alpha)$
$E_4 = -96.39$	$\psi_4 = 0.205\ (\beta\alpha\alpha\alpha + \alpha\beta\alpha\alpha) + 0.677\ (\alpha\alpha\beta\alpha + \alpha\alpha\alpha\beta)$
$E_5 = -97$	$\psi_5 = 1/2\ (\alpha\alpha\beta\alpha - \alpha\alpha\alpha\beta)$
$E_6 = -8.49$	$\psi_6 = 0.971\ \beta\beta\alpha\alpha - 0.119\ (\beta\alpha\beta\alpha + \beta\alpha\alpha\beta + \alpha\beta\beta\alpha + \alpha\beta\alpha\beta) + 0.038\ \alpha\alpha\beta\beta$
$E_7 = -0.39$	$\psi_7 = 0.233\ \beta\beta\alpha\alpha + 0.443\ (\beta\alpha\beta\alpha + \beta\alpha\alpha\beta + \alpha\beta\beta\alpha + \alpha\beta\alpha\beta) - 0.404\ \alpha\alpha\beta\beta$
$E_8 = 0$	$\psi_8 = 1/2\ (\beta\alpha\beta\alpha + \beta\alpha\alpha\beta - \alpha\beta\beta\alpha - \alpha\beta\alpha\beta)$
$E_9 = 0$	$\psi_9 = 1/2\ (\beta\alpha\beta\alpha - \beta\alpha\alpha\beta + \alpha\beta\beta\alpha - \alpha\beta\alpha\beta)$
$E_{10} = 0$	$\psi_{10} = 1/2\ (\beta\alpha\beta\alpha - \beta\alpha\alpha\beta - \alpha\beta\beta\alpha + \alpha\beta\alpha\beta)$
$E_{11} = 4.88$	$\psi_{11} = 0.062\ \beta\beta\alpha\alpha + 0.200\ (\beta\alpha\beta\alpha + \beta\alpha\alpha\beta + \alpha\beta\beta\alpha + \alpha\beta\alpha\beta) + 0.914\ \alpha\alpha\beta\beta$

For this distinct non-first-order case, in addition to the eight transitions of the first two submatrices, corresponding to those of the A_2X_2 case, two combination transitions ($\psi_2 \rightarrow \psi_{11}$ and $\psi_4 \rightarrow \psi_6$) occur in the second submatrix, which were of zero intensity in the first-order case. The frequencies and the square roots of the intensities of the transitions are given as following:

Site I_1	$\psi_1 \rightarrow \psi_2$	$E_2 - E_1 = 94.39$	$I_e^- = 0.944$
Site IV_1	$\psi_1 \rightarrow \psi_4$	$E_4 - E_1 = 101.61$	$I_e^- = 1.764$
	$\psi_4 \rightarrow \psi_6$	$E_6 - E_4 = 87.90$	$I_e^- = 0.030$
Site I_2	$\psi_2 \rightarrow \psi_6$	$E_6 - E_2 = 95.12$	$I_e^- = 1.074$
Site II_2	$\psi_4 \rightarrow \psi_7$	$E_7 - E_4 = 96.00$	$I_e^- = 1.111$
Site II_2^a	$\psi_5 \rightarrow \psi_9$	$E_9 - E_5 = 97.00$	$I_e^- = 1.414$
Site IV_2	$\psi_4 \rightarrow \psi_{11}$	$E_{11} - E_4 = 101.27$	$I_e^- = 1.969$
Site V_2^a	$\psi_3 \rightarrow \psi_8$	$E_8 - E_3 = 103.00$	$I_e^- = 1.414$
Site V_2	$\psi_2 \rightarrow \psi_7$	$E_7 - E_2 = 103.22$	$I_e^- = 1.318$
	$\psi_2 \rightarrow \psi_{11}$	$E_{11} - E_2 = 108.49$	$I_e^- = 0.087$

As expected the combination transitions are of very small intensity (0.0009 and 0.0076). Therefore it is justified to discard them in the derivation of the kinetic exchange matrices. From Fig. 15 and the site frequencies given above, it is seen that the degeneracies of the A_2X_2 case are now removed.

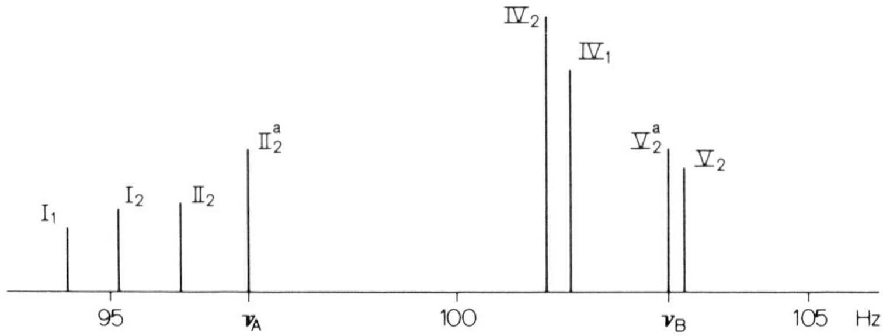

Fig. 15. Transitions corresponding to the change of total spin $F_Z = 2$ to $F_Z = 1$ and $F_Z = 1$ to $F_Z = 0$ for an A_2B_2 spin system with the parameters $\nu_A = 97$ Hz, $\nu_B = 103$ Hz and $J_{AB} = 2$ Hz

Together with the unspecified transitions of the third and fourth submatrix, the sites of the complete A_2B_2 spectrum ordered by increasing frequency are given as I_1, I_2, II_2, II_3, II_2^a, II_3^a, III_4, III_3, IV_2, IV_1, V_2^a, V_3^a, V_2, V_3, VI_3, and VI_4. The only degeneracies still occurring are between the antisymmetrical transitions II_2^a and II_3^a, and between V_2^a and V_3^a.

With the given eigenfunctions and allowed transitions, the prerequisites for the derivation of the kinetic exchange matrices are provided. Since the mathematical procedure [*25, 28*] has been described in the previous section, we will not give details of the calculation here. Although the basis function exchange matrices (cf. matrix XXX for the two-spin case) as well as the eigenfunction exchange matrices (cf. matrix XXXII) are all different for each permutation of the exchange mechanism considered, the kinetic exchange matrices derived for each permutation of the mechanism are identical.

For the first submatrix, *i.e.* transitions I_1 and IV_1, the exchange relations for the A_2X_2 case are given by the kinetic exchange matrices XXXIX (two-pair mechanism) and XL (one-pair mechanism), and the corresponding matrices for our A_2B_2 example are XLI and XLII, respectively.

$$\begin{array}{cc} & \begin{array}{cc} I_1 & IV_1 \end{array} \\ \begin{array}{c} I_1 \\ IV_1 \end{array} & \begin{pmatrix} -k & k \\ k & -k \end{pmatrix} \end{array}$$

XXXIX

$$\begin{array}{cc} & \begin{array}{cc} I_1 & IV_1 \end{array} \\ \begin{array}{c} I_1 \\ IV_1 \end{array} & \begin{pmatrix} -k/2 & k/2 \\ k/2 & -k/2 \end{pmatrix} \end{array}$$

XL

$$\begin{array}{cc} & \begin{array}{cc} I_1 & IV_1 \end{array} \\ \begin{array}{c} I_1 \\ IV_1 \end{array} & \begin{pmatrix} -1.56\,k & 0.83\,k \\ 0.83\,k & -0.44\,k \end{pmatrix} \end{array}$$

XLI

$$\begin{array}{cc} & \begin{array}{cc} I_1 & IV_1 \end{array} \\ \begin{array}{c} I_1 \\ IV_1 \end{array} & \begin{pmatrix} -0.78k & 0.42k \\ 0.42k & -0.22k \end{pmatrix} \end{array}$$

XLII

The matrices resemble the submatrices of the kinetic exchange matrices XXII (AX case) and XXXIII (AB case). Again the diagonal elements contain the intensities (here in fact 1/2 and 1/4 of the square of I_e^-) of the transitions reached by exchange. The exchange behavior of sites I_1 and IV_1 alone cannot be used to differentiate between the two- and one-pair mechanism, since the exchange rate of each of the two sites differs by a factor of two for the two mechanisms. Thus, using 2 k instead of k in the one-pair exchange matrices XL and XLII, the same DNMR band shapes are obtained for the two lines as with the matrices XXXIX and XLI.

However, the different exchange rates of sites I_1 and IV_1 for the two mechanisms do allow mechanistic implications when compared with the exchange behavior of the transitions of the second submatrix. While the two-pair mechanism leads to the kinetic exchange matrices XLIII in the A_2X_2 case and to XLV in the A_2B_2 case, the corresponding matrices for the one-pair mechanism are XLIV and XLVI, respectively.

	I_2	II_2	II_2^a	IV_2	V_2^a	V_2
I_2	$-k$			k		
II_2		$-k$				k
II_2^a			$-k$		k	
IV_2	k			$-k$		
V_2^a			k		k	
V_2		k				$-k$

XLIII

	I_2	II_2	II_2^a	IV_2	V_2^a	V_2
I_2	$-k$	$.25k$	$.25k$		$.25k$	$.25k$
II_2	$.25k$	$-.75k$		$.25k$		$.25k$
II_2^a	$.25k$		$-.75k$	$.25k$	$.25k$	
IV_2		$.25k$	$.25k$	$-k$	$25k$	$.25k$
V_2^a	$.25k$		$.25k$	$.25k$	$-.75k$	
V_2	$.25k$	$.25k$		$.25k$		$-.75k$

XLIV

	I_2	II_2	II_2^a	IV_2	V_2^a	V_2
I_2	$-1.07k$	$-.49k$		$.66k$		$.33k$
II_2	$-.49k$	$-.67k$		$.30k$		$.50k$
II_2^a			$-k$		k	
IV_2	$.66k$	$.30k$		$-.85k$		$.45k$
V_2^a			k		$-k$	
V_2	$.33k$	$.50k$		$.45k$		$-1.33k$

XLV

	I_2	II_2	II_2^a	IV_2	V_2^a	V_2
I_2	$-1.05k$	$.15k$	$.29k$	$.02k$	$.29k$	$.08k$
II_2	$.15k$	$-.81k$	$-.11k$	$.45k$	$-.11k$	$.10k$
II_2^a	$.29k$	$-.11k$	$-.75k$	$.14k$	$.25k$	$.20k$
IV_2	$.02k$	$.45k$	$.14k$	$-.64k$	$.14k$	$.24k$
V_2^a	$.29k$	$-.11k$	$.25k$	$.14k$	$-.75k$	$.20k$
V_2	$.08k$	$.10k$	$.20k$	$.24k$	$.20k$	$-.95k$

XLVI

Since, as mentioned in the previous section, the coefficients of the kinetic exchange matrices of non-first-order spin systems do not represent exchange probabilites, one has not to worry about the minus sign in some off-diagonal elements of the A_2B_2 matrices XLV and XLVI. The AB case, discussed in the previous section, will be used to illustrate why negative off-diagonal elements may occur. Using in this case $-\psi_3$ as eigenfunction instead of ψ_3, all off-diagonal coefficients (0.894) in the kinetic exchange matrix XXXIII will become negative, but two elements of the I_e^- vector will also become negative. Thus as required, the calculated band shapes are independent from multiplication of the eigenfunctions by minus one. In contrast to the AB case, however, it is not possible here to obtain all off-diagonal elements as positive numbers by multiplying some eigenfunctions by minus one.

For our goal to differentiate between the one-pair and two-pair mechanism, we will use only the diagonal elements of the kinetic exchange matrices. There are three major characteristic differences in the exchange behavior of the sites for the two mechanisms. The first one is the faster exchange of site V_2 in the two-pair-mechanism than in the one-pair mechanism, as seen by comparing matrices XLV and XLVI. The other two differences become obvious by comparing the matrices XLV and XLVI with the matrices XLI and XLII. Thus, while site I_1 is exchanging more rapidly than site I_2 in the two-pair mechanism, just the reverse is true for the one-pair mechanism. The third clear difference is seen for the sites IV_1 and IV_2, for which the differential exchange rate is more distinct for the one-pair mechanism, the exchange of site IV_2 being almost three times faster than that of site IV_1, compared to a factor of only 1.9 for the two-pair mechanism.

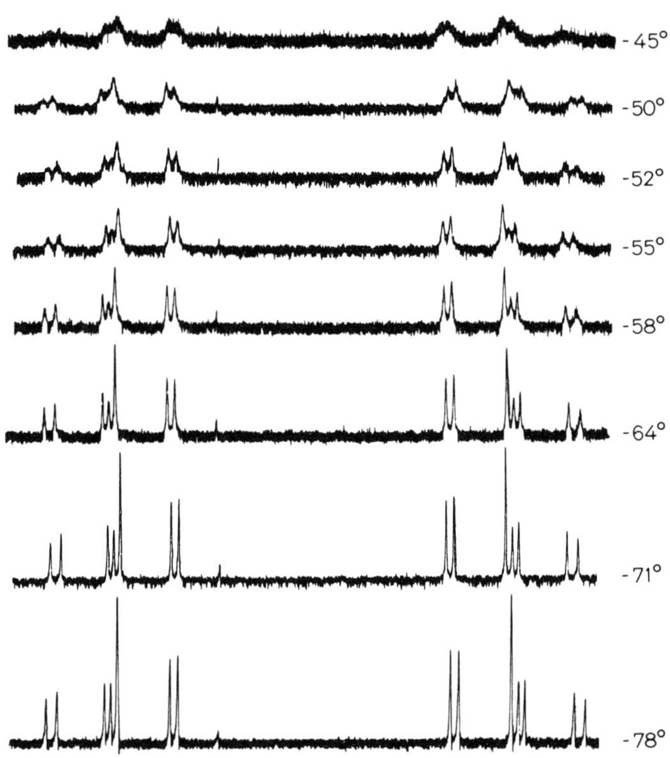

Fig. 16a. Comparison of experimental and calculated ^{19}F-DNMR spectra of SF_4 *2* (Whitesides *et al.* [25]). Experimental spectra of purified SF_4;

These three differences are indeed distinctly seen in the calculated DNMR spectra of SF_4 [*25*], shown in Fig. 16, for which the small non-first-order character of the static ^{19}F-NMR spectrum was amplified by using a weak magnetic field (9.2 MHz). The comparison with the experimental DNMR spectra for highly purified SF_4 clearly shows that only the two-pair mechanism can account for the intramolecular fluorine exchange process, which as the rearrangement of phosphoranes was interpreted to be a Berry rearrangement (cf. Sections 6, 7 and 8.1).

The SF_4 study therefore constitutes a convincing example for the advantageous use of non-first-order character in mechanistic studies. Other mechanistic studies of non-first-order spin systems, such as the polytopal rearrangements of 5-, 6-, 7-, and 8-coordinated transition metal hydrides and of ML_5 type transition metal complexes, have been reviewed recently by Jesson and Muetterties [*71*].

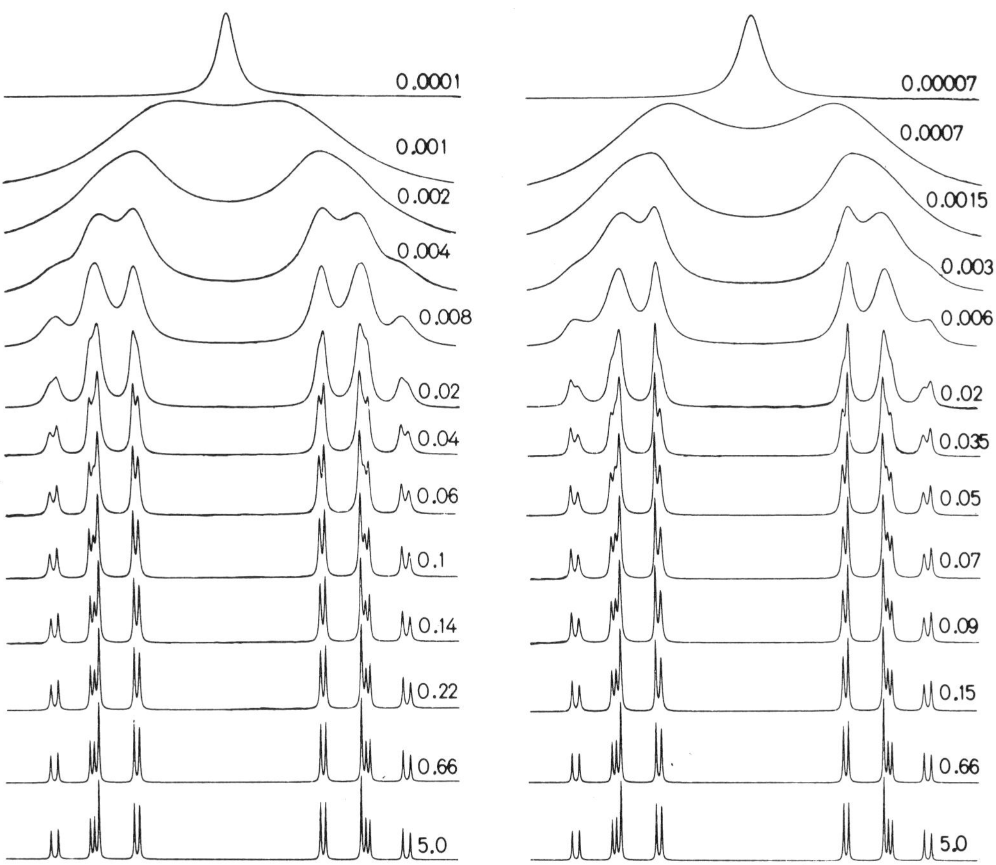

Fig. 16b. Comparison of experimental and calculated ^{19}F-DNMR spectra of SF_4 *2* (Whitesides *et al.* [25]). *Left column:* Spectra calculated for the two-pair exchange; *Right column:* Spectra calculated for the one-pair exchange

10. Intermolecular Exchange Reactions

In this final section some features of the mechanistic analysis of intermolecular exchange reactions will be shown by describing the approach of Chan and Reeves [*72*] which allows the elucidation of complex reaction schemes between several compounds. By this method, in addition to systems in which all reaction components can be seen by NMR, *i.e.* "closed systems", systems comprising reaction components of too small concentration to be observable, *i.e.* "truncated systems", can also be treated. In the band shape equation, of course, only the observable components are specified.

There are some helpful relations concerning the construction of the kinetic exchange matrices. In the case of a system consisting of three observable compounds, each giving rise to only one site, the matrix XLVII is built from pseudo-first-order rate constants (cf. Section 1), which are given capital letters to distinguish them from the specific rate constants. Thus the diagonal elements are given by the rate of magnetization transfer divided by the population of the site, *i.e.* the inverse lifetime of the site, while the corresponding off-diagonal elements equal the rate of magnetization gain from that site divided by its population, *i.e.* the rate of formation of the respective sites divided by the population of the site from which they obtain magnetization.

$$\begin{array}{c} \\ \mathrm{I} \\ \mathrm{II} \\ \mathrm{III} \end{array} \begin{array}{c} \begin{array}{ccc} \mathrm{I} & \mathrm{II} & \mathrm{III} \end{array} \\ \begin{pmatrix} -K_{11} & K_{12} & K_{13} \\ K_{21} & -K_{22} & K_{23} \\ K_{31} & K_{32} & -K_{33} \end{pmatrix} \end{array}$$

XLVII

$$\begin{array}{c} \\ \mathrm{I} \\ \mathrm{II} \\ \mathrm{III} \end{array} \begin{array}{c} \begin{array}{ccc} \mathrm{I} & \mathrm{II} & \mathrm{III} \end{array} \\ \begin{pmatrix} -K_1-K_3 & K_1' & K_3' \\ K_1 & -K_1'-K_2 & K_2' \\ K_3 & K_2 & -K_2'-K_3' \end{pmatrix} \end{array}$$

XLVIII

As a consequence, a balance of the elements of the kinetic exchange matrix is required, as in the previously discussed intramolecular exchange reactions. Since we will follow the formulation given by Chan and Reeves who used the band shape Eq. (2) (Section 2.2), this condition requires the sum of the elements of each matrix column to be zero. In the case of closed systems, there is a further relation, $K_{ij}P_j = K_{ji}P_i$, which connects the off-diagonal elements of the matrix by the corresponding populations of the sites. The physical meaning of this relation is that for any pair of sites, the rate of magnetization gain from each other is the same, provided that the magnetization transfer does not proceed via nonobservable intermediates. For a closed system therefore, the maximum number of independent pseudo-first-order rate constants is $N(N-1)/2$. In our case, using the labeling of matrix XLVIII, these are the elments K_1–K_3. For truncated systems, however, all $N(N-1)$ off-diagonal elements may be independent.

The numerical values of these pseudo-first-order rate constants are determined by band shape simulation of the spectra recorded at different temperatures and for different concentrations of the reaction components. With the determined K values, alternative reaction schemes can be tested by deriving the relations between the pseudo-first-order rate constants and the specific rate constants for each scheme and solving the equations for the specific rate constants. The reaction scheme which leads to consistent values may then be considered to represent the mechanism of the exchange process.

As a specific example, we will describe the halogen exchange between Me_2SnBr_2, Me_2SnBrI, and Me_2SnI_2 studied by Chan and Reeves [*29*]. Although the low abundance in isotopes ^{117}Sn and ^{119}Sn give rise to satellites of the methyl resonances in the ^1H-NMR spectra (cf. Fig. 17), the system can be treated as a three-site problem, since the spin-spin splitting is preserved during the exchange process.

Assuming at first that no further species than the three observed components are present, *i.e.* closed system, the kinetic exchange matrix XLIX can be used to simulate the DNMR spectra. Since a direct exchange between Me_2SnI_2 and Me_2SnBr_2 may be excluded, only two independent pseudo-first-order rate constants, K_1 and K_2, had to be determined. The reaction scheme

$$Me_2SnI_2 + Me_2SnBr_2 \underset{k_{-1}}{\overset{k_1}{\rightleftharpoons}} 2\ Me_2SnIBr$$

$$Me_2SnI_2 + Me_2SnIBr \xrightarrow{k_2} Me_2SnIBr + Me_2SnI_2$$

$$Me_2SnBr_2 + Me_2SnIBr \xrightarrow{k_3} Me_2SnIBr + Me_2SnBr_2$$

allows to relate the derived K values with the specific rate constants by the following equations

$$K_1 = k_1[Me_2SnBr_2] + k_2[Me_2SnIBr]$$

$$K_2 = k_3[Me_2SnBr_2] + k_{-1}[Me_2SnIBr]$$

Attempts to solve the two equations for k_1, k_{-1}, k_2, and k_3, using the simulated K values obtained for several sample compositions and for different temperatures, were not successful, implying that the proposed reaction scheme does not account for the observed exchange process. The inclusion of a third independent pseudo-first-order rate constant, K_3, in the kinetic exchange matrix XLIX led to no change, since the simulated K_3 values were very small if not zero, confirming the assumption that no direct exchange between Me_2SnI_2 and Me_2SnBr_2 occurs.

$$\begin{array}{l} \\ Me_2SnI_2 \\ Me_2SnIBr \\ Me_2SnBr_2 \end{array} \begin{array}{c} \begin{array}{ccc} Me_2SnI_2 & Me_2SnIBr & Me_2SnBr_2 \end{array} \\ \begin{pmatrix} -K_1 & K_1' & 0 \\ K_1 & -K_1'-K_2 & K_2' \\ 0 & K_2 & -K_2' \end{pmatrix} \end{array}$$

XLIX

$$\begin{array}{l} \\ Me_2SnI_2 \\ Me_2SnIBr \\ Me_2SnBr_2 \end{array} \begin{array}{c} \begin{array}{ccc} Me_2SnI_2 & Me_2SnIBr & Me_2SnBr_2 \end{array} \\ \begin{pmatrix} -K_1 & K_3 & 0 \\ K_1 & -K_2-K_3 & K_4 \\ 0 & K_2 & -K_4 \end{pmatrix} \end{array}$$

L

Another experimental fact cannot be accounted for by the reaction scheme above. Thus it was observed that traces of iodine retard the halogen exchange, while addition of tetrabutylammonium bromide leads to an acceleration of the exchange.

These observations suggested the occurrence of a truncated system. Accordingly, matrix L was used to simulate the experimental DNMR spectra. In Fig. 17 the excellent agreement between the experimental spectra and the band shapes calculated by use of the depicted values K_1–K_4 is shown for a mixture of Me_2SnI_2 (0.6 M) and Me_2SnBr_2 (0.4 M) in toluene at 29.5 °C without and with addition of iodine (0.005 M). The equilibrium concentrations for the three observable compounds Me_2SnI_2, Me_2SnIBr, and Me_2SnBr_2 in both cases are 0.376, 0.448, and 0.176 mole/l, respectively.

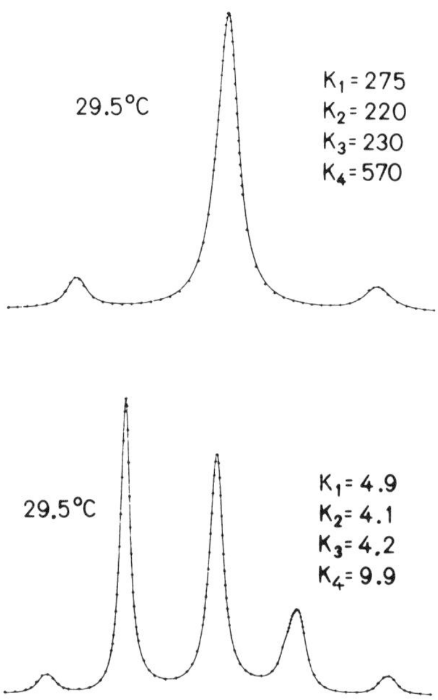

Fig. 17. Simulation of the experimental ^{1}H-DNMR spectra (characterized by the spectral points) for a Me_2SnI_2/Me_2SnBr_2 mixture (molar ratio 0.6 : 0.4) (Reeves *et al.* [29]).
Above: Mixture not containing iodine;
Below: Mixture containing 0.005 M iodine

The new reaction scheme which accounts for the experimental observations includes ionization reactions and exchange reactions of the following type

$$Me_2SnBr_2 \underset{a_{-2}}{\overset{a_2}{\rightleftharpoons}} Me_2SnBr^+ + Br^-$$

$$Me_2SnBr_2 + I^- \underset{b_{-2}}{\overset{b_2}{\rightleftharpoons}} Me_2SnIBr + Br^-$$

The relations between the pseudo-first-order rate constants K_1–K_4 and the specific rate constants now contain also the concentrations of the respective halide ions such as

$$K_2 = a_2([Me_2SnBr_2]/[Me_2SnIBr]) + b_{-2}\,[Br^-]$$

Thus the iodine effect (Fig. 17) can be explained by a reduction of the halide ion concentrations by formation of I_3^- and I_2Br^-. On the other hand, the addition of tetrabutylammonium bromide is seen to accelerate the exchange.

Similar classical type approaches permitted to elucidate proton transfer processes, which have been reviewed recently by Grunwald and Ralph [*73*]. Mechanistic studies of intermolecular exchange reactions, however, are not limited to classical methods. In fact, the early quantum mechanical theories of exchange of coupled spins were exemplified for intermolecular exchange reactions [*20, 21*]. Since coupled spin systems have already been treated in detail in Sections 8 and 9, we will conclude the section by only mentioning two recent important contributions in this field. The first is the permutation of indices method developed by Kaplan and Fraenkel in 1972 [*74*], and the second work is that of Meakin, English, and Jesson in 1976 [*75, 76*], who analyzed intermolecular exchange reactions of metal complexes by use of permutational analysis.

11. References

1. Gutowsky, H. S.: Time-dependent magnetic perturbations. In: Dynamic Nuclear Magnetic Resonance Spectroscopy. Jackman, L. M., Cotton, F. A. (eds.) New York: Academic Press 1975, p. 1
2. Bloembergen, N., Purcell, E. M., Pound, R. V.: Phys. Rev. **73**, 679 (1948)
3. Ernst, R. R., Anderson, W. A.: Rev. Sci. Inst. **37**, 93 (1966)
4. Levy, G. C., Holak, T., Steigel, A.: J. Am. Chem. Soc. **98**, 495 (1976)
5. Binsch, G.: Band-Shape Analysis. In: Dynamic Nuclear Magnetic Resonance Spectroscopy. Jackman, L. M., Cotton, F. A. (eds.). New York: Academic Press 1975, p. 45
6. Jackman, L. M., Cotton, F. A. (eds.): Dynamic Nuclear Magnetic Resonance Spectroscopy. New York: Academic Press 1975
7. Loewenstein, A., Connor, T. M.: Ber. Bunsenges. Physik. Chem. **67**, 280 (1963)
8. Binsch, G.: Topics Stereochem. **3**, 97 (1968)
9. Gutowsky, H. S., Holm, C. H.: J. Chem. Phys. **25**, 1228 (1956)
10. Binsch, G.: J. Am. Chem. Soc. **91**, 1304 (1969)
11. Binsch, G., Kleier, D. A.: Program 165, Quantum Chemistry Program Exchange, Indiana University 1970
12. Jaeschke, A., Muensch, H., Schmid, H. G., Friebolin, H., Mannschreck, A.: J. Mol. Spectr. **31**, 14 (1969)
13. Shoup, R. R., Becker, E. D., McNeel, N. L.: J. Phys. Chem. **76**, 71 (1972)
14. Phillips, W. D.: J. Chem. Phys. **23**, 1363 (1955)
15. Jackman, L. M.: Rotation about partial double bonds in organic molecules. In: Dynamic Nuclear Magnetic Resonance Spectroscopy. Jackman, L. M., Cotton, F. A. (eds.). New York: Academic Press 1975, p. 203
16. Drakenberg, T., Dahlquist, K. I., Forsen, S.: Acta Chem. Scand. **24**, 694 (1970)
17. Steigel, A., Sauer, J., Kleier, D. A., Binsch, G.: J. Am. Chem. Soc. **94**, 2770 (1972)
18. Hahn, E. L., Maxwell, D. E.: Phys. Rev. **88**, 1070 (1952)
19. McConnell, H. M.: J. Chem. Phys. **28**, 430 (1958)
20. Kaplan, J. I.: J. Chem. Phys. **28**, 278 (1958)
21. Alexander, S.: J. Chem. Phys. **37**, 967, 974 (1962)
22. Sack, R. A.: Mol. Phys. **1**, 163 (1958)
23. Johnson jr., C. S.: Advan. Magn. Res. **1**, 33 (1965)

24. Reeves, L. W., Shaw, K. N.: Can. J. Chem. **48**, 3641 (1970)
25. Klemperer, W. G., Krieger, J. K., McCreary, M. D., Mutterties, E. L., Traficante, D. D., Whitesides, G. M.: J. Am. Chem. Soc. **97**, 7023 (1975)
26. Johnson jr., C. S., Moreland, C. G.: J. Chem. Educ. **50**, 477 (1973)
27. Meakin, P., Mutterties, E. L., Tebbe, F. N., Jesson, J. P.: J. Am. Chem. Soc. **93**, 4701 (1971)
28. Krieger, J. K., Deutch, J. M., Whitesides, G. M.: Inorg. Chem. **12**, 1535 (1973)
29. Chan, S. O., Reeves, L. W.: J. Am. Chem. Soc. **95**, 673 (1973)
30. Gordon, R. G., McGinnes, R. P.: J. Chem. Phys. **49**, 2455 (1968)
31. Dyer, D. S., Ragsdale, R. O.: J. Am. Chem. Soc. **89**, 1528 (1967)
32. Steigel, A., Brownstein, S.: J. Inorg. Nucl. Chem., Supplement **1976**, p. 145
33. Cotton, F. A., George, J. W., Waugh, J. S.: J. Chem. Phys. **28**, 994 (1958)
34. Muetterties, E. L., Phillips, W. D.: J. Am. Chem. Soc. **81**, 1084 (1959)
35. Chen, M. M. L., Hoffmann, R.: J. Am. Chem. Soc. **98**, 1647 (1976)
36. Klemperer, W. G.: Delineation of nuclear exchange processes. In: Dynamic Nuclear Magnetic Resonance Spectroscopy. Jackman, L. M., Cotton, F. A. (eds.). New York: Academic Press 1975, p. 23
37. Saunders, M.: Tetrahedron Letters **1963**, 1699
38. Saunders, M.: Measurement of rates of fast reactions using magnetic resonance. In: Magnetic Resonance in Biological Systems. Ehrenberg, A., Malmström, B., Vänngard, T. (eds.). Oxford: Pergamon Press 1967, p. 85
39. Whitesides, G. M., Fleming, J. S.: J. Am. Chem. Soc. **89**, 2855 (1967)
40. Cotton, F. A.: Stereochemical nonrigidity in organometallic compounds. In: Dynamic Nuclear Magnetic Resonance Spectroscopy. Jackman, L. M., Cotton, F. A. (eds.). New York: Academic Press 1975, p. 45
41. Mislow, K., Raban, M.: Topics in Stereochem. **1**, 1 (1967)
42. Jennings, W. B.: Chem. Rev. **75**, 307 (1975)
43. Anet, F. A. L., Bourn, A. J. R., Lin, Y. S.: J. Am. Chem. Soc. **86**, 3576 (1964)
44. Bushweller, C. H., Anderson, W. G., Stevenson, P. E., O'Neil, J. W.: J. Am. Chem. Soc. **97**, 4338 (1975)
45. Bushweller, C. H., O'Neil, J. W., Bilofsky, H. S.: J. Am. Chem. Soc. **93**, 542 (1971)
46. Bushweller, C. H., Anderson, W. G., Stevenson, P. E., Burkey, D. L., O'Neil, J. W.: J. Am. Chem. Soc. **96**, 3892 (1974)
47. Hummel, J. P., Gust, D., Mislow, K.: J. Am. Chem. Soc. **96**, 3679 (1974)
48. Mislow, K.: Acc. Chem. Res. **9**, 26 (1976)
49. Muetterties, E. L.: J. Am. Chem. Soc. **91**, 1636 (1969)
50. Gutowsky, H. S., McCall, D. W., Slichter, C. P.: J. Chem. Phys. **21**, 279 (1953)
51. Berry, R. S.: J. Chem. Phys. **32**, 933 (1960)
52. Gielen, M., Vanlautem, N.: Bull. Soc. Chim. Belg. **79**, 679 (1970)
53. Musher, J. I.: J. Am. Chem. Soc. **94**, 5662 (1972)
54. Hässelbarth, W., Ruch, E.: Theor. Chim. Acta (Berl.) **29**, 259 (1973)
55. Whitesides, G. M., Bunting, W. M.: J. Am. Chem. Soc. **89**, 6801 (1967)
56. Whitesides, G. M., Eisenhut, M., Bunting, W. M.: J. Am. Chem. Soc. **96**, 5398 (1974)
57. Musher, J. I.: Inorg. Chem. **11**, 2335 (1972)
58. Eaton, S. S., Eaton, G. R.: J. Am. Chem. Soc. **95**, 1825 (1973)
59. Holm, R. H.: Stereochemically nonrigid metal chelate complexes. In: Dynamic Nuclear Magnetic Resonance Spectroscopy. Jackman, L. M., Cotton, F. A. (eds.). New York: Academic Press 1975, p. 317
60. Pignolet, L. H.: Topics Curr. Chem. **56**, 91 (1975)
61. Gillespie, P., Hoffman, P., Klusacek, H., Marquarding, D., Pfohl, S., Ramirez, F., Tsolis, E. A., Ugi, I.: Angew. Chem. **83**, 691 (1971); Angew. Chem. Inter. Ed. **10**, 687 (1971)
62. Altmann, J. A., Yates, K., Csizmadia, I. G.: J. Am. Chem. Soc. 98, 1450 (1976)
63. Jesson, J. P., Meakin, P.: J. Am. Chem. Soc. **96**, 5760 (1974)
64. Meakin, P., English, A. D., Ittel, S. D., Jesson, J. P.: J. Am. Chem. Soc. **97**, 1254 (1975)
65. Whitesides, G. M., Mitchell, H. L.: J. Am. Chem. Soc. **91**, 5384 (1969)

66. Eisenhut, M., Mitchell, H. L., Traficante, D. D., Kaufman, R. J., Deutch, J. M., Whitesides, G. M., J. Am. Chem. Soc. **96**, 5385 (1974)
67. Furtsch, T. A., Diersdorf, D. S., Cowley, A. H.: J. Am. Chem. Soc. **92**, 5760 (1970)
68. Moreland, C. G., Doak, G. O., Littlefield, L. B.: J. Am. Chem. Soc. **95**, 225 (1973)
69. Moreland, C. G., Doak, G. O., Littlefield, L. B., Walker, N. S., Gilje, J. W., Braun, R. W., Cowley, A. H.: J. Am. Chem. Soc. **98**, 2161 (1976)
70. Wiberg, K. B., Nist, B. J.: The interpretation of NMR Spectra. New York: Benjamin Press 1962, p. 319
71. Jesson, J. P., Muetterties, E. L.: Dynamic molecular processes in inorganic and organometallic compounds. In: Dynamic Nuclear Magnetic Resonance Spectroscopy. Jackman, L. M., Cotton, F. A. (eds.). New York: Academic Press 1975, p. 253
72. Chan, S. O., Reeves, L. W.: J. Am. Chem. Soc. **95**, 670 (1973)
73. Grunwald, E., Ralph, E. K.: Proton transfer processes. In: Dynamic Nuclear Magnetic Resonance Spectroscopy. Jackman, L. M., Cotton, F. A. (eds.) New York: Academic Press 1975, p. 621
74. Kaplan, J. I., Fraenkel, G.: J. Am. Chem. Soc. **94**, 2907 (1972)
75. Meakin, P., English, A. D., Jesson, J. P.: J. Am. Chem. Soc. **98**, 414 (1976)
76. English, A. D., Meakin, P., Jesson, J. P., J. Am. Chem. Soc. **98**, 422 (1976)

Received July 26, 1976

Rotation of Molecules and Nuclear Spin Relaxation

Hans Wolfgang Spiess

Institut für Physikalische Chemie der Johannes-Gutenberg-Universität Mainz, D-6500 Mainz

Contents

Preface

"Nuclear spin relaxation" to many scientists stands for relaxation either because of the dipolar coupling between the nuclear spins or the quadrupolar interaction for nuclei with spin $I > 1/2$. There is a simple reason for this; until the late 1960's most of the relaxation studies were performed on abundant nuclei with spin $I = 1/2$; protons, in particular, where the dipolar relaxation generally predominates in condensed phases; and on nuclei with quadrupole moment. For the $I = 1/2$ nuclei of low natural abundance (*e.g.*, ^{13}C) of great interest for studying the dynamic behavior of organic molecules, the dipolar interaction between ^{13}C and ^{1}H likewise represents by far the most important relaxation mechanism. Additionally, two other relaxation mechanisms, anisotropic shielding (anisotropic chemical shift) and spin-rotation interaction can give appreciable contributions to relaxation rates and offer new possibilities for using nuclear spin relaxation to study the reorientation of molecules. In this contribution we will deal with the theoretical background necessary to exploit the various relaxation mechanisms and demonstrate their advantages by a number of experimental examples, restricting ourselves to rotational motions in solids and in liquids of low viscosity.

Chapter 4 is of primary importance. In that chapter we give a detailed description of how to calculate line shapes for slow rotational jump motions in solids, and a uniform derivation of the spin relaxation rates in liquids for all the intramolecular couplings, and completely anisotropic rotation of the mocules in liquids using irreducible tensor calculus. This part has been written for a reader not familiar with irreducible tensors; therefore, the few theorems and results from group theory actually needed are reproduced (Chapter 2 and the appendices). For reader convenience, the results are collected in a number of tables. It is hoped those tables giving the relaxation rates $1/T_1$ and $1/T_2$ for all the intramolecular couplings of an asymmetric molecule and completely anisotropic rotational diffusion will be appreciated by individuals wanting to analyze relaxation data.

The relaxation data, in many cases, can be analyzed only in combination with NMR on rigid solids, although it need not necessarily be the same experimenter in both cases. A student, even engaged in relaxation studies of liquids only, nevertheless will need a good deal of knowledge about solid state NMR and it is hoped the corresponding sections in Chapters 3 and 5 will be of help in this area. Similarly, such students will need some theoretical knowledge concerning magnetic couplings; in particular, about the close relationship between magnetic shielding and spin-rotation interaction, since the strength of the latter coupling for many compounds of interest is not known from an experiment but has to be calculated from the corresponding shielding tensor. The formulae needed are given in Chapter 5.

Theoretical derivations and discussions of experimental results have been separated. For readers interested in developments in nuclear spin relaxation but not so interested in the details, it will probably be more advantageous to undertake an easy reading of Chapters 5 and 6 rather than try to work through Chapters 2 through 4. The literature on nuclear spin relaxation is so vast this article will not attempt to cover it completely. Representative examples have been selected instead where, naturally, I was personally engaged in many of the experiments. I apologize to all who may have contributed more clear-cut and beautiful examples I didn't mention.

Most of the work this contribution is based comes from my stay at the Max-Planck-Institute, Department of Molecular Physics, in Heidelberg, West Germany. It is a great pleasure to thank Prof. K. H. Hausser for his generous support of our work and for his special interest and encouragement of relaxation studies. Many of the results were obtained only through the very productive collaboration in our group. I wish to thank all of my colleagues who have been actively engaged in our program. Above all it has been a wonderful experience for me to be able to work so closely with U. Haeberlen, who really has become a friend during this time. I also greatly am indebted to him for working through the manuscript in detail and initiating many improvements. For the same reason I also owe special thanks to Prof. H. Sillescu and, moreover, his encouragement throughout the writing of the manuscript is highly appreciated. Without his interest and patience I would have hardly found the time for preparing this contribution. Last, but not least, I want to thank the editors of this series for their encouragement and help in publishing this volume. I gratefully acknowledge the patience and accuracy with which Mrs. M. Janssen prepared the final manuscript, including the typing of the numerous tables. Finally, I would like to thank my wife, who not only kept me in good humor during the long time of writing but also helped directly by typing the first draft.

1. Introduction

Nuclear spin relaxation has been developed as a standard method for studying molecular motions in liquids, solids, polymers, and–to a lesser extent–gases, staring with the pioneering work of Bloembergen, Purcell, and Pound [*1*]. Of the great variety of molecular motions possible (*e.g.*, translations, rotations, vibrations) *rotations* are particularly important for nuclear spin relaxation. Conversely, nuclear spin relaxation can be especially successful if information about rotational motions is desired. In this case nuclear spin relaxation can yield quantitative information over an extraordinary wide range of characteristic frequencies, from about 1 Hz to 10^{14} Hz. It shoud be noted that, typically, the nuclear spin relaxation *times* actually observed are much longer than the characteristic times of rotation of molecules. Therefore, a *theory* is needed which links the time constants of nuclear spin relaxation to the correlation functions describing the rotational motion, and so a considerable part of this volume is devoted to this goal. For rapid motions in liquids of low viscosity nuclear spin relaxation thus in principle, is inferior to scattering experiments (*e.g.*, light and neutron scattering) which allow the determination of correlation *functions*.

For *slow* rotational motions in *solids* or highly viscous systems with characteristic frequencies comparable with splittings in the NMR spectra, however, we can determine correlation functions directly. The effect slow molecular motion has on the NMR spectra is similar to spin exchange, well-known from high resolution NMR in liquids [*2*]. While in liquids spin exchange usually results from chemical exchange, in solids spin exchange can also result from rotational motions since the magnetic and electric interactions experienced by the nuclear spins are anisotropic and, therefore, are angular dependent. In conventional wide line NMR in solids, however, the line shape is determined by the dipolar interaction of a given spin with all other spins in the solid. The result is a rather structureless line and a motional narrowing is monitored mainly through the width of a symmetric line [*3*]. This is the reason why line shype analyses to be described here have not been demonstrated experimentally in NMR earlier. If the unselective dipolar interaction is suppressed (*e.g.*, by pulse techniques [*4, 5*] or simply by applying high magnetic fields of superconducting magnets), the line shape often is dominated by the magnetic shielding tensor[1] resulting in a characteristic powder pattern. As an example, Fig. 1.1 shows ^{31}P spectra of solid white phosphorus at room temperature and at 25 K [*6*]. Since in the P_4 tetrahedra the ^{31}P nuclei occupy positions related to each other by cubic symmetry, rotational jump motion of the molecule averages out the anisotropic coupling. As a result at high temperatures we find a sharp symmetric line at the centre of gravity of the powder spectrum observed at low temperatures. Clearly there must be an intermediate region with characteristic line shapes which can be exploited to yield information about the type and time scale of the motion.

[1] The reason for the phenomenon "chemical shift" is the shielding of the external magnetic field by the electrons surrounding the nucleus; therefore, the term "magnetic shielding" is preferable to the phenomenological term "chemical shift". Nevertheless, since the latter is commonly used, especially in chemical literature, it will also be used here alternatively.

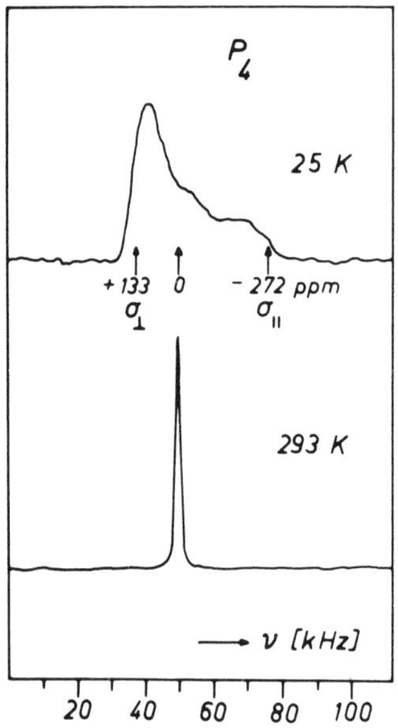

Fig. 1.1. ^{31}P spectra of solid white phosphorus P_4 at 92 MHz [*6*].

The general procedure to calculate such powder line shapes under presence of motion of the spins is treated in detail in the first half of Chapter 4. As background material, we first discuss the spin Hamiltonians pertinent to NMR in Chapter 2. The spin Hamiltonians are introduced in the form most suited with rotations in both ordinary space and spin space (*i.e.,* as irreducible tensor operators). Since we will make use of the transformation properties of the spin Hamiltonians later, some basic properties of irreducible tensor operators are also repeated in Chapter 2. A brief discussion of solid state spectra is given in Chapter 3 because first we must understand the spectra in the rigid case before we can calculate the effects of motion. Solid state spectra have been treated in more detail in the two monographs published recently on high resolution NMR in solids [*7, 8*]

The other half of this volume is devoted to nuclear spin relaxation in *liquids* of low viscosity where NMR also offers considerable advantages. Nuclear spin relaxation is very selective; therefore, we are not restricted to only determining a single correlation time for a given molecule. Instead, anisotropic rotational motion of molecules in liquids can be detected by determining correlation times for different nuclei in the molecule. These selective experiments have now become standard mainly because of the development of the Fourier transform technique [*9*]. This technique highly increased the sensitivity of NMR and made ^{13}C and ^{15}N accessible to the chemist. These nuclei primarily offer high selectivity since because of the large chemical shifts relaxation times for different carbon positions in the molecule can be determined simultaneously. Additionally for these nuclei, several relaxation mechanisms give important contributions and can be exploited to yield dynamic information.

The most important relaxation mechanism for these nuclei is the *intra*molecular dipole-dipole interaction with protons of the same molecule. These nuclei, therefore, are particularly well suited for studying rotational motions in liquids. In this respect,

^{13}C or ^{15}N relaxation yields information similar to that obtained from quadrupolar relaxation of spin $I > 1/2$ nuclei. In addition to the dipolar interaction, for spin $I = 1/2$ nuclei (other than protons) we find appreciable contributions to the relaxation rate from spin-rotation interaction and anisotropic shielding. The spin-rotation interaction offers unique information since it allows the determination of the correlation time of angular momentum (τ_J), which, together with the correlation time of reorientation (τ_c), allows a rather detailed description of the rotation of molecules in liquids. Relaxation through anisotropic shielding proves to be especially valuable for studying anisotropic rotational motions of planar molecules in particular.

A uniform derivation of nuclear spin relaxation rates for all intramolecular relaxation mechanisms is given in the second half of Chapter 4. Many details of the derivation are given so that the interested reader can follow the calculations closely and make use of the results given in condensed form in a number of tables. Our derivation of the relaxation rates closely follows Abragam [*3*] and for some readers may seam outdated in view of the new theoretical approach towards spin relaxation using Mori's formalism [*10*]. My intention was, however, to present explicit formulae directly applicable to the analysis of relaxation data rather than to contribute to the basic understanding of relaxation phenomena.

For the analysis of relaxation data, the coupling constants for the respective relaxation mechanisms have to be known. Only in the case of the dipolar interaction, the strength of the coupling can be calculated classically from the molecular geometry. In all other cases the coupling constants have to be determined experimentally, in particular by solid state measurements. Since we are dealing with anisotropic couplings described by second rank tensors we need to know not only the absolute values of the coupling parameters but also the orientation of the principal axes system relative to the molecular frame.

Therefore in Chapter 5 we give a survey of different experimental methods for determining these coupling parameters, where considerable emphasis is placed on shielding tensors, which have become accessible to experiment in recent years only. In addition the theory of the coupling tensors is reviewed briefly where again the shielding and the spin-rotation interaction are treated most extensively. The theoretical expressions derived there enable us to understand the special properties of the shielding tensor and the orientation of its unique axis, in particular, which makes this quantity especially useful to study anisotropic rotational motions. The close relationship between magnetic shielding and spin-rotation interaction is stressed and explicit relations between the respective tensor elements are derived. This is of considerable importance for relaxation applications, since it offers a means to calculate spin-rotation constants, hard to get otherwise, from the shielding tensor of the same nucleus.

Finally, in Chapter 6, a number of experimental examples are presented. After a brief survey of the current knowledge of shielding tensors we give a comprehensive account of experimental examples where NMR line shape analysis has been used to obtain information about rotation of molecules in solids. As far as liquids are concerned we illustrate the possibilities offered by NMR relaxation studies by discussing a number of representative examples where the various relaxation mechanisms have been exploited to gather detailed information about rotation of molecules in liquids.

2. Nuclear Spin Hamilton Operators

The nuclear spins of a sample placed into a magnetic field experience a number of interactions which we want to classify as *external* and *internal* ones. Both are important if we want to study the dynamic behaviour of molecules in condensed matter. Without *external* couplings, detection of NMR signals would be impossible, but only the *internal* couplings of the nuclear spins to their surroundings provide the basis for using nuclear spins to study molecular motions. Therefore, we will mainly be concerned with internal interactions; restricting ourselves to *diamagnetic nonconducting* materials.

The Hamiltonians for the different couplings of the spins to their surroundings are part of the total Hamiltonian of the system which, besides these terms, contains a large number of other operators. In principle, we could solve the total Hamiltonian exactly and then sort out those parts of the energy which depend on nuclear spin variables. This rather tedious approach–remember that averaging, *e.g.*, over vibrational states etc. would have to be performed–is circumvented in magnetic resonance by using the concept of "spin Hamilton operators" [*1, 2*]. The interactions between the nuclear spins and their surroundings is described by phenomenological coupling parameters. The complete spin Hamiltonian must contain the nuclear spin operators in such a way that all observable spectra and relaxation phenomena can be described correctly. Of this requirement is fulfilled, the coupling parameters can be determined *experimentally*. In order to *explain* these coupling parameters, we have to express them in terms of eigenfunctions of the system. A pertubation treatment is usually adequate since the nuclear spin-dependent energies of magnetic and electric interactions are small compared with the total energy of the system. The calculation of theoretical values for the coupling parameters, therefore, is a task of theoretical chemistry or solid state physics rather than of magnetic resonance.

The spin Hamiltonians pertinent to NMR in diamagnetic nonconducting materials can be found in almost every monograph on magnetic resonance (*e.g.,* Abragam [*2*]; a particularly clear and consistent treatment is given by Haeberlen [*3*]; see also Mehring [*4*]). The magnetic dipole moment of a given nucleus i, $\vec{\mu}_i$, couples with the external magnetic field $\mathbf{B}_o$, the magnetic moments of the other nuclei in the sample, $\vec{\mu}_j$, and the total angular momentum of the molecule $\mathbf{J}$. Nuclei with spin $I > 1/2$, in addition, can experience the electric quadrupole interaction between their quadrupole moment and the electric field gradient at the nuclear site. All of these interactions can be described as coupling of two vectors $\mathbf{V}$ and $\mathbf{U}$ by a general cartesian tensor of second rank :

$$\mathcal{H} \sim \mathbf{V}\mathfrak{R}\mathbf{U}. \tag{2.1}$$

Of course, many of the following considerations apply equally well to electron spin Hamiltonians. On the other hand we will occasionally use arguments which hold for nuclear spins only (*e.g.*, the dipolar interaction between two localized nuclear spins can be treated classically, contrary to the dipolar interaction between two electron spins).

We will mainly deal with molecular rotations in condensed phases in this review and, consequently, we are interested especially in the transformation properties of the spin Hamilton operators on rotations. This does not mean only the transformation properties of the complete Hamiltonians but, in particular, those of the different constituents since the spin Hamiltonian itself must be invariant under rotations. The transformation properties of operators are known from group theory; being especially simple if we formulate the spin Hamiltonians in irreducible tensor notation as treated in the books of Rose [*5*], Edmonds [*6*], Tinkham [*7*], Brink and Satchler [*8*], and Heine [*9*]. Therefore, the spin Hamiltonians are introduced as irreducible tensor operators in the following sections. Some parts of these might look rather formal. In the following chapters it should become clear, however, that a lot of calculations can be avoided by using results from group theory. These results, however, can be used only if we formulate our problems in the "language group theory understands". Therefore, detailed recipes for constructing the irreducible tensor operators we have to deal with in NMR are given here. Naturally, for readers familiar with irreducible tensor calculus, these sections will not contain new information; for others, however, it might be useful to define irreducible tensor operators and repeat some of their properties in the next section.

2.1. Irreducible Tensor Operators

2.1.1. Definition

We can describe the rotation of molecules either by a coordinate transformation through a real orthogonal–more general a unitary–matrix $\mathfrak{R}$:

$$\mathbf{x}' = \mathfrak{R}\mathbf{x}, \tag{2.2}$$

($\mathbf{x}'$: new coordinates) or by rotation operators acting on the function $f(\mathbf{x})$. The operator $\mathfrak{I}_R$, which corresponds to $\mathfrak{R}$, is defined by

$$\mathfrak{I}_R f(\mathfrak{R}\mathbf{x}) = f(\mathbf{x}), \tag{2.3}$$

equivalently

$$\mathfrak{I}_R f(\mathbf{x}) = f(\mathfrak{R}^{-1}\mathbf{x}).$$

(cf. Ref. [7], p. 32).

This means the operator $\mathfrak{I}_R$ changes the function f to $\mathfrak{I}_R f = f'$ in such a way that

$$f'(\mathbf{x}') = f(\mathbf{x}). \tag{2.3a}$$

In order to obtain the transformation properties of operators, we apply these relations to the function:

$$g(\mathbf{x}) = \mathcal{O}\psi(\mathbf{x}), \tag{2.4}$$

where $\mathcal{O}$ is an operator acting on the function ψ. On rotation, we obtain from Eq. (2.3)

$$g(\mathbf{x}) = \mathcal{T}_R g(\mathbf{x}') = \mathcal{T}_R \mathcal{O}\psi(\mathbf{x}'), \tag{2.5a}$$

and from Eqs. (2.3a) and (2.4):

$$g(\mathbf{x}) = g'(x') \equiv \mathcal{O}'\psi'(x') = \mathcal{O}'\,\mathcal{T}_R\,\psi(\mathbf{x}'). \tag{2.5b}$$

Equations (2.5a) and 2.5b) must hold for arbitrary values of $\psi(\mathbf{x}')$. From this we get the operator equation:

$$\mathcal{O}'\,\mathcal{T}_R = \mathcal{T}_R\,\mathcal{O}, \tag{2.6}$$

or:

$$\mathcal{O}' = \mathcal{T}_R \mathcal{O} \mathcal{T}_R^{-1}.$$

Irreducible tensor operators have especially simple transformation properties under unitary transformations. One *defines* the irreducible tensor operator of l-th rank (T_l) to be an operator with $(2\,m + 1)$ components T_{lm} which transform according to

$$\mathcal{T}_R\, T_{lm}\, \mathcal{T}_R^{-1} = \sum_{m'=-l}^{l} T_{lm'} \mathcal{D}^{(l)}_{m'm}(\mathfrak{R}). \tag{2.7}$$

Thus the components of an irreducible tensor operator of l-th rank transform under rotations into a linear combination of components of the same tensor operator T_{lm}. $\mathcal{D}^{(l)}_{m'm}(\mathbf{R})$ are components of the Wigner rotation matrix $\mathcal{D}^{(l)}(\alpha, \beta, \gamma)$ where the coordinate transformation $\mathbf{R}$ is described by the eulerian angles α, β, γ, see *e.g.* [7] p. 101. The derivation of explicit formulaefor the matrix elements $\mathcal{D}^{(l)}_{m'm}(\alpha, \beta, \gamma)$ can be found in the text books on group theory [5–9]. The definition of the eulerian angles, however, is not uniform; therefore we give the Wigner rotation matrices used here as an appendix. The discussion of the rotation matrices closely follows the introduction of the spin Hamilton operators (Section 2.3.4 and Appendix C).

2.1.2. Construction of Irreducible Tensor Operators from Vector Operators

As will become clear shortly, in NMR of diamagnetic compounds we have to deal with tensors of rank $l \leqslant 2$ only. The case $l = 0$ is trivial. The sum (2.7) contains only a single term T_{00} (a scalar) invariant under rotations.

For $l = 1$ we can also give an example directly: the vector operator $\mathbf{V} \sim \mathbf{r}$. The cartesian components V_x, V_y, and V_z themselves, however, are not irreducible components T_{lm} since they do not fulfill Eq. (2.7).[2] Linear combinations of the cartesian components, on the other hand, yield the irreducible components, often also referred to as spherical components:

$$T_{1\pm1} = V_{\pm1}, \quad T_{10} = V_0,$$

with: (2.8)

$$V_{\pm1} = \mp \frac{1}{\sqrt{2}} (V_x \pm i V_y), \quad V_0 = V_z.$$

Polar vectors and axial vectors have identical transformation properties under rotation. Under inversion, the axial vectors remain unchanged and the polar ones change sign.

From irreducible tensor operators of rank 1, which are obtained from vector operators according to Eq. (2.8) we can now construct tensor operators of arbitrary rank. The direct product of two irreducible tensor operators T_{l_1} and T_{l_2} is given by:

$$T_{l_1} \times T_{l_2} = \sum_{L=|l_1-l_2|}^{l_1+l_2} T_L \tag{2.9}$$

[5–9]. The notation of Eq. (2.9) is often used in group theory if one is interested in knowing only which values of L are possible at all by forming the different products $T_{l_1 m_1} T_{l_2 m_2}$. Accordingly, we can construct irreducible tensors of higher rank from the components of irreducible tensor operators of lower rank. When formulating the spin Hamiltonians, we will find it convenient to combine two vector operators to resulting tensor operators of rank 0 to 2 (see below). This problem is treated in detail in the monographs on group theory [5–9]. Therefore, only the result is given here. If we form the direct product of two tensor operators T_{l_1} and T_{l_2}, according to Eq. (2.9) we obtain for the components of the resulting tensor operators:

$$T_{LM} = \sum_m (l_1 m l_2 M-m | l_1 l_2 L M) \, T_{l_1 m} \, T_{l_2 M-m}. \tag{2.10}$$

The coefficients in Eq. (2.10) are called *Wigner- or Clebsch-Gordon* coefficients, which we have given here in the notation of Edmonds [6]. Most frequently, we will be interested in the coupling of two vector operators for which $l_1 = l_2 = 1$ and, according to Eq. (2.9) $L = 0, 1, 2$. The corresponding Wigner coefficients are collected in Table 2.1 for convenience. For $l_1 + l_2 \leqslant 4$, these Wigner coefficients are given in terms of square roots of rational numbers in an appendix to Heine's book [9]. For

[2] In cartesian representation we have, likewise, $\mathscr{P}_R V_i \mathscr{P}_R^{-1} = \sum_{j=1}^{3} a_{ij} V_j$ but the elements of the transformation matrix $\mathfrak{a}$ are not identical with $\mathscr{D}^{(1)}$.

Table 2.1. Wigner coefficients $(1\ m\ 1\ M-m\mid 1\,1\,L\,M)$, $m = 0, \pm 1$; $-L \leqslant M \leqslant +L$. For $L = 0$, the coefficients, according to Eq. (2.11a), are given in parentheses

$M-m$ / L	1	0	−1
0	$\frac{1}{\sqrt{3}}$ (−1)	$-\frac{1}{\sqrt{3}}$ (1)	$\frac{1}{\sqrt{3}}$ (−1)
1	$-\frac{1}{2}\sqrt{(1+M)(2-M}$	$\frac{1}{\sqrt{2}}M$	$\frac{1}{2}\sqrt{(1-M)(2+M)}$
2	$\frac{1}{2\sqrt{3}}\sqrt{(1+M)(2+M}$	$\frac{1}{\sqrt{6}}\sqrt{(2-M)(2+M)}$	$\frac{1}{2\sqrt{3}}\sqrt{(1-M)(2-M)}$

a discussion of the different notations used and the relation to Wigner's 3 *j* symbols, the reader is referred to the literature [*5–9*].

According to Eq. (2.10) and Table 2.1 we can construct irreducible tensor operators T_{00}, T_{1m}, and T_{2m} from two vector operators V_0, $V_{\pm 1}$ and U_0, $U_{\pm 1}$, which are obtained from their cartesian components according to Eq. (2.8)

$$T_{00} = \frac{1}{\sqrt{3}}\,(V_{+1}\,U_{-1} + V_{-1}\,U_{+1} - V_0\,U_0), \tag{2.11}$$

$$\begin{aligned} T_{10} &= \frac{1}{\sqrt{2}}(V_{+1}\,U_{-1} - V_{-1}\,U_{+1}), \\ T_{1\pm1} &= \frac{1}{\sqrt{2}}\,(\pm\,V_{\pm 1}\,U_0 \mp V_0\,U_{\pm 1}), \end{aligned} \tag{2.12}$$

$$T_{20} = \frac{1}{\sqrt{6}}\,(V_{+1}\,U_{-1} + V_{-1}\,U_{+1} + 2\,V_0\,U_0) = \frac{1}{\sqrt{6}}\,(3\,V_0\,U_0 - \mathbf{V}\cdot\mathbf{U}), \tag{2.13}$$

$$T_{2\pm1} = \frac{1}{\sqrt{2}}\,(V_{\pm 1}\,U_0 + V_0\,U_{\pm 1}), \quad T_{2\pm2} = V_{\pm 1}\,U_{\pm 1}\,.$$

The *expression* (2.11) for T_{00} which is obtained from Eq. (2.10), differs from the scalar product of the two vectors **V** and **U** by the factor $-1/\sqrt{3}$. Generally, the scalar product of two irreducible tensor operators V_{lm} and U_{lm} is given by:

$$T_{00} = \sum_{m=-l}^{l} (-1)^m\, V_{lm}\,U_{l-m} \equiv \mathbf{V}\cdot\mathbf{U}, \tag{2.11a}$$

which differs from Eq. (2.10) by the factor $-1/\sqrt{2l+1}$. In the following we will only use Eq. (2.11a) when forming irreducible tensor operators T_{00} so that T_{00} is identical with the scalar product (see *e.g.* [7], p. 90).

2.2. Nuclear Spin Operators[3]

We now can discuss the general form of the spin Hamiltonians, since the sum of products of irreducible tensor operators is the most general form of such Hamiltonians. According to Eq. (2.9) such a sum represents nothing other than a sum of irreducible tensor operators. The spin Hamiltonian itself must be invariant under rotations; therefore, it must be a tensor operator T_{00}. Using this property of the spin Hamiltonians, we can immediately clarify what sort of interactions nuclear spins can experience by combining the operators depending only on variables of the nucleus I to an irreducible tensor operator T^I_{lm}.

The operators T^I_{lm} act upon the eigenfunctions of nucleus I corresponding to the different spin states of the nucleus. In NMR, we work with nuclei in well defined states. Accordingly, a given tensor operator T^I_{lm} can only be responsible for a coupling between the nucleus and its surroundings if it has nonvanishing matrix elements over the eigenfunctions of nucleus I. Stated differently, the operator T^I_{lm} must correspond to a nonvanishing multipole moment of the nucleus. As a consequence of the large number of l – values that could, in principle, lead to couplings, only two ($l = 1$ and $l = 2$) are of practical importance in NMR. Since parity is well defined for nuclei, l must be even for all electric[4] and odd for all magnetic multipole moments. Furthermore, l has an upper limit depending upon the total spin of the nucleus I. According to the Wigner-Eckart theorem for the matrix elements of irreducible tensor operators [5–9] T^I_{lm} can only have nonvanishing matrix elements for $l \leqslant 2I$. This leaves for nuclei with spin $I = 1/2$ like ^{1}H, ^{19}F, ^{31}P, ^{13}C, ^{15}N (to mention those most important in NMR) $l \leqslant 1$ only and T^I_{1m} as the only tensor operator of such nuclei. Nuclei with spin $I \geqslant 1$ can have, in addition, a permanent electric quadrupole moment corresponding to T^I_{2m}. The multipole moments for higher values of l [magnetic octupole moment ($l = 3$) and electric hexadecupole moment ($l = 4$)] are without any practical importance in NMR; therefore, we can restrict ourselves to $l = 1$ and $l = 2$ in the following.

We can immediately write down explicit expressions for the operators T^I_{lm} in terms of nuclear spin operators since, according to the Wigner Eckart theorem, the matrix elements of different T^I_{lm} over basis functions (belonging to a given eigenstate of the nucleus I) are proportional to each other. For T^I_{1m} we obtain from Eq. (2.8):

$$T^I_{10} = I_0 = I_z$$

$$T^I_{1\pm 1} = \mp \frac{1}{\sqrt{2}} (I_x \pm iI_y). \tag{2.14}$$

Similarly, we construct T^I_{2m}, with the aid of Eq. (2.13) noting that here both vector operators **V** and **U** are given by the *same* vector operator **I**:

---

[3] The nuclear spin operator should not be confused with the nuclear spin Hamilton operator.

[4] $l = 0$ corresponds to the spherically symmetric part of the nuclear charge distribution and, therefore, does not correspond to couplings different for different spin substates.

$$T^I_{20} = \frac{1}{\sqrt{6}} [3 I_0^2 - I(I+1)]$$

$$T^I_{2\pm1} = \frac{1}{\sqrt{2}} (I_0 I_{\pm1} + I_{\pm1} I_0) \tag{2.15}$$

$$T^I_{2\pm2} = I^2_{\pm1}.$$

Finally, we have to determine the factors of proportionality connecting the operators T^I_{lm} with the magnetic dipole- and the electric quadrupole moment of the nucleus, respectively. These are obtained from the conventional definitions of the dipole moment μ_I and the quadrupole moment eQ_I of the nucleus as expectation values of the corresponding operators μ^I_{Op} and Q^I_{Op} for the nuclear state $I_z = I$:

$$\mu_I = \langle II | \mu^I_{Op} | II \rangle$$

$$= \alpha \langle II | T^I_{10} | II \rangle = \alpha I, \tag{2.14a}$$

or:

$$\mu^I_{Op} = \frac{\mu_I}{I} T^I_{1m} = \gamma_I \hbar T^I_{1m},$$

where γ_I is the magnetogyric ratio of the nucleus I. Similarly:

$$eQ_I = e \langle II | Q^I_{Op} | II \rangle$$

$$= \beta \langle II | T^I_{20} | II \rangle = \beta \frac{1}{\sqrt{6}} [3 I^2 - I(I+1)], \tag{2.15a}$$

or:

$$eQ^I_{Op} = \frac{eQ_I}{I(2I-1)} \sqrt{6}\, T^I_{2m}.$$

2.3. Nuclear Spin Hamilton Operators $\mathcal{H}_I$

2.3.1. General Form of $\mathcal{H}_I$

Since we have introduced the nuclear spin operators as irreducible tensor operators T^I_{1m} and T^I_{2m}, we can–with the aid of group theory–easily give all spin Hamiltonians for diamagnetic nonconducting materials, as far as their symmetry properties are concerned. $\mathcal{H}_I$ can be written as a general tensor coupling:

$$\mathcal{H}_I \sim T^I_l \times \sum_{l'} T^R_{l'}, \tag{2.16}$$

cf. discussion of Eq. (2.9), where we have combined all the operators that do not depend on the spin variables of nucleus I to irreducible tensor operators $T^R_{l'}$. For magnetic couplings $l = 1$ and for electric couplings $l = 2$. (cf. Section 2.2). According to Eq. (2.9), we obtain a rotationally invariant Hamiltonian $\mathcal{H}_I \sim T_{00}$ only for $l' = l$.

Forming the scalar product, according to Eq. (2.11a) yields

$$\mathcal{H}_I \sim T_{00} = \sum_l \sum_{m=-l}^{l} (-1)^m \, T^I_{lm} \, T^R_{l-m}. \tag{2.17}$$

Equation (2.17) expresses the well-known fact that the vector nuclear dipole moment ($l = 1$) can couple only with another vector (*i.e.*, magnetic field) while the nuclear quadrupole moment ($l = 2$) can only interact with a tensor of second rank (*i.e.*, the electric field gradient tensor at the nuclear site). It should be noted, however, that this statement holds for the symmetry properties of the corresponding operators only, it does not imply that only external magnetic fields are involved.

2.3.2. Magnetic Interactions

Let us now discuss the different magnetic couplings T^R_{1m}. The first one we think of is the external magnetic field $\mathbf{B}_0$. Additional magnetic fields are caused by other nuclear magnetic moments leading to the dipolar coupling. For a pair of spins i and j, the classical expression for the magnetic field $\mathbf{B}_{ij}$, generated by spin j at the site of spin i, is given[5]

$$\mathbf{B}_{ij} = -\frac{\vec{\mu}_j}{r^3_{ij}} + 3\,\frac{(\vec{\mu}_j\,\mathbf{r}_{ij})\,\mathbf{r}_{ij}}{r^5_{ij}}. \tag{2.18}$$

Here $\mathbf{r}_{ij}$ is the internuclear vector of nuclei i and j. From the components of $\mathbf{B}_0$ and $\mathbf{B}_{ij}$, we can now obtain the tensor operators $T^{B_0}_{1m}$ and $T^{B_{ij}}_{1m}$ [see Eq. (2.8)] and construct the spin Hamiltonians according to Eq. (2.17). In the case of $\mathbf{B}_0$, this is exactly what we do. If $\mathbf{B}_0$ is homogeneous, we have:

$$\mathbf{B}_0 = (0, 0, B_0) \tag{2.19}$$

$$\mathbf{B}_{\pm 1} = 0, \quad B_0 = B_0,$$

where we have chosen the z axis parallel to $\mathbf{B}_0$.

Together with Eqs. (2.17) and (2.14), we obtain for the spin Hamiltonian of the Zeeman interaction of the nuclear dipole moment with the external magnetic field:

$$\hbar\mathcal{H}_Z \sim \gamma_I \hbar I_0 B_0. \tag{2.20}$$

Since we have used only symmetry considerations, scalar factors (including the sign) are not obtained in this way. For the Zeeman interactions, they are well-known as:

$$\hbar\mathcal{H}_Z = -\gamma_I \hbar I_0 B_0. \tag{2.21}$$

In principle, we could handle the dipolar coupling in exactly the same way. The operator $T^{B_{ij}}_{1m}$ would then contain, however, both space- and spin variables of nucleus j.

[5] A clear elementary derivation of Eq. (2.28) is given by Sillescu [*10*].

Since we will be interested in the transformation properties of the operators under rotations in both ordinary space and spin space, we will find it much more appropriate to express the spin Hamiltonians as products of irreducible tensor operators T_{lm} and R_{lm}; the operators R_{lm} being independent of spin operators. The operators T_{lm}, on the other hand, are built according to Eqs. (2.11a)–(2.13) from the two vector operators corresponding to the coupling to be described. Therefore, at least one of these vector operators is the nuclear spin operator. For the Zeeman interaction we have $R^Z_{lm} = R^Z_{00} = 1$. For the dipolar coupling R^D_{lm} depends only on space coordinates of the two nuclei i and j. The tensor operators T^D_{lm} and R^D_{lm}, for the dipolar coupling of two nuclear spins i and j, are obtained directly, if in:

$$\hbar \mathcal{H}_D = \frac{\vec{\mu}_i \vec{\mu}_j}{r^3_{ij}} - 3\,\frac{(\vec{\mu}_i\ \mathbf{r}_{ij})(\vec{\mu}_j\ \mathbf{r}_{ij})}{r^5_{ij}}, \tag{2.22}$$

we insert for $\vec{\mu}_i$, $\vec{\mu}_j$, and $\mathbf{r}_{ij}$ the sperical operators. As an illustration, this is worked out explicitly in Appendix A. The well-known result is:

$$\mathcal{H}_D = -\frac{3}{r^5_{ij}} \sum_{m=-2}^{2} T^D_{2m}\, R^D_{2-m}, \tag{2.23}$$

where T^D_{2m} and R^D_{2m} are obtained from I^i_m, I^j_m, and r_{1m}, respectively, according to Eq. (2.13) (cf. Appendix A). Equation (2.23) expresses the fact that the Hamiltonian for the dipolar coupling of two spins can be written as the scalar product of two *irreducible tensors* of *second rank*.

Besides these classical couplings in NMR, we observe a number of additional magnetic interactions of the nuclei with their surroundings. We will introduce these couplings phenomenologically here and, in particular, consider their symmetry properties. Only if the nuclear spin interacts with a well-defined vector, such a coupling can cause line shifts or splittings in a spectrum. In addition to direct classic interactions, we now also allow for *indirect* couplings involving the electrons of the molecules.

In diamagnetic compounds, the total orbital and spin angular momentum operators of the electrons have vanishing expectation values. A detailed discussion of this important fact is given in Section 5.2. As vectors that can couple with the nuclear spin, we are, therefore, left with

i) the external magnetic field $\mathbf{B}_0$,
ii) the other nuclear spins of the sample $\vec{\mu}_j$, or
iii) the total angular momentum of the molecule $\mathbf{J}$
(in the gas phase only, as far as *spectra* are concerned, with regard to relaxation also in liquids and even solids).

This enables us to give the general form of spin Hamiltonians $\mathcal{H}_\lambda$ for the different couplings λ:

$$\mathcal{H}_\lambda = C^\lambda \sum_{l=0}^{2} \sum_{m=-l}^{l} (-1)^m\, T^\lambda_{lm}\, R^\lambda_{l-m}. \tag{2.24}$$

The values $l = 0$, 1, and 2 follow from Eq. (2.9) since the T_{lm}^{λ} are built out of two vectors ($l' = 1$). The scalar factors C^{λ} contain the factors of proportionality, cf. Eq. 2.14a). Explicit expressions for the different T_{lm}^{λ} and R_{lm}^{λ} are given below (see Tables 2.3 and 2.4).

The coupling tensors R_{lm}^{λ} which have been introduced here in a rather formal way, correspond to well-known spectral parameters of NMR, microwave, and molecular beam spectroscopy [*2*]. They are collected together with their conventional labels in Table 2.2. It is seen that to each of the classical couplings of **I** with $\mathbf{B}_0$, $\mathbf{I}_j$, and **J** (in the last case, however, a classical coupling does not exist), there corresponds an indirect coupling. Likewise, the theoretical explanation of these indirect couplings

Table 2.2. Magnetic couplings of nuclear spins

λ	Coupling vectors	Conventional label	Allowed values of l
Z	$\mathbf{I}, \mathbf{B}_0$	Zeeman interaction	$l = 0$
CS	$\mathbf{I}, \mathbf{B}_0$	Chemical shift Magnetic shielding	$l = 0, 1, 2$
D	$\mathbf{I}^i, \mathbf{I}^j$	Dipole-dipole coupling	$l = 2$
J	$\mathbf{I}^i, \mathbf{I}^j$	*J*-coupling; indirect spin-spin coupling	$l = 0, 1, 2$
SR	$\mathbf{I}, \mathbf{J}$	Spin-rotation interaction	$l = 0, 1, 2$

through the interaction of the nuclear spin with the orbital motion and the spin of the electrons of the molecules is very similar, as discussed in detail in Section 5.2.

The values of l given in Table 2.2 are given according to symmetry consideration only. The actual importance of the contributions for different l values are discussed in relation to spectra in Chapter 3 and in relation to spin relaxation in Chapter 4.

2.3.3. Electric Interactions

In NMR, we have to consider only a single spin dependent electric interaction, namely the quadrupole coupling of the nuclear quadrupole moment with the electric field gradient tensor at the nuclear site. The nuclear spin operator, in this case, is the irreducible tensor operator T_{2m}^{I} [cf. Eq. (2.15)]. Therefore, we can immediately write down the corresponding spin Hamiltonian:

$$\mathcal{H}_Q \sim \sum_{m=-2}^{2} (-1)^m \, T_{2m}^{I} \, R_{2-m}^{Q}; \tag{2.25}$$

no summation over l is necessary. The tensor operator R_{2m}^{Q} can be interpreted classically as the tensor of the second derivatives of the electric potential at the nuclear site (field gradient tensor). $\mathcal{H}_Q$ corresponds, therefore, to a direct coupling analogous to the coupling of the magnetic dipole moment with the external field.

2.3.4. Relation Between Cartesian and Irreducible Tensors

The spin dependent operators T_{lm}^{λ} can be constructed from vector operators according to Eqs. (2.11)–(2.13), (cf. Table 2.4) below. The coupling tensors R_{lm}^{λ}, on the other hand, have been introduced phenomenologically and, therefore, cannot be constructed a priori from known vectors. Being general second rank tensors, the coupling tensors can, however, be described and determined experimentally as matrices $\mathfrak{R}^{\lambda}$ in an arbitrary axes system. Often we will find a molecular fixed (or crystal fixed) axes system to be most appropriate; therefore, we have to know the relation between R_{ab}^{λ} $(a, b = x, y, z)$ and R_{lm}^{λ}. The following considerations are the same for all interactions and the superscript λ is omitted accordingly. A general second rank tensor can be described by 9 matrix elements in a cartesian coordinate system. It can be decomposed, however, into a sum of irreducible tensors of rank $l = 0, 1, 2$ (see [6–9]):

$$\mathfrak{R} = \mathfrak{R}^{(0)} + \mathfrak{R}^{(1)} + \mathfrak{R}^{(2)},$$

$$\begin{aligned} &l = 0: R_{ab}^{(0)} = \frac{1}{3}\, Tr\, \mathfrak{R}\,\delta_{ab} = R\,\delta_{ab} && \text{(the isotropic part)},\\ &l = 1: R_{ab}^{(1)} = \frac{1}{2}(R_{ab} - R_{ba}) && \text{(the antisymmetric part)},\\ &l = 2: R_{ab}^{(2)} = \frac{1}{2}(R_{ab} + R_{ba}) - R\,\delta_{ab} && \text{(the traceless symmetric part)}. \end{aligned} \tag{2.26}$$

The cartesian components $R_{ab}^{(1)}$ should not be confused with the R_{lm}.

Only the tensors with $l = 0$ and $l = 2$ are of general importance in NMR. For each symmetric tensor of second rank $(\mathfrak{R}^{(0)} + \mathfrak{R}^{(2)})$ there exists a *principal axes system* (X, Y, Z) for which the matrix $\mathfrak{R}^{(2)}$ is diagonal. The diagonal elements are called *principal elements*. We denote them by $\rho_{XX}^{(2)}, \rho_{YY}^{(2)}, \rho_{ZZ}^{(2)}$ in order to distinguish them from the components $R_{ab}^{(2)}$. The principal axes system often is fixed by molecular symmetry. For low symmetries, however, different interactions can have different principal axes systems. The principal axes are labeled according to the following convention[6]:

$$|\rho_{ZZ}^{(2)}| \geqslant |\rho_{XX}^{(2)}| \geqslant |\rho_{YY}^{(2)}|. \tag{2.27}$$

Since $\vec{\vec{\rho}}^{(2)}$ is a traceless tensor, it is sufficient to characterize it by the principal value:

$$\delta = \rho_{ZZ}^{(2)}$$

[6] This convention for the X and Y axes differs from that of Abragam [*2*]; generally used in quadrupole resonance. Our convention, which has also been used by Haeberlen [*3*], assures that either $\rho_{XX} \leqslant \rho_{YY} \leqslant \rho_{ZZ}$ or $\rho_{XX} \geqslant \rho_{YY} \geqslant \rho_{ZZ}$ and, therefore, is more convenient to characterize shielding tensors.

and the asymmetry parameter:

$$\eta = \frac{\rho_{YY}^{(2)} - \rho_{XX}^{(2)}}{\delta}, \qquad 0 \leqslant \eta \leqslant 1 \tag{2.28}$$

rather than by three principal elements.

The antisymmetric part $\mathfrak{R}^{(1)}$ cannot be brought into diagonal form; therefore, we find it convenient to connect the cartesian components $R_{ab}^{(1)}$ and the spherical components R_{lm} in the principal axes system of the symmetric part $\mathfrak{R}^{(2)}$. Since we call the tensor elements in this principal axes system $\rho_{ab}^{(1)}$, we denote the corresponding spherical components by ρ_{lm}. We then can obtain the R_{lm} in an axes system that is rotated with respect to the principal axes system (X, Y, Z) according to Eq. (2.7):

$$R_{lm} = \sum_{m'=-l}^{l} \rho_{lm'} \mathfrak{D}_{m'm}^{(1)}(\alpha, \beta, \gamma). \tag{2.29}$$

Here, the eulerian angles $((\alpha, \beta, \gamma) \equiv \Omega)$ describe the transformation of functions from the principal axes system into the axes system in which we want to know the R_{lm}. A detailed definition is given in Appendix C. As mentioned above, the principal axes systems can be different for different interactions λ; then the eulerian angles also are different for different λ.

In the principal axes system of $\mathfrak{R}^{(2)}$ the tensor $\mathfrak{R} = \vec{\vec{\rho}}$ is given in cartesian representation by:

$$\begin{aligned} \vec{\vec{\rho}} &= R \begin{pmatrix} 1 & 0 & 0 \\ 0 & 1 & 0 \\ 0 & 0 & 1 \end{pmatrix} + \begin{pmatrix} 0 & \rho_{XY} & \rho_{XZ} \\ -\rho_{XY} & 0 & \rho_{YZ} \\ -\rho_{XZ} & -\rho_{YZ} & 0 \end{pmatrix} \\ &= \qquad \vec{\vec{\rho}}^{(0)} \qquad + \qquad \vec{\vec{\rho}}^{(1)} \\ &\quad + \delta \begin{pmatrix} -1/2\,(1+\eta) & 0 & 0 \\ 0 & -1/2\,(1-\eta) & 0 \\ 0 & 0 & 1 \end{pmatrix} \\ &\quad + \qquad \vec{\vec{\rho}}^{(2)}. \end{aligned} \tag{2.30}$$

From these cartesian components we can determine the ρ_{lm}. This is worked out in detail in Appendix B. The result is[7]:

$$\begin{aligned} &\rho_{00} = R, \\ &\rho_{10} = -i\sqrt{2}\,\rho_{XY}; \quad \rho_{1\pm1} = \rho_{XZ} \pm i\rho_{YZ}; \\ &\rho_{20} = \sqrt{\frac{3}{2}}\,\delta; \quad \rho_{2\pm1} = 0; \quad \rho_{2\pm2} = -\frac{1}{2}\delta\eta. \end{aligned} \tag{2.31}$$

[7] The nomenclature is in accordance with that of Haeberlen [3]. Unfortunately the minus sign in $\rho_{2\pm2} = -1/2\,\delta\eta$ is missing in Eq. (2.19) of Ref. [3] and in some further equations.

The spin Hamiltonians which we have already given in general form in Eq. (2.24) are completely characterized through the factors C^λ, the operators T_{lm}^λ, the principal elements ρ_{lm}^λ and the eulerian angles $\Omega^\lambda = (\alpha^\lambda, \beta^\lambda, \gamma^\lambda)$.

The C^λ are chosen in accordance with the conventional definitions of the corresponding coupling tensors (cf. Table 2.3), due to convention $C^Z \neq C^{CS} \neq C^{SR}$ and $C^D \neq C^J$ only. Besides that, Table 2.3 contains the conventional symbols used for

Table 2.3. Factors C^λ and conventional symbols of coupling parameters [cf. Eqs. (2.26)–(2.28), (2.33)]

Interaction	λ	C^λ	R^λ	δ_λ	η_λ	Antisymmetric constituents
Zeeman	Z	$-\gamma_I$	1	–	–	–
Chemical shift	CS	γ_I	σ_{iso}	$2/3\ \Delta\sigma$	η_σ	Possible
Dipole-dipole	D	$-2\gamma_{Ii}\gamma_{Ij}$	0	r_{ij}^{-3}	0	–
J-coupling	J	1	J_{iso}	$2/3\ \Delta J$	η_J	Possible
Spin-rotation	SR	1	c_{iso}	$2/3\ \Delta c$	η_c	Possible
Quadrupole	Q	$\dfrac{eQ}{2I(2I-1)\hbar}$	0	eq	η_Q	–

the different coupling tensors: $\overleftrightarrow{\sigma}$ tensor, $\mathfrak{J}$ tensor, and $\mathfrak{c}$ tensor. Instead of δ^λ one often uses the anisotropy of the corresponding coupling, in particular in order to characterize the $\overleftrightarrow{\sigma}$ tensor

$$\Delta\sigma = \sigma_{ZZ} - \frac{1}{2}(\sigma_{XX} + \sigma_{YY}), \tag{2.32}$$

which reduces, in the case of axial symmetry, to

$$\Delta\sigma = \sigma_{ZZ} - \sigma_{XX,YY} = \sigma_{\parallel} - \sigma_{\perp}. \tag{2.32a}$$

Therefore, the relation between the anisotropy of the coupling λ and δ^λ is given also. For the antisymmetric parts, we will keep the general labels ρ_{ab}^λ. Actually, the $l = 1$ terms can be interpreted as components of a vector **A** with components $A_x = \rho_{YZ}$, $A_y = -\rho_{XZ}$, $A_z = \rho_{XY}$, and one could specify that vector by its length and its polar angles in the principal system of the $l = 2$ tensor. In order to avoid confusion of the different angles, we will not do this.

The complete list of Table 2.3 does not, of course, imply that all coupling parameters δ^λ, η^λ, and the antisymmetric components are different from zero for arbitrary point symmetry. If the point symmetry of the *nuclear site* or the *internuclear vector* (in case of J-coupling) is known, however, group theory tells which parts will vanish. These questions are treated in detail for the symmetric part of the tensors in a review article by Weil *et al.* [*11*]. Antisymmetric constituents are included in the tables given by Buckingham *et al.* [*12, 13*]; therefore, we can omit a discussion of these questions here. Explicit expressions for the spin dependent operators T^{λ}_{lm} are collected in Table 2.4. Depending on the specific problems we are interested in, we can freely choose the coordinate system for the T^{λ}_{lm} and thus

Table 2.4. Spin dependent operators T^{λ}_{lm} in the laboratory system (z axis $\parallel \mathbf{B}_0$)

λ	T_{00}	T_{10}	$T_{1\pm1}$	T_{20}	$T_{2\pm1}$	$T_{2\pm2}$
Z			–	–	–	–
	$I_0 B_0$	–				
CS			$\frac{1}{\sqrt{2}}(\pm I_{\pm1} B_0)$	$\sqrt{\frac{2}{3}}\, I_0 B_0$	$\frac{1}{\sqrt{2}} I_{\pm1} B_0$	–
D	–	–	–			
				$\frac{1}{\sqrt{6}}(3I^i_0 I^j_0 - \mathbf{I}^i \cdot \mathbf{I}^j)$	$\frac{1}{\sqrt{2}}(I^i_{\pm1} I^j_0 + I^i_0 I^j_{\pm1})$	$I^i_{\pm1} I^j_{\pm1}$
J	$\mathbf{I}^i \cdot \mathbf{I}^j$	$\frac{1}{\sqrt{2}}(I^i_{+1} I^j_{-1} - I^i_{-1} I^j_{+1})$	$\frac{1}{\sqrt{2}}(\pm I^i_{\pm1} I^j_0 \mp I^i_0 I^j_{\pm1})$			
SR	$\mathbf{I} \cdot \mathbf{J}$	$\frac{1}{\sqrt{2}}(I_{+1} J_{-1} - I_{-1} J_{+1})$	$\frac{1}{\sqrt{2}}(\pm I_{\pm1} J_0 \mp I_0 J_{\pm1})$	$\frac{1}{\sqrt{6}}(3 I_0 J_0 - \mathbf{I} \cdot \mathbf{J})$	$\frac{1}{\sqrt{2}}(I_{\pm1} J_0 + I_0 J_{\pm1})$	$I_{\pm1} J_{\pm1}$
Q	–	–	–	$\frac{1}{\sqrt{6}}[3I_0^2 - I(I+1)]$	$\frac{1}{\sqrt{2}}(I_{\pm1} I_0 + I_0 I_{\pm1})$	$I^2_{\pm1}$

the "axis of quantization", to make the calculation as convenient as possible ("principle of spectroscopic stability", cf. Van Vleck [*14*]). Most frequently, we will be interested in the T^{λ}_{lm} in the *laboratory* system whose z axis is parallel to the magnetic field since the Zeeman interaction for nuclei with spin $I = 1/2$ dominates over the internal couplings in the magnetic fields usually employed in NMR ($B_0 > 1\ T$). Therefore, in Table 2.4 the T^{λ}_{lm} are given in the laboratory frame. If the T^{λ}_{lm} are needed in another coordinate system, they can be constructed according to Eqs. (2.11a), (2.12), and (2.13).

By combining Eqs. (2.24) and (2.29), we finally obtain the spin Hamiltonians in the desired form:

$$\mathcal{H}_{\lambda} = C^{\lambda} \sum_{l=0}^{2} \sum_{m=-l}^{l} (-1)^m\, T^{\lambda}_{lm} \sum_{m'=-l}^{l} \rho^{\lambda}_{lm'}\, \mathfrak{D}^{(1)}_{m'-m}(\alpha^{\lambda}, \beta^{\lambda}, \gamma^{\lambda}). \tag{2.33}$$

In order to get explicit expressions from Eq. (2.33) together with Tables 2.3 and 2.4, we also need the Wigner rotation matrices which, therefore, are reproduced in Appendix C for $l = 0, 1, 2$.

In Eq. (2.33), we have now written the spin Hamiltonians as sums over products of three factors each[8]: $T^{\lambda}_{lm}, \rho^{\lambda}_{lm}, \mathfrak{D}^{(l)}_{m'-m}(\Omega^{\lambda})$. This form of the spin Hamiltonians

[8] Note that "cross products" with different values of l do not appear.

is suited especially for the treatment of rotations of molecules since, typically, T^{λ}_{lm} and ρ^{λ}_{lm} are then time independent and the rotation can be described through the Wigner rotation matrices only.

3. NMR Spectra of Solids

NMR spectra of solids differ markedly from those of liquids or dissolved substances. In an isotropic liquid, all anisotropic constituents of the spin Hamiltonians are averaged out because of rapid translational and rotational motion of the molecules. As a consequence, the only terms of Eq. (2.33) that "survive" are those with $l = 0$, the scalar or isotropic parts of the couplings. Furthermore, in a liquid a molecule cannot have a well-defined total angular momentum; therefore, out of the variety of coupling terms (cf. Tables 2.3 and 2.4) we are left with only two: the *isotropic chemical shift* and the *scalar-J-coupling*. The spectra resulting from these two terms can nevertheless be extremely complex. Their discussion is far beyond the scope of this review and we must refer to the literature instead (see, *e.g.*, [*1*].

Here, we are interested in obtaining dynamic information through the anisotropic parts of the spin Hamiltonians ($l = 1, 2$). Therefore, we are also interested in spectra from which we can determine these anisotropic parts directly; in particular, NMR spectra of solids (for a brief discussion of liquid crystals, see Section 5.1.1). The discussion of solid state spectra is very similar for the different interactions, since we always have to deal with coupling tensors of second rank. For $\lambda = CS, SR, Q$, we have single particle interactions as far as the nuclear spin is concerned. The nucleus being observed will be labelled I. For $\lambda = D, J$, on the other hand, we have pair interactions between two nuclear spins. If the two spins are like spins, we denote them by I^i and I^j; unlike spins, in particular those of different isotopes, are denoted by I and S, the I spins being observed. If different interactions λ are treated together, I and I^i are interchangable. In case of $\lambda = D, J$, we must, in addition, sum over all I^j or S if we want to obtain the spectrum of the I^i or I spin, respectively. For all other interactions we will, in most cases, only have to deal with intramolecular contributions. In detail, we will treat only the case where all internal couplings are so small, compared with the Zeeman interaction, that the spectra can be described adequately by first order perturbation theory with respect to the internal coupling. This condition is fulfilled, in general, for all couplings except the quadrupole coupling, if we work in magnetic fields $B_0 > 1\ T$ common in NMR (see Section 3.1).

In Section 3.2, we will give a brief discussion of strong quadrupole coupling which not only has the well-known consequences for the spectrum of the nucleus with a quadrupole moment [*2*], but can also strongly affect the spectrum of a spin $I = 1/2$ which is coupled via the pair interactions $\lambda = D, J$ to a nucleus with quadrupole moment. This case is treated in some detail.

3.1. Solid State Spectra Dominated by the Zeeman Interaction

3.1.1. Truncation of T^λ_{lm}

In most cases, especially for $I = 1/2$ nuclei, the Zeeman interaction of the nuclear spins with the external field is completely predominant. It is then convenient to devide the spin Hamiltonian according to:

$$\mathcal{H} = \mathcal{H}_Z + \sum_{\lambda \neq Z} \mathcal{H}_\lambda, \tag{3.1}$$

the solid state spectra now can be calculated using perturbation theory, starting with the eigenfunctions of $\mathcal{H}_Z$ and treating the operators $\mathcal{H}_\lambda$ as perturbing Hamiltonians. The eigenfunctions of $\mathcal{H}_Z$ are eigenfunctions of $I_0 = I_z$ (cf. Table 2.4). We denote them by $|I, m_z\rangle$.

Restricting ourselves to first order perturbation theory for *all internal Hamiltonians* $\mathcal{H}_\lambda$, we need consider only those parts of the T^λ_{lm} which have nonvanishing diagonal elements $\langle I, m_z | T^\lambda_{lm} | I, m_z\rangle$. The interaction between two spins—through dipolar or J-coupling—calls for special attention. If the two spins I^i and I^j are like spins, the complete operator $\mathbf{I}^i \cdot \mathbf{I}^j = I^i_0 I^j_0 - I^i_{+1} I^j_{-1} - I^i_{-1} I^j_{+1}$ leads to diagonal elements, whereas, for unlike spins I, S, only the term $I_0 \cdot S_0$ can have diagonal elements.

This simplyfies the T^λ_{lm} operators considerably, as shown by comparison of Tables 3.1 and 2.4. This simplification is often called "restricting to secular terms"

Table 3.1. Spin dependent operators T^λ_{lm} in the laboratory system valid in first order perturbation theory (I^i, I^j like spins; I, S unlike spins)[9]

λ	T_{00}	T_{10}	$T_{1\pm1}$	T_{20}	$T_{2\pm1}$	$T_{2\pm2}$
CS	$I_0 B_0$	–	–	$\sqrt{\frac{2}{3}}\, I_0 B_0$	–	–
D	–	–	–	$\frac{1}{\sqrt{6}}(3 I^i_0 I^j_0 - \mathbf{I}^i \cdot \mathbf{I}^j)$; $\sqrt{\frac{2}{3}}\, I_0 S_0$	–	–
J	$\mathbf{I}^i \cdot \mathbf{I}^j$; $I_0 S_0$					
Q	–	–	–	$\frac{1}{\sqrt{6}}[3 I_0^2 - I(I+1)]$	–	–

[9] The spin-rotational interaction can manifest itself as lines splitting in a spectrum only for molecules in well defined rotational states. This interaction, therefore, need not be considered here. Even in solid hydrogen spin-rotation, interaction cannot be observed in the spectrum [*2*].

or "truncation of internal Hamiltonians". We again emphasize that the condition $\mathcal{H}_Z \gg \mathcal{H}_\lambda$ must be fulfilled simultaneously *for all* λ in order that the T^λ_{lm} of Table 3.1 can be used.

From the T^λ_{lm} of Table 3.1, together with Eq. (2.33) we can now already make some general statements about first order NMR spectra without any further calculations:

i) The energy values and, as a consequence, the line positions in the spectra are influenced only by the symmetric parts of the coupling tensors. Even if $\rho^\lambda_{1m} \neq 0$, they have no effect on the first order spectra since the corresponding T^λ_{1m} vanish.

ii) The spectra for dipole-dipole, *J*-, and quadrupole coupling must be symmetric about the frequency determined by the corresponding isotropic part since m_z can take the values $-I \leqslant m_z \leqslant +I$. This is true, in general, for the line *positions;* whereas, the intensities can be different at very low temperatures.

3.1.2. Angular Dependence of the Spectra

According to Table 3.1, the spin dependent operators are given by T^λ_{00} and T^λ_{20} alone in first order. The T^λ_{00} terms give isotropic contributions only and angular dependence of the line position in a NMR spectrum can result only from terms in $\mathcal{H}_\lambda$ containing T^λ_{20} as a factor: therefore, the angular dependence is alike for all interactions λ. From Eq. (2.33) we obtain:

$$\begin{aligned}\mathcal{H}_\lambda &= C^\lambda T^\lambda_{20} \sum_{m'=-2}^{2} \rho^\lambda_{2m'} \mathfrak{D}^{(2)}_{m'0}(\alpha^\lambda, \beta^\lambda, \gamma^\lambda) \\ &= C^\lambda T^\lambda_{20}\left[-\tfrac{1}{2}\delta_\lambda\eta_\lambda\left(\mathfrak{D}^{(2)}_{-20} + \mathfrak{D}^{(2)}_{20}\right) + \sqrt{\tfrac{3}{2}}\,\delta_\lambda \mathfrak{D}^{(2)}_{00}\right] \qquad (3.2) \\ &= C^\lambda T^\lambda_{20}\,\frac{1}{2}\sqrt{\frac{3}{2}}\,\delta_\lambda\left[3\cos^2\beta^\lambda - 1 - \eta_\lambda \sin^2\beta^\lambda \cos 2\alpha^\lambda\right].\end{aligned}$$

As expected, the $\mathcal{H}_\lambda$ and, consequently, the energy values depend on *two* eulerian angles only; α^λ and β^λ. These are the two polar angles specifying the direction of the magnetic field vector $\mathbf{B}_0$ in the principal axes system of the interaction tensor $\vec{\rho}^{(2),\lambda}$ (cf. Appendix C).

Clearly, the common angular dependence of $\mathcal{H}_\lambda$, expressed in Eq. (3.2), does not imply that the spectra for different interactions are identical if one uses a normalised δ_λ. Within our first order perturbation theory treatment in the basis $|I, m_z\rangle$, the selection rule for spectral lines is $|h\nu_m| = |E_{m_z} - E_{m'_z}|$; $\Delta m = |m'_z - m_z| = 1$; therefore, the I spectrum consists of a single line for $\lambda = CS$ but of $2I$ lines for $\lambda = Q$. For $\lambda = D, J$ we have in the I^i spectrum $2I^j + 1$ or $2S + 1$ lines, respectively, symmetric about their average position. Of course, summation has to be performed over all I^j and S spins in the solid. The angular dependence for a given λ, however, is the same for different spectral lines and has the same form for different λ. It is, therefore, sufficient to consider the angular dependence of $\mathcal{H}_{CS}$ only in the following.

3.1.3. NMR Spectra of Single Crystals

From powder spectra, one normally can determine the principal values of the coupling tensors but not the orientation of the principal axes system (cf. Section 5.1.2). If we want to obtain dynamic information through the internal couplings, however, the complete determination of the coupling tensors (including the angular information about the direction of the principal axes) is of special importance. This holds true, in particular, if only little is known about an anisotropic coupling, as was the case for $\overleftrightarrow{\sigma}$ tensors only a few years ago. Besides, the experimental determination of $\overleftrightarrow{\sigma}$ tensors by single crystal measurements is important since their theoretical calculation is difficult and the directions of the principal axes, in many cases, cannot be predicted easily. From such single crystal measurements, we know that the unique directions of $\overleftrightarrow{\sigma}$ tensors frequently are different from those of the more common couplings $\lambda = D, Q$. We will show in Chapter 6 that, for this reason, we can obtain especially interesting information about molecular rotations through $\overleftrightarrow{\sigma}$ tensors; therefore, we will give a short description of single crystal spectra here.

In single crystals, we can follow directly the angular dependence of the NMR lines expressed in Eq. (3.2). We will restrict ourselves to the case where the angular dependence is dominated by a single interaction. This might seem a rather severe restriction but, in NMR, this can often be achieved (cf. Chapter 5). More detailed descriptions of solid state spectra and many references to original papers can be found in a number of books and review articles (for dipole-dipole and quadrupole coupling, see Abragam [*2*]; for quadrupole coupling see also Cohen and Reif [*3*] and Weil *et al.* [*4*]; for the $\overleftrightarrow{\sigma}$ tensor see Haeberlen [*5*] and Mehring [*6*]). Typically, one determines the angular dependence of the spectra from rotation patterns where the crystal is turned about an axis perpendicular to $\mathbf{B}_0$. The angular dependence of $\mathcal{H}_\lambda$ then is given simply by:

$$\mathcal{H}_\lambda \sim A \cos 2 (\phi - \phi_0) + B, \qquad (3.3)$$

even if $\eta_\lambda \neq 0$. The derivation of Eq. (3.3) can be found in Ref. [*6*] and will not be repeated here. The angular dependence of the NMR spectra for such rotation patterns can accordingly always be described by simple cosine functions having different phase angles ϕ_0 and being centered about different mean values B. For cylindrical symmetry of the coupling tensor ($\eta_\lambda = 0$), Haeberlen [*5*] gives explicit formulae for A, B, and ϕ_0. It should be mentioned that the angular dependence of NMR lines can no longer be written in the simple form (3.3) if the axis of rotation, $\mathbf{N}$, and $\mathbf{B}_0$ are not perpendicular to each other [*4–6*]. If, for example, one wants to suppress the dipolar interaction between two spins with especially short internuclear distance, one can mount the crystal in such a way that the rotation axis is parallel to the specific internuclear vector and choose the angle $\beta = 54°44'$ between $\mathbf{N}$ and $\mathbf{B}_0$. This assures $\mathcal{H}_D \sim 3 \cos^2\beta - 1 = 0$ for that selected pair of spins, cf. [*5*], p. 153.

Another important question that can be answered using Eq. (3.3) is, how many of such rotation patterns are needed to completely determine a coupling tensor. We will not go into details here and refer the reader to Ref. [4] instead. The general symmetric tensor of second rank $\mathfrak{R}^{(0)} + \mathfrak{R}^{(2)}$ that we want to determine is characterized by 6 parameters [cf. Eq. (2.30)]: R, δ, η, and the three eulerian angles specifying the principal axes system in a molecule or crystal fixed axes system. The angular dependence of $\mathcal{H}_\lambda$, Eq. (3.3) tells us a single rotation pattern can yield 3 parameters at most. The detailed discussion shows that, in general, rotations about three mutually perpendicular axes are needed to determine completely the symmetric coupling tensor. In principle, up to 5 parameters can be obtained from a single rotation pattern if the axis of rotation $\mathbf{N}$ and $\mathbf{B}_0$ are not perpendicular [4–6], but the accuracy is rather limited. Analysis of the spectra can be simplified considerably if the axes of rotation are selected carefully with respect to the crystal symmetry and to the orientation of the molecules in the unit cell.

3.1.4. Powder Spectra

A powder sample consists of small crystallites whose crystal axes are distributed randomly in an ideal powder. The line shape resulting from the angular dependence of $\mathcal{H}_\lambda$ [Eq. (3.2)] is obtained by a superposition of the spectra for all values possible for the eulerian angles $\alpha^\lambda, \beta^\lambda$ describing the orientation of $\mathbf{B}_0$ in a given principal axes system *X, Y, Z.* Again we can refer to the literature, in particular to Haeberlen [5], who has discussed powder line shapes for rigid solids in detail.[10] Here we only want to discuss the case of axially symmetric coupling tensors in some detail, the reason being that in Chapter 4 we will calculate how such powder spectra change under the influence of slow rotational jump motions of the molecules in solids. Having this already in mind, we introduce the concept of "curves of constant frequency" shown in Fig. 3.1. Let us consider the hatched area in Fig. 3.1. For all values ϑ, φ that are within this area, the corresponding frequencies have values between ω_a and ω_b. Here ϑ and φ are the polar angles[11] of $\mathbf{B}_0$ in the principal axes system *X, Y, Z.* For axial symmetry of the coupling, these curves of constant frequency simply are small circles ϑ = const about the *Z* axis. Such curves of constant frequency can be determined experimentally in single crystals if one can orient the crystal arbitrarily with respect to the magnetic field by use of a two circle goniometer [9]. Here, we want to employ the concept of curves of constant frequency in order to calculate the line shape $f(\omega)$. The calculation of the intensity for $\omega_a \leqslant \omega \leqslant \omega_b$ then reduces to calculating the area between the two curves of constant frequency ω_a and ω_b:

$$\int_{\omega_a}^{\omega_b} f(\omega)\, d\omega = c \int_{\vartheta_a}^{\vartheta_b} \int_{\varphi_a}^{\varphi_b} \sin\vartheta \, d\vartheta \, d\varphi. \tag{3.4}$$

---

[10] Reading the corresponding section in Ref. [5] will probably be more fruitful than going back to the original papers by Bloembergen and Rowland [7, 8], who have calculated these powder line shapes more than 20 years ago.

[11] Because of the equivalence of the eulerian angles $\beta^\lambda, \alpha^\lambda$ with the polar angles ϑ, φ of $\mathbf{B}_0$ in the *X, Y, Z* system, we will use the more familiar polar angles here.

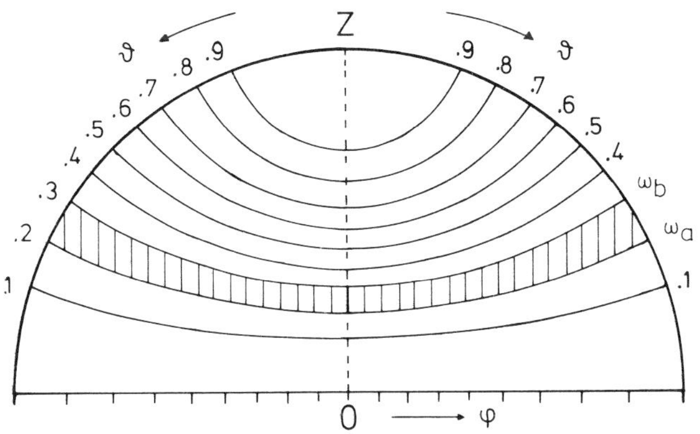

Fig. 3.1. Curves of constant frequency for an axially symmetric shielding tensor in stereographic projection.

Here, $\vartheta_a, \vartheta_b, \varphi_a, \varphi_b$ are the values of ϑ and φ, respectively, for the curves of constant frequency ω_a and ω_b. As the frequency is independent of φ, the integration over φ is trivial and can be taken into account by properly choosing the normilization constant c.

According to Eq. (3.2), we have for the normalized frequency:

$$\omega = \frac{\Delta}{2}(3\cos^2\vartheta - 1) \tag{3.5}$$

Here, we have introduced a *positive* frequency $\Delta \sim |\delta|$ and have set $\omega = 0$ equal to the central frequency.[12] We then obtain:

$$\int_{\vartheta_a}^{\vartheta_b} \sin\vartheta\, d\vartheta = \int_{\omega_a}^{\omega_b} \frac{1}{\sqrt{3}\,\Delta} \frac{d\omega}{\sqrt{\frac{2\omega}{\Delta} + 1}}. \tag{3.6}$$

Comparison of Eqs. (3.4) and (3.6) yields:

$$f(\omega) = \frac{1}{\sqrt{3}\,\Delta} \frac{1}{\sqrt{1 + \frac{2\omega}{\Delta}}}. \tag{3.7}$$

According to Eq. (3.5) ω can take the values $-\frac{\Delta}{2} \leqslant \omega \leqslant \Delta$. The characteristic powder line shape $f(\omega)$ with the singularity at $\omega = -\frac{\Delta}{2}$ is plotted in Fig. 3.2a. The frequencies $\omega_Z = \omega_\parallel$ and $\omega_{X,Y} = \omega_\perp$ correspond to $\mathbf{B}_0$ being, respectively, parallel and perpendicular to the unique direction of the coupling tensor.

[12] The introduction of Δ is necessary since we have defined δ_λ conventionally and, therefore, not uniformly (cf. Table 2.3)

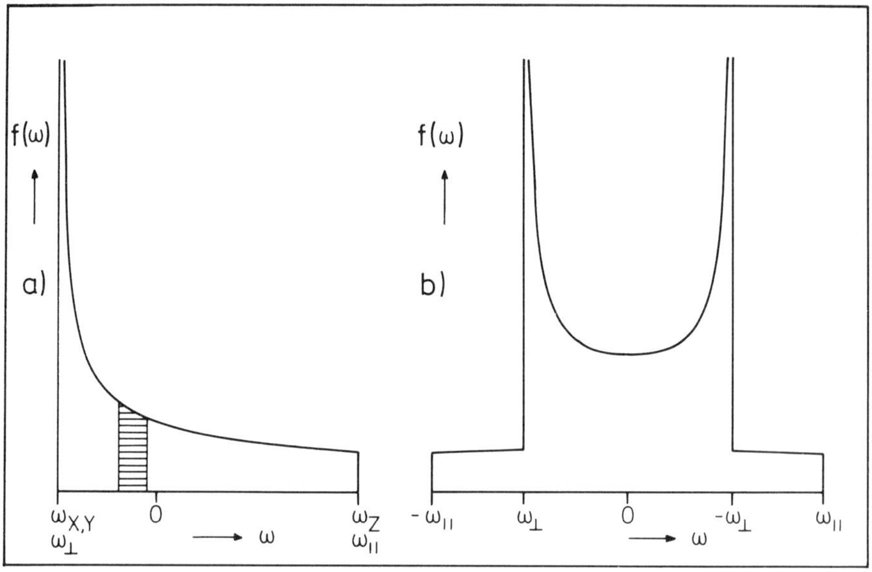

Fig. 3.2a and b. Powder spectra for axially symmetric coupling tensors. (a) Shielding tensor, $I = 1/2$, the hatched area corresponds to that in Fig. 3.1; (b) dipolar coupling between two $I = 1/2$ spins or quadrupolar coupling $I = 1$

For $\lambda = CS$, this already is the complete powder spectrum; experimental examples will be given in Chapters 5 and 6. For $\lambda = D, J, Q$ the powder spectrum is a superposition of single components with line shapes according to Eq. (3.7) and Fig. 3.2a. Each component has a counterpart according to the spin dependent operators T_{20}^{λ} so that the total spectrum is symmetric about $\omega = 0$. If there is an odd number of single components (total spin half integer), we have, in addition, a δ function at $\omega = 0$. For the dipolar coupling of two spins $I = 1/2$ we obtain in this manner the Pake diagram [*2, 10*]. Formally identical with this is the powder spectrum for $I = 1$ in the presence of quadrupole coupling (Fig. 3.2b).

The general case $\eta \neq 0$ can be treated in exactly the same way; a detailed description can be found in [*5*]. Therefore only the result will be given here. With our convention about the principal axes, Eq. (2.27), for the normalized frequencies (Δ-positive!), we always have $\omega_X \leqslant \omega_Y \leqslant \omega_Z$, as shown in Fig. 3.3. The normalized line shape $f(\omega)$ is given by:

a) for $\omega_X \leqslant \omega \leqslant \omega_Y$:

$$f(\omega) = \frac{1}{\pi\sqrt{(\omega_Z - \omega)(\omega_Y - \omega_X)}} \quad F\left(\sqrt{\frac{(\omega_Z - \omega_Y)(\omega - \omega_X)}{(\omega_Z - \omega)(\omega_Y - \omega_X)}},\ \frac{\pi}{2}\right), \tag{3.8}$$

and b) for $\omega_Y \leqslant \omega \leqslant \omega_Z$:

$$f(\omega) = \frac{1}{\pi\sqrt{(\omega_Z - \omega_Y)(\omega - \omega_X)}} \quad F\left(\sqrt{\frac{(\omega_Z - \omega)(\omega_Y - \omega_X)}{(\omega_Z - \omega_Y)(\omega - \omega_X)}},\ \frac{\pi}{2}\right).$$

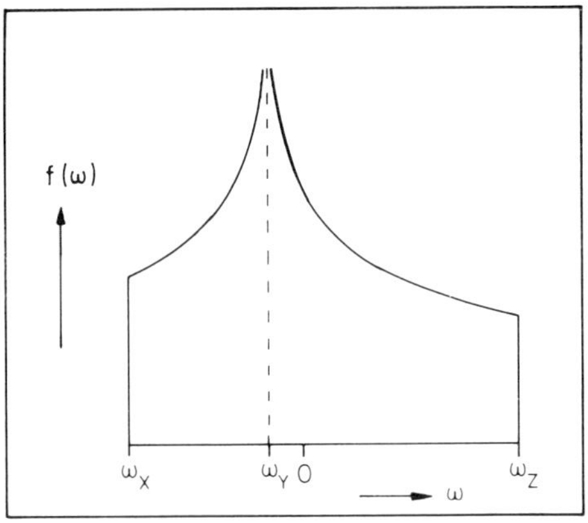

Fig. 3.3. Powder spectrum for a shielding tensor with asymmetry parameter $\eta = 2/3$

The incomplete elliptic integral of the first kind $F(k, \varphi)$ is tabulated. In both cases, $\eta = 0$ and $\eta \neq 0$, the coupling parameters (δ and η) can be determined directly from the frequencies ω_X, ω_Y, and ω_Z : $\delta \sim \omega_Z$; $\eta = (\omega_Y - \omega_X)/\omega_Z$. From the position of the central frequency one obtains the isotropic part R, often relative to a reference.

Powder spectra are useful especially if the compound under investigation contains a nuclear species in a single position only, (*e.g.*, ^{15}N in pyridine) or if the molecule is highly symmetric, making different nuclear positions equivalent (*e.g.* ^{13}C in benzene). If, on the other hand, a compound contains a nuclear species in different positions (*e.g.* ^{13}C in aromatic rings *and* carbonyl groups), it will often not be possible to reliably analyse the resulting superposition of overlapping powder spectra.

3.1.5. Partially Ordered Solids

In solids or solid polymers, one often observes partial ordering of the molecules. This can be a desired property of the sample (*e.g.*, in a drawn polymer). In other cases, however, partial ordering results from the failure of preparing a random sample (*e.g.*, by freezing a liquid in the NMR sample tube). NMR and ESR line shapes resulting in these cases have recently been treated in detail [*11*]. A particularly simple expression for the line shape is obtained if both the orientational distribution and the coupling tensor are axially symmetric:

$$\begin{aligned} f(\omega) &= 8\,\pi^2 \sum_{l=0,2,4\ldots} P_{l00}\, P_l(\cos\theta)\, P_l(\cos\beta_0)\, f_l(\omega) \\ &= 8\,\pi^2 \sum_{l=0,2,4\ldots} a_l f_l(\omega). \end{aligned} \tag{3.9}$$

Here θ is the angle between the Z axis of the coupling tensor and the z axis of the molecular system used to specify the orientation of the individual molecules in the sample system. The z' axis of this sample system, in turn, is defined by the direction of partial ordering (*e.g.* the draw direction). The angle between the z' axis of the sample system and $\mathbf{B}_0$ is β_0 (for details cf. [*11*]). For a given l, therefore, the two Legendre polynomials $P_l(\cos\theta)$ and $P_l(\cos\beta_0)$ have fixed values for a given sample in a fixed orientation to $\mathbf{B}_0$. The line shape function $f(\omega)$ is thus obtained as a weighted superposition of "subspectra" $f_l(\omega)$, given by

$$f_l(\omega) = \frac{1}{\Delta\sqrt{3}} \frac{1}{\sqrt{1+\frac{2\omega}{\Delta}}} P_l[x(\omega)]$$

with

$$x(\omega) = \frac{1}{\sqrt{3}} \sqrt{1+\frac{2\omega}{\Delta}}. \tag{3.10}$$

The weighting factor a_l, besides depending on the angular factors, contains the factor $\bar{P}_{l00}$, the l-th moment of the orientational distribution. For further details see [*11*]. Such subspectra $f_l(\omega)$ are drawn for some values of l in Fig. 3.4. For an isotropic sample, the usual powder spectrum [Eq. (3.7)] is obtained as the special case

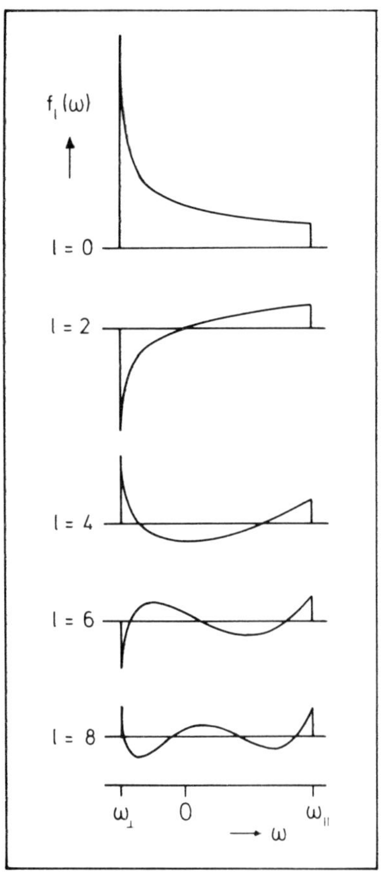

Fig. 3.4. Subspectra $f_l(\omega)$ for composing line shapes of partially ordered solids according to Eq. (3.9) [*11*]

$l = 0$ for which $P_l = 1$. Theoretical spectra of partially ordered systems are shown in Fig. 3.5 as illustrative examples. Each was obtained by a superposition of three subspectra; the ordinary powder spectrum ($l = 0$) and two others with higher values of l. For the spectra of Fig. 3.5a we have chosen $a_0 = 1, a_2 = \pm 1/2, a_4 = \pm 1/2$; and for Fig. 3.5b, likewise, $a_0 = 1, a_4 = \pm 1/2, a_6 = \pm 1/2; a_l = 0$ for all other values of l. The +-sign corresponds to the upper spectra, and the −-sign to the lower ones, respectively. The upper spectra are more appealing if one is interested in determining the degree of partial ordering of a sample. The lower ones can be "dangerous" if the partial ordering is not a desired property but is encountered accidentally, since the deviations from the usual powder spectrum are not so evident. Determining $\Delta\sigma$ from a spectrum of the type of Fig. 3.5b, not realizing that it stems from a partially ordered sample, therefore can result in serious errors for that coupling parameter.

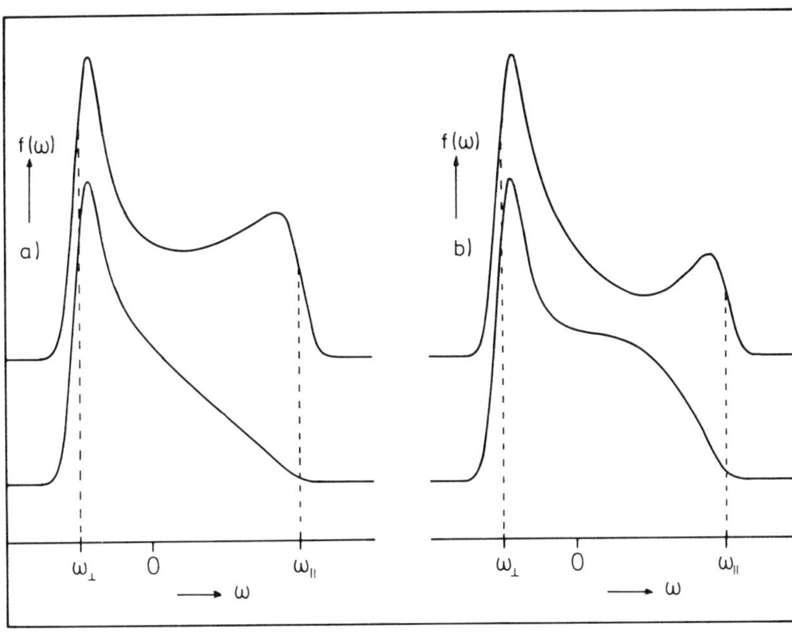

Fig. 3.5a and b. Powder spectra of partially ordered solids calculated according to Eq. (3.9). (a) $a_0 = 1, a_2 = \pm 1/2, a_4 = \pm 1/2$; (b) $a_0 = 1, a_4 = \pm 1/2, a_6 = \pm 1/2; a_l = 0$ for all other values of l. The + and − signs correspond to the upper and lower spectra, respectively (Courtesy of R. Hentschel)

The subspectra shown in Fig. 3.4 should be especially helpful, if one seeks a qualitative understanding of the deviations from the random powder spectrum caused by partial ordering without elaborate calculations.

3.2. Solid State Spectra in the Presence of Strong Quadrupole Coupling

The simple truncated T^{λ}_{lm} operators of Table 3.1 can be used only if the $\mathcal{H}_\lambda$ *for all* internal couplings are so small compared with the Zeeman interaction that it is permissible to restrict to first order perturbation theory $\mathcal{H}_\lambda \ll \mathcal{H}_Z$. This condition generally is fulfilled for $\lambda = CS, D, J$ or at least it can be fulfilled by applying strong magnetic fields. For $\lambda = Q$, on the other hand, it can be fulfilled, in general, only for nuclei with small quadrupole moments, such as ^{2}D, or for almost cubic symmetry of the surroundings (*e.g.*, in metals [*12*]). In other cases, such as the halogen nuclei and ^{14}N in organic compounds, the Zeeman interaction is only of the same order or even small compared with the quadrupole coupling [*13, 14*].

3.2.1. Spectra for Nuclei with $I > 1/2$

The spectra of the quadrupolar nuclei will not be dealt with in detail (cf. Refs. [*2, 13, 14*]). It should be noted, however, that shielding tensors $\vec{\vec{\sigma}}$ can also be determined for such nuclei even in the presence of large quadrupole coupling. In most cases, however, one will only be able to determine the diagonal elements of $\vec{\vec{\sigma}}$ in the principal axes system of the field gradient tensor q [*15*]. In general, the matrix of the Hamiltonian $\mathcal{H}_Z + \mathcal{H}_{CS} + \mathcal{H}_Q$ (Eq. (2.33) for $l = 0,2$) must be diagonalized numerically [*15*]. Typically, one also has to use single crystals in order to perform an analysis as described in [*15*].

Half integer spins lead to a special case of some practical importance for exceedingly large quadrupole coupling. In order to ease the description, we will use, for the moment, the Z^Q axis of the field gradient tensor as axis of quantization; labelling the spin substates by M_Z. For zero magnetic field, the $M_Z = \pm 1/2$ levels are degenerate. This degeneracy is lifted by applying an external magnetic field. The transition between the two split levels can be observed as a NMR transition (cf. Fig. 3.6). For $\eta_Q = 0$ its frequency varies between the Larmor frequency ω_L for

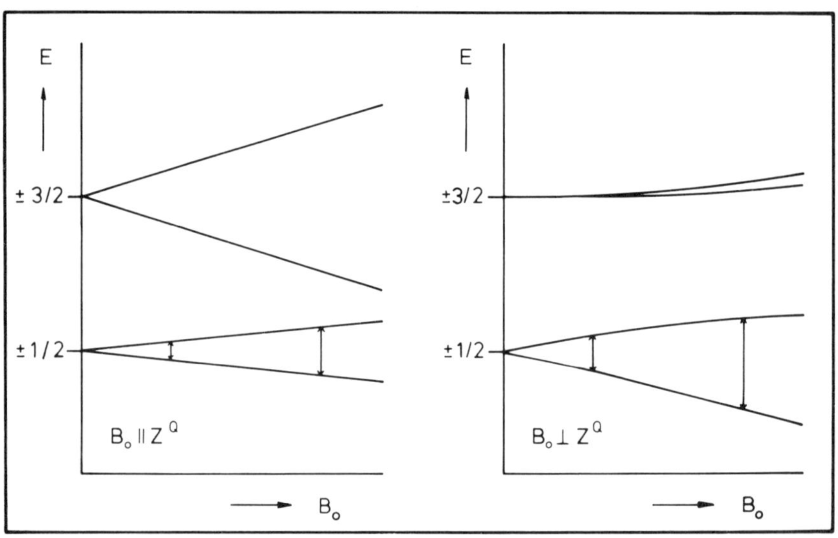

Fig. 3.6. Zeeman splitting of the $\pm M_Z$ levels (degenerate in zero field) for $I = 3/2$ and large quadrupole coupling $\mathcal{H}_Z \ll \mathcal{H}_Q$

$\mathbf{B}_0 \parallel Z^Q$ and $(I + 1/2) \cdot \omega_L$ for $\mathbf{B}_0 \perp Z^Q$; it is independent of δ_Q in the limit[13] $\omega_L \ll C^Q \delta_Q/\hbar$. The analysis of this transition, therefore, yields information about the shielding tensor in a rather direct way. Single crystal spectra are treated in [*17*]. powder spectra in [*18*].

3.2.2. Dipolar Coupling of a Spin I = 1/2 with a Spin S > 1/2

As mentioned already in the introduction to this chapter, strong quadrupole coupling also manifests itself in the spectrum of a spin $I = 1/2$ if it is coupled–through dipole-dipole or J-coupling–with a spin $S > 1/2$. Calculation of the dipolar coupling, in this case, seems to be rather timely since the dipolar interaction recently is used again in order to get structural information (cf. Section 5.1.2). In molecules of sufficiently high symmetry the unique axis of the field gradient tensor and the internuclear vector often will be parallel. For this special case, the contribution to the second moment of the dipolar coupling to the S spin has been calculated before [*19, 20*] and single crystal spectra for ^{13}C–have been observed recently [*21*]. We will give a slightly more general treatment here.

By strong quadrupole coupling, we mean the eigenfunctions of the S spin are no longer eigenfunctions of $\mathcal{H}_Z$. The eigenfunctions of the I spins, however, are still eigenfunctions of $\mathcal{H}_Z$ to a high degree of approximation. Therefore, we only need consider those spin operators in Table 2.4 which are diagonal in the $|I, m_z\rangle$ representation. For the dipolar coupling there are only two terms in addition to those of Table 3.1:

$$T_{2\pm1} = \frac{1}{\sqrt{2}} I_0 S_{\pm 1}. \tag{3.11}$$

For J-coupling, there are, in addition, two terms $T_{1\pm1} = \mp \frac{1}{\sqrt{2}} I_0 S_{\pm1}$, identical with $T_{2\pm1}$ except for the sign change. This case is of some theoretical interest since it offers the possibility that the antisymmetric part of a coupling tensor could manifest itself in a spectrum which involves first order perturbation theory for the I spin only. We nevertheless will leave this term aside here since anisotropic J-coupling is of no importance in solid state NMR.

For the dipolar coupling we obtain from Eq. (2.33) together with Eq. (3.11):

$$\begin{aligned} \mathcal{H}_D &= C^D \sqrt{\frac{3}{2}}\, \delta_D \{-[T_{21}\mathfrak{D}^{(2)}_{0-1} + T_{2-1}\mathfrak{D}^{(2)}_{01}] + T_{20}\mathfrak{D}^{(2)}_{00}\} \\ &= \frac{-\gamma_I\gamma_S\hbar}{r^3_{IS}} \{(3\cos^2\beta - 1) I_0 S_0 + \frac{3}{2\sqrt{2}} \sin 2\beta\, I_0 (S_{+1} e^{i\gamma} - S_{-1} e^{-i\gamma})\}. \end{aligned} \tag{3.12}$$

[13] This statement holds for very low values of ω_L only. In fact, one can determine, for example, δ_Q and η_Q for $^{77,79}Br$ or ^{127}I by studying the magnetic field dependence of that transition without having a spectrometer to directly observe the quadrupole transitions at much higher frequencies [*16*].

The eulerian angles β and γ describe the laboratory system with $z \parallel \mathbf{B}_0$ in the principal axes system of the dipolar coupling between I and S, *i.e.*, $Z \parallel \mathbf{r}_{IS}$ (see Fig. 3.7).

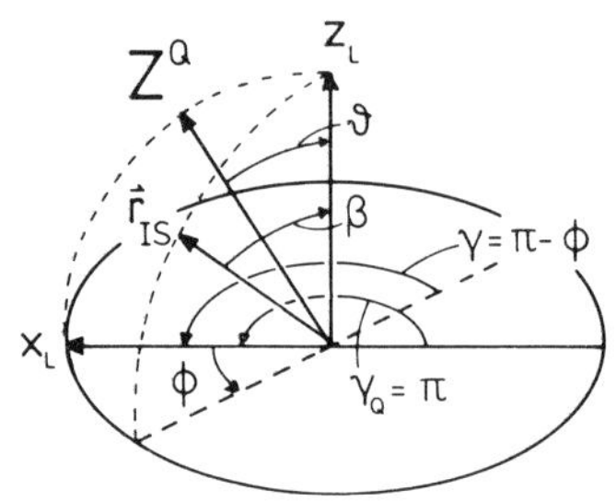

Fig. 3.7. Dipolar coupling between I and S when S experiences large quadrupolar coupling; definition of angles

The operator for the two major interactions of the S spin can be expressed similarly:

$$\mathcal{H}_S = \mathcal{H}_Z + \mathcal{H}_Q. \tag{3.13}$$

Here, we denote the eulerian angle β by ϑ in order to avoid confusion with β of Eq. (3.12). Furthermore we will assume an axially symmetric field gradient tensor. The eulerian angle γ^Q can then be chosen freely. If we choose $\gamma^Q = \pi$ we get the usual expression for $\mathcal{H}_S$ [*2, 3*]:

$$\mathcal{H}_S = -\gamma_S B_0 S_0 + \frac{e^2 q Q}{8\,I(2\,I-1)} \{(3\cos^2\vartheta - 1)[3\,S_0^2 - S(S+1)] + \frac{3\sqrt{2}}{2} \sin 2\,\vartheta\,[S_0(-S_{+1} + S_{-1}) + (-S_{+1} + S_{-1})S_0] + 3\sin^2\vartheta\,[S_{+1}^2 + S_{-1}^2]\}. \tag{3.14}$$

By this choice $\gamma^Q = \pi$, the position of the x axis of the laboratory frame is also fixed. It lies in the plane spanned by $\mathbf{B}_0$ and Z^Q in such a way that, seen from the laboratory frame, Z^Q has polar angles $\vartheta' = \vartheta$ and $\varphi' = 0$ (see Fig. 3.7). This also fixes the eulerian angle γ in Eq. (3.12). If we denote the polar angles of $\mathbf{r}_{IS}$ in the laboratory system by $\theta = \beta$ and ϕ, then $\gamma = \pi - \phi$ (cf. Fig. 3.7).

The eigenfunction of the S spin ψ_S^i, belonging to a given value of the energy E_i of the Hamiltonian (3.14) is a linear combination of the eigenfunctions $|k\rangle$ of S_Z:

$$\psi_S^i = \sum_{k=-S}^{S} a_k^i |k\rangle. \tag{3.15}$$

As the operator (3.14) is real, the coefficients a_k^i can also be chosen as real. The change in energy of a given I spin level caused by dipolar interaction with a certain spin state i of the S spin can be calculated by standard first order pertubation theory with $\mathcal{H}_D$ Eq. (3.12) and the eigenfunctions (3.15). For each nondegenerate state i the energy shift of the state $|m_I\rangle$ is given by:

$$\Delta E_{m_I,i} = \frac{-\gamma_I \gamma_S \hbar}{r_{IS}^3} m_I \Big[(3\cos^2\beta - 1) \sum_{k=-S}^{S} k\,(a_k^i)^2 + \frac{3}{2}\sin 2\beta \cos\gamma \sum_{k=-S+1}^{S} \sqrt{S(S+1) - k(k+1)}\; a_k^i a_{k-1}^i\Big]. \tag{3.16}$$

Since the selection rule $|\Delta m_I| = 1$ remains valid, $\Delta E_{m_I,i}$ directly gives the line position in the I spin spectrum due to the dipolar coupling with the S spin. The dipolar coupling is obtained as a sum of two terms of different angular dependence. Only if the internuclear vector $\mathbf{r}_{IS}$ is in the plane perpendicular to both Z^Q and $\mathbf{B}_0$, the familiar $(3\cos^2\beta - 1)$ dependence is preserved; the strength of the coupling is modified, however. If $\mathbf{r}_{IS}$, $\mathbf{B}_0$, and Z^Q all are in the same plane, on the other hand, the contribution of the $\sin^2\beta$ term is maximum and can even dominate the angular dependence. Spectra of the I spin calculated for this case are shown in Fig. 3.8 for different values of α the angle between $\mathbf{r}_{IS}$ and Z^Q, which has a fixed value for a given pair of spins in a single crystal. The parameters for the quadrupole coupling and the Zeeman interaction are those of transdiiodoethylene. For this compound, the dipolar coupling between ^{1}H and ^{127}I was of interest in order to explain the multipulse spectra [*22*]. The Zeeman interaction for the spectra in Fig. 3.8b–d corresponds to a proton resonance frequency of 90 MHz. Because of the exceedingly large quadrupole coupling of ^{127}I in organic halides, these spectra are typical for the case $\omega_L \ll \omega_Q$.

The main differences to the familiar dipolar coupling ($\omega_Q = 0$, Fig. 3.8a) are:

i) The splitting in the I spin spectrum is *not symmetric* about the central frequency. This asymmetry is even more pronounced if $\omega_L \approx \omega_Q$, (see *e.g.* [*21*]. The average of the different frequencies remains constant, however.

ii) There is no "magic angle" for which the dipolar coupling vanishes simultaneously for all S spin states.

Comparison of Fig. 3.8a and 3.8b–d shows that the $T_{2\pm1}$ terms clearly do not just give small corrections to the dipolar spectra. In fact, calculation of the dipolar coupling between I and S, without taking into account that S experiences a strong quadrupole coupling, would lead to meaningless results.

4. Molecular Motions

In this chapter, we will investigate the influence of molecular motions, rotational motions in particular, on NMR parameters (*i.e.*, line shapes and relaxation times). We will concentrate on two limiting cases:

i) Slow rotational jump motions in solids which change the solid state spectra (dealt with in Chapter 3) in a characteristic fashion. Here we will discuss, in detail, the theoretical procedures for calculating line shapes (Section 4.1).

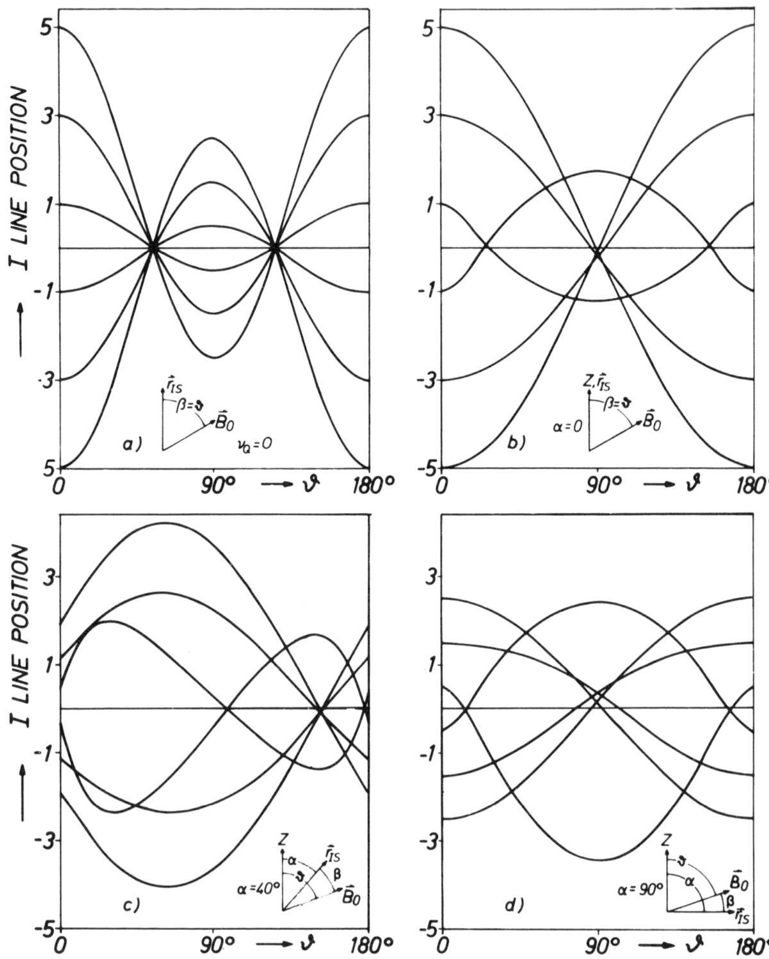

Fig. 3.8a–d. Dipolar coupling of a spin $I = 1/2$ and $S = 3/2$. (a) $\nu_Q^S = 0$; (b)–(d) transdidiodoethylene $\nu_Q^S = 227$ MHz, $\nu_L^S = 18$ MHz [*22*]. [From Spieß, H. W., *et al.*: J. Magn. Reson. **25**, 55 (1977)]

ii) Rapid rotational motions in liquids, whereby the anisotropic couplings are completely averaged out in first order. These couplings determine, however, the relaxation rates–provided oxygen or other paramagnetic impurities are absent–and, therefore, the natural widths of the NMR lines. Here we will give a uniform derivation of the longitudinal and transverse relaxation rates for the different interactions, including antisymmetric constituents of the coupling tensors, and taking into account anisotropic motion of the molecules in the liquid (Section 4.2).

The transition region, probably most interesting, where the anisotropic couplings are not completely averaged out is encountered (*e.g.*, in polymer melts). A detailed description of both the recent experimental and theoretical activity in this field is beyond the scope of this review.

4.1. Rotational Jumps in Solids

When discussing NMR spectra of solids in Chapter 3, we have tacitly assumed that the nuclei are rigidly fixed to their lattice sites. For a given nucleus, we, therefore, had a principal axes system for each interaction and, in this system the magnetic field $\mathbf{B}_0$ had a fixed orientation. We know, on the other hand, that in solids, molecular motions do take place. We will show in this section that analysis of solid state NMR spectra offers a rather direct way to study reorientational motions in solids. In fact, the narrowing of the proton NMR spectrum, due to the rotation of the benzene molecules in the solid phase, had been investigated by Andrew and Eades in 1953 [*1*] already. In regular broad line proton resonance, however, the line shape is determined by the dipolar coupling to all nuclear spins in the solid. While an isolated spin pair leads to a simple powder spectrum, as discussed in Chapter 3 (cf. Fig. 3.2b), the additional splitting, due to the remaining spins, produces a rather structureless line shape (often describable by a gaussian) and one is limited to determine its second moment–in exceptional cases, higher moments also (see, *e.g.*, Abragam [*2*]).

If the resolution of the solid state spectra is increased, however, by one of the methods mentioned in Chapter 5, powder spectra of the simple type (as shown in Figs. 3.2 and 3.3) are obtained and, therefore, we will describe how the line shape changes in the presence of motions of the spins (experimental examples will be given in Chapter 6). For simplicity we will restrict ourselves to axially symmetric tensors and furthermore, to the limit of strong Zeeman interaction $\mathcal{H}_Z \gg \mathcal{H}_\lambda$. Then the discussion is completely alike for different couplings, and we will again use the shielding tensor $\overset{\rightrightarrows}{\sigma}$ as example. Such line shapes, however, are also of considerable interest for quadrupole coupling of ^{2}D where the condition $\mathcal{H}_Z \gg \mathcal{H}_Q$ can be met easily.

4.1.1. Calculation of the Line Shape

We will restrict ourselves to *intra*molecular couplings. Then the coupling tensors R^λ_{lm} are stationary in a molecule fixed axes system. Let us consider, as an example, the nuclear spins at the corner of an octahedron. First, we note that then R^λ_{2m} will be axially symmetric and R^λ_{1m} will vanish since the nuclear site has a threefold symmetry C_{3v}. Let us discuss the spectrum for an octahedron fixed in space first. For simplicity we will use the $\overset{\rightrightarrows}{\sigma}$ tensor as an example. The frequencies ω_i for the nuclei 1,1′; 2,2′, 3,3′ then will coincide, respectively, and the NMR spectrum will consist of the three discrete frequencies as shown in Fig. 4.1.

If the molecule rotates, the eulerian angles α^λ and β^λ (describing the orientation of $\mathbf{B}_0$ in the principal axes system of a given nuclear spin) will be time dependent. Probably the easiest reorientational motion one can imagine is a jump process, whereby the octahedron as a whole keeps its orientation but the nuclei are interchanged. Thus each of the individual nuclei during the jump process will take on *all* six positions of the octahedron. It is immaterial for our consideration that the motion of the nuclei is correlated as long as we do not take into account the coupling between

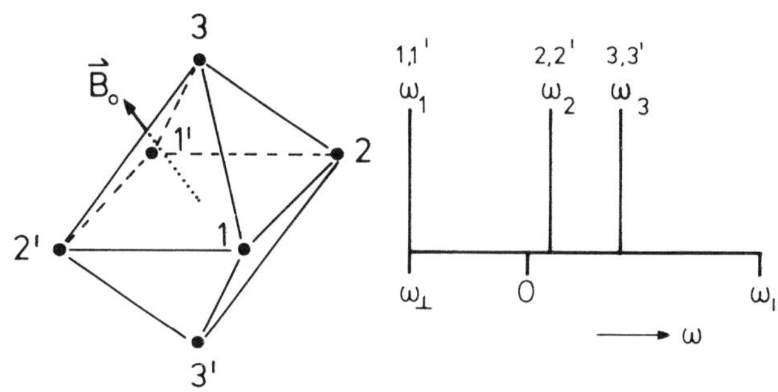

Fig. 4.1. NMR spectrum due to anisotropic shielding for an octahedron fixed in space

the nuclear spins. Then all six nuclear spins will lead to identical spectra which can be calculated according to the well-known theory of spin exchange [*3, 2*].

We will now show how to calculate the line shape for an arbitrary molecule undergoing a jump process. For each orientation of $\mathbf{B}_0$ relative to a molecular frame in which $\mathbf{B}_0$ has polar angles, ϑ and φ the spectrum of the nuclear spins is given by an intensity function $f(\omega, \vartheta, \varphi)$ which depends parametrically on the frequencies $\omega_1, \omega_2, \dots \omega_n$ for the different nuclear positions for that orientation of $\mathbf{B}_0$. To be more precise, ω_i are the frequencies corresponding to the positions the nuclei of a molecule can take on between two jumps of the molecules; thus they need not necessarily correspond to the nuclear positions of a molecule fixed in space. The spectrum $f(\omega, \vartheta, \varphi)$ is the Fourier transform of the correlation function $g(t, \vartheta, \varphi)$, the real and imaginary parts of which can be detected as the components of the nuclear magnetization $m_x(t)$ and $m_y(t)$ in the rotating frame (*i.e.*, the free induction signal)

$$f(\omega, \vartheta, \varphi) = Re \int_0^\infty g(t, \vartheta, \varphi)\, e^{-i\omega t}\, dt. \tag{4.1}$$

We will label correlation function and spectrum of a single molecule by $g(t)$ and $f(\omega)$ and of the powder spectrum by $G(t)$ and $F(\omega)$, respectively. If the molecule is fixed in space, $g(t, \vartheta, \varphi)$ is simply given by:

$$g(t, \vartheta, \varphi) = \sum_{\alpha=1}^{n} g_\alpha(t, \vartheta, \varphi) = \sum_{\alpha=1}^{n} e^{i\omega_\alpha(\vartheta,\varphi)t}, \tag{4.2}$$

where, for simplicity, we have neglected other relaxation processes. Let us now select a nuclear spin having a frequency ω_α during the time $t_1 \leqslant t \leqslant t_2$. If at time t_2 a rapid jump of the molecule occurs, changing the orientation of $\mathbf{B}_0$ in the principal axes system of that nuclear spin, its frequency will change so that for times $t \geqslant t_2$ it will be ω_β. If the jump process can be described by a stationary Markov process, the change in frequency can also be described by a stationary Markov process. By this we mean the probability that the frequency has a certain value ω_β at time $t' = t + \Delta t$, if we *know* that it had the value ω_α at time t, is

i) independent of the values of ω before the time t.

ii) dependent only on the time difference $\Delta t = t' - t$ and can accordingly be written as $P(\omega_\alpha | \omega_\beta, \Delta t)$.

Introducing the transition probability $\pi(\omega_\alpha, \omega_\beta)$, we obtain for small time intervals Δt:

$$P(\omega_\alpha | \omega_\beta, \Delta t) = \delta_{\omega_\alpha \omega_\beta} + \pi(\omega_\alpha, \omega_\beta) \cdot \Delta t, \tag{4.3}$$

where

$$\sum_{\beta=1}^{n} \pi(\omega_\alpha, \omega_\beta) = 0. \tag{4.4}$$

In order to simplify the notation, one introduces a vector **g** with the components g_α, the n-dimensional matrix $i\overset{\rightrightarrows}{\omega}$ with diagonal elements $i\omega_\alpha$, and combines the transition probabilities to a matrix $\overset{\rightrightarrows}{\pi}$ with elements $\pi_{\alpha\beta} = \pi(\omega_\alpha, \omega_\beta)$. We then obtain the compact expression [2]:

$$\frac{d\mathbf{g}}{dt} = \mathbf{g}(i\overset{\rightrightarrows}{\omega} + \overset{\rightrightarrows}{\pi}), \tag{4.5}$$

which can be integrated immediately as:

$$\mathbf{g}(t) = \mathbf{g}(0)\, e^{(i\overset{\rightrightarrows}{\omega} + \overset{\rightrightarrows}{\pi})t}. \tag{4.6}$$

In order to calculate the correlation function $g(t)$, one has to sum over the componets g_α. This summation can also be written as the scalar product with the n-dimensional vector $\mathbf{1}$ whose components are all equal to unity:

$$g(t) = \mathbf{g}(0) \cdot e^{(i\overset{\rightrightarrows}{\omega} + \overset{\rightrightarrows}{\pi})t} \cdot \mathbf{1}. \tag{4.7}$$

Since we want to calculate the absorption $f(\omega)$ as the real part of the Fourier transform of $g(t)$ (cf. Eq. (4.1), the vector $\mathbf{g}(0)$, describing the initial condition, also has to be real. Typically, $\mathbf{g}(0)$ will be proportional to the n-dimensional vector $\mathbf{1}$ introduced above.

The expression (4.7) for $g(t)$ holds for a certain orientation (ϑ, φ) of the magnetic field relative to the molecule. In order to calculate the powder spectrum, one must integrate over ϑ and φ. For this integration, we have two obvious possibilities:

i) One first calculates the spectrum for a given orientation $f(\omega, \vartheta, \varphi)$ and then integrates over ϑ and φ [4]

$$F(\omega) = \int\int f(\omega, \vartheta, \varphi) \sin \vartheta \, d\vartheta \, d\varphi. \tag{4.8a}$$

ii) One first calculates the correlation function for the complete powder spectrum

$G(t)$ by integrating over ϑ and φ:

$$G(t) = \iint g(t, \vartheta, \varphi) \sin \vartheta d\vartheta d\varphi \tag{4.8b}$$

and afterwards calculates the Fourier transform numerically [5].

Let us now discuss the difficulties encountered in the numerical calculation of the line shape $F(\omega)$, especially in the integration according to Eq. (4.8a), concentrating on cubic systems, octahedra, and tetrahedra.

4.1.2. Numerical Calculation of $F(\omega)$

For a given orientation of $\mathbf{B}_0$, we obtain from Eq. (4.1)

$$f(\omega, \vartheta, \varphi) = Re \int_0^\infty g(t)\, e^{-i\omega t} dt. \tag{4.9}$$

The line shape $f(\omega)$ does not depend explicitly upon the angles ϑ and φ, only implicitly through $\mathrm{g}(0)$ and $i\vec{\vec{\omega}}$. Therefore, we will keep the arguments ϑ, φ only in $f(\omega, \vartheta, \varphi)$. Inserting the expression (4.7) for $g(t)$ yields:

$$f(\omega, \vartheta, \varphi) = Re \int_0^\infty \mathrm{g}(0) \cdot e^{[i(\vec{\vec{\omega}} - \omega \mathfrak{E}) + \vec{\vec{\pi}}]t} \cdot \mathbf{1}\, dt$$

$$\equiv Re \int_0^\infty \mathrm{g}(0) \cdot e^{\mathfrak{A} t} \cdot \mathbf{1}\, dt. \tag{4.10}$$

Here $\mathfrak{E}$ is the n-dimensional unit matrix and the matrix $\mathfrak{A}$ is given according to Eq. (4.10) by

$$\mathfrak{A} = i(\vec{\vec{\omega}} - \omega\mathfrak{E}) + \vec{\vec{\pi}}. \tag{4.10a}$$

Equation (4.10) can be solved by two standard procedures.

The direct straightforward integration yields:

$$f(\omega, \vartheta, \varphi) = -Re\, [\mathrm{g}(0) \cdot \mathfrak{A}^{-1} \cdot \mathbf{1}]. \tag{4.11}$$

For each orientation ϑ, φ the n-dimensional matrix $\mathfrak{A}$ has to be inverted for each value of the frequency ω, needed to characterize the line shape. This method, therefore, will be useful only if the matrix inversion can be performed in closed form (see below). As demonstrated first by Gordon and McGynnis [6], we can save a considerable amount of computer time by first diagonalizing the matrix $i\vec{\vec{\omega}} + \vec{\vec{\pi}}$ numerically using the *QR* algorithm [7]. In this method, one also has to determine, however, the matrix $\mathfrak{S}$ which transforms the matrix $i\vec{\vec{\omega}} + \vec{\vec{\pi}}$ into a diagonal matrix $\vec{\vec{\lambda}}$ according to:

$$\vec{\vec{\lambda}} = \mathfrak{S}^{-1}(i\vec{\vec{\omega}} + \vec{\vec{\pi}})\mathfrak{S}. \tag{4.12}$$

Thus both eigenvalues and eigenvectors of $i\vec{\vec{\omega}} + \vec{\vec{\pi}}$ have to be calculated. Then the Fourier transform is performed easily as:

$$f(\omega, \vartheta, \varphi) = Re\, [g(0) \cdot \mathfrak{S}\, (\int_0^\infty e^{(\vec{\vec{\lambda}} - i\omega \mathfrak{E})t} dt) \mathfrak{S}^{-1} \cdot 1]$$

$$= -Re\, [g(0) \cdot \mathfrak{S}\, (\vec{\vec{\lambda}} - i\omega \mathfrak{E})^{-1} \mathfrak{S}^{-1} \cdot 1]. \tag{4.13}$$

Inversion of the diagonal matrix $\vec{\vec{\lambda}} - i\omega\,\mathfrak{E}$ is trivial and $f(\omega, \vartheta, \varphi)$ is obtained as a simple sum by:

$$f(\omega, \vartheta, \varphi) = -Re \sum_{j=1}^{n} \frac{(g(0) \cdot \mathfrak{S})_j\, (\mathfrak{S}^{-1} \cdot 1)_j}{\lambda_j - i\omega}. \tag{4.14}$$

The main advantage of this method, instead of using Eq. (4.11), results from the fact that the transformation matrix $\mathfrak{S}$ has to be calculated only once for a given orientation, regardless of how many values of the frequency ω we want to use.

In both cases, however, the spatial integration is performed over $f(\omega, \vartheta, \varphi)$ (*i.e.*, after the Fourier transform). This integration can be done analogous to the calculation of the powder spectrum for an axially symmetric tensor as described in Section 3.1.4. There we have introduced the curves of constant frequency and thereby have reduced the calculation of the line shape to the calculation of the areas between such curves of constant frequency. Let us consider an octahedron, for which, in Fig. 4.2a, curves of constant frequency, ω_1 = const, ω_2 = const, ω_3 = const, are plotted. Because of the symmetry of the octahedron, one has to integrate over $0 \leqslant \vartheta \leqslant \pi/2$ and $0 \leqslant \varphi \leqslant \pi/4$ only. The curves of constant frequency divide the surface of the sphere into triangles, the sides of which, in general, are not geodesic lines. Each of the triangles corresponds to a certain spectrum. In order to calculate $F(\omega)$, we calculate $f(\omega, \vartheta, \varphi)$ for the values of ϑ, φ corresponding to the centre of the specific triangle and weigh this spectrum according to the area of that triangle. Use of this division of the sphere guarantees a smooth transition from low jump frequencies to the rigid case.

For an octahedron, this procedure can, in fact, be used without much difficulty and offers a particularly simple way to calculate the line shape. But already for a tetrahedron the division of the sphere through the curves of constant frequency becomes prohibitively complicated (cf. Fig. 4.2b). We now have to deal with general polygons of quite different sizes. Although the areas of these polygons still can be calculated analytically [*4*], this does not offer a practical procedure. Figure 4.2b should make clear, however, that numerical instabilities are encountered if the integration is performed numerically [*4, 5*], especially in the limit of slow exchange.

If we integrate numerically over $f(\omega, \vartheta, \varphi)$, it is necessary to sample the frequency ω by equal increments $\Delta\omega$ and this leads to the complicated division of the surface of the sphere. If the spatial average is calculated in the time domain already [*i.e.*, integration over $g(t, \vartheta, \varphi)$, Eq. (4.8b)], there is no necessity to allow only values $\omega_\alpha = \omega_\perp + \kappa \cdot \Delta\omega$. Integration over $g(t, \vartheta, \varphi)$, however, is possible only if the

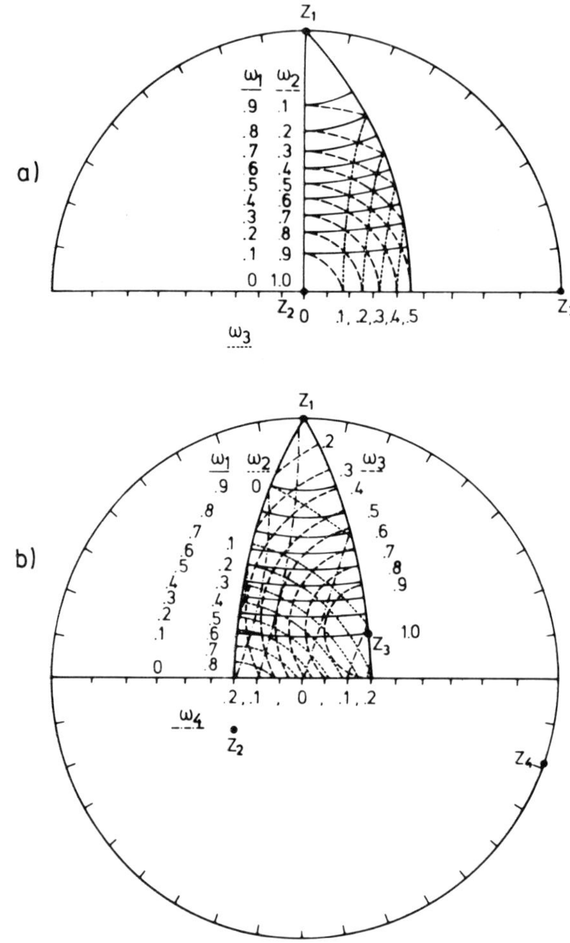

Fig. 4.2a and b. Curves of constant frequency for an octahedron (a) and a tetrahedron (b) in stereographic projection. The unique axes of the $\vec{\vec{\sigma}}$ tensors for the different positions in the molecule and the corresponding frequencies are labelled Z_i and ω_i, respectively. The frequencies $\omega_i = \omega_\perp + \kappa\Delta\omega$ are labelled after κ ranging from 0 to 1 [4]

transformation matrix **S** is known, which diagonalizes the matrix $i\vec{\vec{\omega}} + \vec{\vec{\pi}}$, cf. Eq. (4.12). Only with knowledge of $\mathfrak{S}$, we can actually calculate $g(t, \vartheta, \varphi)$ according to:

$$g(t, \vartheta, \varphi) = g(0) \cdot \mathfrak{S}\, e^{-\vec{\vec{\lambda}} t}\, \mathfrak{S}^{-1} \cdot 1. \tag{4.15}$$

After spatial integration over ϑ and φ, according to Eq. (4.8b) the line shape of the powder pattern $F(\omega)$ is obtained through a numerical complex Fourier transform, according to Eq. (4.1). In general, the transformation matrix $\mathfrak{S}$ must be determined numerically. In fact, this procedure, until now, has been used only for the exchange between two sites [5], where $\mathfrak{S}$ can easily be obtained analytically. But for more complicated systems, also, this method seems to be most promising, because it is generally applicable.

4.1.3. Line Shape for an Octahedron

For an octahedron, we have to deal with three axially symmetric tensors only. The unique axes of these three tensors, moreover, are mutually perpendicular. This makes the division of the surface of the sphere by the curves of constant frequency simple enough; the integration over $f(\omega, \vartheta, \varphi)$ can be performed easily. The inversion of the matrix $\mathfrak{A}$, Eq. (4.10a), moreover, can be performed in closed form. Therefore, this simple case will be treated here in some detail since the essential characteristics of the line shapes are obtained already. For an octahedron, the matrix $\mathfrak{A}$ is given by

$$\mathfrak{A} = \begin{pmatrix} i(\omega_1-\omega)-2\,\Omega & \Omega & \Omega \\ \Omega & i(\omega_2-\omega)-2\,\Omega & \Omega \\ \Omega & \Omega & i(\omega_3-\omega)-2\,\Omega \end{pmatrix}. \tag{4.16}$$

In terms of τ^{-1} (the mean jump frequency of the octahedron), we have $\Omega = 1/3\ \tau^{-1}$. This matrix can be inverted elementarily through formation of the adjunct matrix [8]

$$\mathfrak{A}^{-1} = \frac{1}{Det\,\mathfrak{A}} \begin{pmatrix} \begin{matrix}-(\omega_2-\omega)(\omega_3-\omega)+3\,\Omega^2 \\ -2\,i\Omega(\omega_2+\omega_3-2\,\omega)\end{matrix} & \begin{matrix}3\,\Omega^2 \\ -i\Omega(\omega_3-\omega)\end{matrix} & \begin{matrix}3\,\Omega^2 \\ -i\Omega(\omega_2-\omega)\end{matrix} \\ \begin{matrix}3\,\Omega^2 \\ -i\Omega(\omega_3-\omega)\end{matrix} & \begin{matrix}-(\omega_1-\omega)(\omega_3-\omega)+3\,\Omega^2 \\ -2\,i\Omega(\omega_1+\omega_3-2\omega)\end{matrix} & \begin{matrix}3\,\Omega^2 \\ -i\Omega(\omega_1-\omega)\end{matrix} \\ \begin{matrix}3\,\Omega^2 \\ -i\Omega(\omega_2-\omega)\end{matrix} & \begin{matrix}3\,\Omega^2 \\ -i\Omega(\omega_1-\omega)\end{matrix} & \begin{matrix}-(\omega_1-\omega)(\omega_2-\omega)+3\,\Omega^2 \\ -2\,i\Omega(\omega_1+\omega_2-2\omega)\end{matrix} \end{pmatrix}. \tag{4.17}$$

Since the mean value of the frequencies ω_1, ω_2, and ω_3 is independent of the orientation of the octahedron, we can set $\omega_1 + \omega_2 + \omega_3 = 0$. Introducing the abbreviations

$$\begin{aligned} C &= (\omega_1-\omega)(\omega_2-\omega) + (\omega_2-\omega)(\omega_3-\omega) + (\omega_1-\omega)(\omega_3-\omega) \\ D &= (\omega_1-\omega)(\omega_2-\omega)(\omega_3-\omega), \end{aligned} \tag{4.18}$$

we obtain

$$Det\,\mathfrak{A} = 2\,\Omega C - i(D + 9\omega\Omega^2). \tag{4.19}$$

The initial condition $\mathbf{g}(0)$ is given by the vector $\mathbf{1}$ and this gives for $f(\omega)$

$$\begin{aligned} f(\omega, \vartheta, \varphi) &= -Re\,\frac{-C + 27\,\Omega^2 + i\,18\,\Omega\omega}{2\,\Omega C - i(D + 9\,\omega\Omega^2)} \\ &= \frac{2\,\Omega C(C - 27\,\Omega^2) + 18\,\Omega\omega\,(D + 9\,\omega\Omega^2)}{4\,\Omega^2 C^2 + (D + 9\,\omega\Omega^2)^2}. \end{aligned} \tag{4.20}$$

We, therefore, have an analytical expression for the line shape $f(\omega, \vartheta, \varphi)$ for a given orientation and the numerical calculation of $f(\omega, \vartheta, \varphi)$, according to Eq. (4.20) is rapid indeed. In order to calculate the areas between the curves of constant frequency, cf. Fig. 4.2a, we approximate small circles, which are not described by ϑ = const, by great circles. We can check whether any serious errors are introduced in this manner by calculating the powder spectrum in the rigid case $\Omega = 0$, using that division of the surface of the sphere, and comparing the result with the theoretical powder spectrum calculated from Eq. (3.7). The excellent agreement is shown in Fig. 4.3.

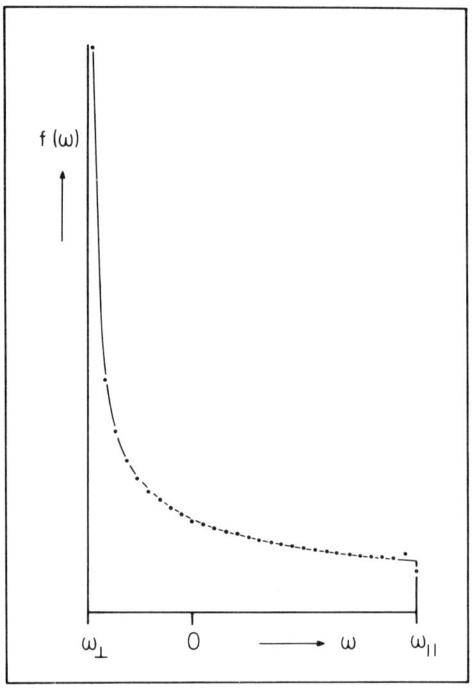

Fig. 4.3. Powder spectrum for an octahedron in the rigid limit obtained through integration using the division of the surface of the sphere of Fig. 4.2a (...); theoretical powder spectrum (——)

The results of the numerical calculation of the line shape for octahedral symmetry are depicted in Fig. 4.4. In the limit of rapid exchange $\Delta\omega \cdot \tau \ll 1$, only a sharp single line in the centre of the spectrum $\omega = 0$ is obtained since, for such a system, all anisotropic couplings are averaged out. As might have been expected, the most characteristic line shapes are observed if the mean jump frequency is comparable with the width of the powder spectra in the rigid case (*i.e.*, $\Delta\omega \cdot \tau \approx 1$). The pronounced shoulder at $\omega = \omega_\perp$, even for relatively high values of the jump frequency, seems to be especially noteworthy. For $\Delta\omega \cdot \tau \approx 5$, this leads to the characteristic double peak spectra. Experimental examples of solid state spectra exhibiting these features are presented in Chapter 6.

In the region $\Delta\omega \cdot \tau \approx 1$ the line shape for solids, where the jumps occur through fixed angles, can be distinguished clearly from the line shapes for viscous liquids,

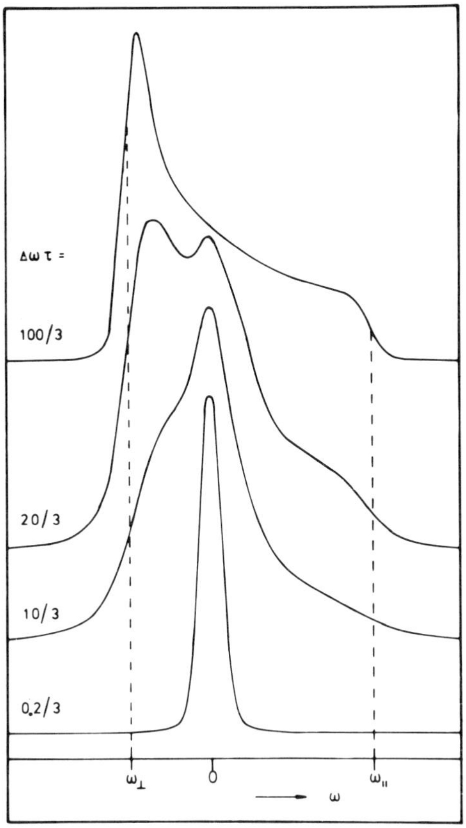

Fig. 4.4. Line shape of an octahedron for various values of the jump frequency τ^{-1}, $\Delta\omega = \omega_{\|} - \omega_{\perp}$, line width of single component comprising the rigid powder spectrum: $2\,\delta = \Delta\omega/15$

which have been investigated by Sillescu [*9, 10*] and Freed and coworkers [*11, 12*]. Contrary to the situation in solids, where the molecule, as a whole, maintains its orientation and only the nuclei are interchanged; in liquids, the orientation of the molecule changes also. Whereas in the rotational jump model for solids exchange occurs between a small number of discrete frequencies only, in an isotropic liquid, each nuclear spin will take on all frequencies between $\omega_{\perp}$ and $\omega_{\|}$ in the process of reorientation. Two models for the reorientation of molecules in liquids have been treated by Sillescu [*9*], Debye's rotational diffusion model [*13*], and Eyring's rotational jump model [*14*]. Typical line shapes [*9*] for slow motions in liquids are shown, for comparison, in Fig. 4.5. The prominent features of the powder line shapes of solids, especially in the region between $\omega_{\perp}$ and $\omega = 0$, are not obtained for isotropic reorientation in liquids. The differences to the solid state spectra are less pronounced for highly anisotropic motion in liquids [*10*].

Last but not least, the reader is referred to the elegant work of Alexander, Baram, and Luz [*15*] who calculated line shapes taking into account both correlated solid like jumps and rotational diffusion.

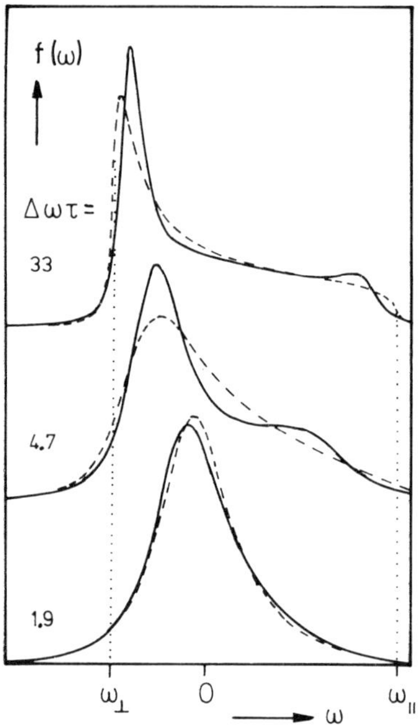

Fig. 4.5. Line shapes for highly viscous liquids and axially symmetric coupling tensor (Sillescu [9]); —— rotational diffusion model, - - - rotational jump model

4.1.4. Line Shapes for Noncubic Systems

In cubic systems, the anisotropic coupling is averaged out completely in the limit of rapid exchange and the powder spectrum degenerates to a single line. For noncubic systems, on the other hand, an anisotropic coupling remains even for rapid exchange; its strength being reduced compared with the value it has in the rigid case. Since we have restricted ourselves to the case $\mathcal{H}_Z \gg \mathcal{H}_\lambda$, the orientational dependence of $\mathcal{H}_\lambda$ is given solely by:

$$\mathcal{H}_\lambda = C^\lambda T_{20}^\lambda R_{20}^\lambda \tag{4.21}$$

(cf. Section 3.1.1). The jump process does not influence the spin dependent operators T_{20}^λ; therefore, the time average of $\mathcal{H}_\lambda$, $\overline{\mathcal{H}}_\lambda$, determining the spectra in the limit of rapid exchange, is given by:

$$\overline{\mathcal{H}}_\lambda = C^\lambda T_{20}^\lambda \overline{R_{20}^\lambda}. \tag{4.22}$$

Thus the angular dependence of the spectra is completely analogous to the situation in the rigid case; the only difference being that R_{20}^λ must be replaced by its time average

$$\overline{R_{20}^\lambda} = \sum_{i=1}^{n} a_i R_{20}^\lambda (\alpha_i^\lambda, \beta_i^\lambda). \tag{4.23}$$

Between two jumps, the molecule has a certain orientation i, the eulerian angles of $\mathbf{B}_0$ for a given principal axes system of a coupling λ are α_i^λ and β_i^λ. The $R_{20}^\lambda(\alpha_i^\lambda, \beta_i^\lambda)$ can, but need not, be different for each orientation. By introducing weighting factors a_i, we have also taken into account that the mean residence time for different orientations may be different. The $R_{20}^\lambda(\alpha_i^\lambda, \beta_i^\lambda)$ are obtained from the ρ_{2m}^λ (the irreducible components in the corresponding principal axes system) according to Eq. (2.29). Since $\overline{R_{20}^\lambda}$ is time independent for equivalent nuclear spins of a molecule, there also exists a *molecule fixed* principal axes system for which $\overline{\rho_{2m}^\lambda}$ also is time independent. The orientation of that principal axes system cannot be changed by the jump process. Two cases of importance are:

i) Rotation about a fixed axis (example: rotation of the benzene molecule about its C_6 axis) and

ii) 180° jumps which interchange nuclei of a planar molecule (example: water molecule in a solid hydrate).

For Case (i) the averaged tensor is axially symmetric if the axis of rotation has at least threefold symmetry. Such planar jump processes have been studied recently by Baram, Alexander, and Luz [*16*]. For Case (ii) the averaged tensor will in general be nonaxially symmetric, even if the coupling tensors are axially symmetric in the rigid case [*4, 5*].

We will consider a rather simple case here: the exchange between two axially symmetric tensors with mutually perpendicular axes. Such an exchange corresponds to either a 180° jump between two positions or a rotation about a fourfold axis.[14] Accordingly, the averaged coupling tensor is axially symmetric; the coupling parameter δ, however, is reduced to half its value and has opposite sign, as in the rigid case. The corresponding spectra are shown in Fig. 4.6. The exchange between two axially symmetric tensors, whose unique axes form the tetrahedral angle with each other, has been treated by Mehring [*5*]. In this case a totally asymmetric tensor ($\eta = 1$) results in the rapid exchange limit. Thus changing the angle between the unique axes from 90° to 109.5°, corresponds to a variation of the asymmetry parameter η (from $\eta = 0$ to $\eta = 1$). Line shape analyses of this kind for H_2O in solid hydrates should allow a very accurate determination of the H-O-H angle or D-O-D angle [*17*].

4.2. Hindered Rotations in Liquids, Relaxation

In liquids, the reorientation of the molecules is so fast that the anisotropic constituents of the internal couplings do not directly cause line shifts and splittings in the spectrum. These couplings determine, however, the longitudinal relaxation time T_1 and the transverse relaxation time T_2 [*2*]. We will discuss only those relaxation times measured in the laboratory frame. Starting from the pioneer work of Bloembergen *et al.* [*18*] and Kubo and Tomita [*19*], explicit expressions for the various relaxation

[14] An example would be the ^{13}C resonance in a tetragonal metal carbonyl. The $\overleftrightarrow{\sigma}$ tensor then is approximately axially symmetric since it is dominated by the electrons of the CO ligand (cf. Chapter 6)

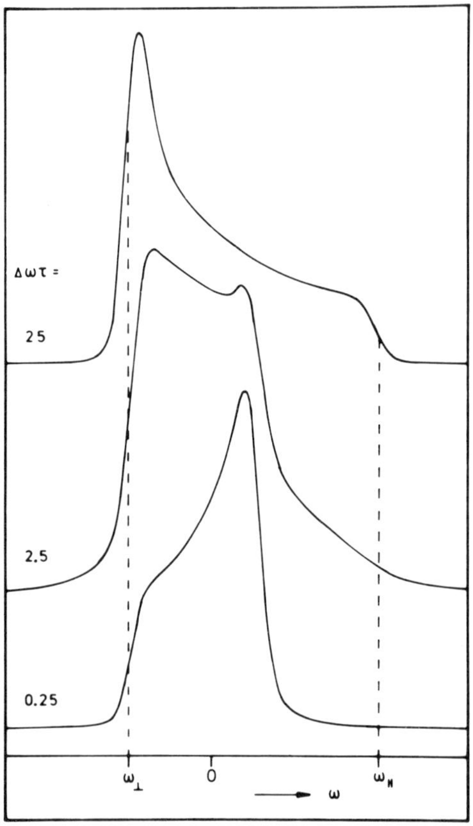

Fig. 4.6. Line shape for exchange between two axially symmetric $\vec{\vec{\sigma}}$ tensors with mutually perpendicular unique axes. Line width $2\delta = \Delta\omega/15$ (cf. Fig. 4.4)

mechanisms have been derived already by Abragam [*2*]. His restriction to isotropic diffusional motion of the molecules in the liquids has since been removed in a number of papers taking into account anisotropic rotational diffusion [*20–25*]. These papers concentrate on the calculation of autocorrelation functions of the operators R^{λ}_{2m}, depending on space variables. Here, we will use the transformation properties of the spin dependent operators T^{λ}_{lm} being irreducible tensor operators in order to obtain a uniform treatment of the different relaxation mechanisms. Such a derivation has been given already for the dipolar coupling between like spins by Haeberlen and Waugh [*26*]. If the transformation properties of the T^{λ}_{lm} are exploited, the contribution of the antisymmetric constituents of the $\vec{\vec{\sigma}}, \mathfrak{J}$, and c tensors to the relaxation rates are obtained without extra difficulty (cf. also [*27–28*]). Moreover the conditions under which the relaxation process can be described by Bloch equations [*2*] can be discussed more easily if irreducible tensor calculus is used. Finally, the differences for different relaxation mechanisms are seen more clearly.

We will treat each mechanism independent of all the others. This means we write for the total relaxation rate of a nuclear species I, $1/T^I_1$

$$\frac{1}{T^I_1} = \sum_{\lambda} \frac{1}{T^{\lambda}_1}. \tag{4.24}$$

Interference of the different couplings, leading to deviations from Eq. (4.24) have been discussed in detail by Blicharski [*29*]. In liquids of low viscosity these deviations are exceedingly small and treating the mechanisms separately is justified. Similarly, we will treat the nuclear spins in the liquid as being independent of each other since the coupling between the spins is weak. Even in the case of correlated motion (*e.g.*, the protons in a methyl group), the deviations from the expressions derived, neglecting the correlated motion, are extraordinarily small [*2*]. This statement holds for simple nonviscous liquids and nonselective pulse experiments only. For methyl groups in polymers, on the other hand, deviations from Eq. (4.24) are indeed expected [*30*].

For selective experiments where individual lines of a multiplet are irradiated and their time constants determined, cross correlation terms must be considered (cf. below, Refs. [*43–46*]). For the majority of relaxation studies where nonselective pulses are used and the time constants are obtained mainly from the initial decay or built-up, respectively, of magnetization, our treatment is directly applicable. Most of our derivations can also be used to describe selective experiments, except, of course, for some final simplifications (*e.g.*, where averages over the spin states, corresponding to the individual lines of a multiplet, are taken or cross terms are neglected).

4.2.1. Classification of Relaxation Mechanisms

The derivation of spin relaxation rates for liquids is a typical example for a semi-classical treatment of a problem. The spin dependent parts of the coupling Hamiltonians are treated quantum mechanically, whereas the remaining parts of the operators, depending on the space variables, are treated classically. This semi-classical treatment seems but natural from our introduction of spin Hamiltonians. There we had introduced the space dependent parts of the operators through phenomenological coupling parameters which now become time dependent. The brownian motion of the molecules thus renders the spin Hamiltonians time dependent ($\mathcal{H}_\lambda \to \mathcal{H}_\lambda(t)$).

The starting point for our derivation of the relaxation rates is the equation of motion of the density matrix in the rotating frame ρ^* (cf. [*2*] Chapter VIII, Eq. (33)):

$$\frac{d\rho^*}{dt} = -\int_0^\infty \langle[\mathcal{H}_\lambda^*(t), [\mathcal{H}_\lambda^*(t-\tau), \rho^*(t)]]\rangle \, d\tau, \tag{4.25}$$

where $\langle ... \rangle$ indicates average over an ensemble of spin systems. The conditions for the validity of this master equation have been discussed in detail by Abragam [*2*] and later (*e.g.*, by Deutch [*31*]) and we will not repeat their arguments here. Within the semi-classical treatment, the fact that the spin system reaches thermal equilibrium is normally introduced as an "ad hoc" assumption (see, however, Deutch [*31*]). Therefore we have to replace $\rho^*(t)$ by $\rho^*(t) - \rho_0^*$ in all linear equations for $\rho^*(t)$. Here, ρ_0^* is the density matrix in thermal equilibrium (cf. [*2*], p. 277).

The density matrix ρ^* and the internal Hamiltonians $\mathcal{H}_\lambda^*$ in the rotating frame are given by:

$$\rho^*(t) = e^{i\mathcal{H}_Z t}\rho(t)\,e^{-i\mathcal{H}_Z t}, \quad \mathcal{H}_\lambda^*(t) = e^{i\mathcal{H}_Z t}\,\mathcal{H}_\lambda(t)\,e^{-i\mathcal{H}_Z t}, \tag{4.26}$$

where ρ and $\mathcal{H}_\lambda$ are defined in the laboratory frame. By inserting the expression (2.24) for $\mathcal{H}_\lambda$, we obtain:

$$\mathcal{H}_\lambda^*(t) = C^\lambda e^{i\mathcal{H}_Z t}\Big(\sum_{l=0}^{2}\sum_{m=-l}^{l}(-1)^m\,T_{lm}^\lambda(t)\,R_{l-m}^\lambda(t)\Big)\,e^{-i\mathcal{H}_Z t}$$

$$= C^\lambda \sum_{l=0}^{2}\sum_{m=-l}^{l}(-1)^m e^{i\mathcal{H}_Z t}\,T_{lm}^\lambda(t)\,e^{-i\mathcal{H}_Z t}R_{l-m}^\lambda(t), \tag{4.27}$$

since $R_{lm}^\lambda(t)$ is independent of spin variables.

Time dependent Hamiltonians $\mathcal{H}_\lambda^*(t)$ can result from two basically different mechanisms and this calls for the following classification of the different relaxation mechanisms depicted schematically in Fig. 4.7.

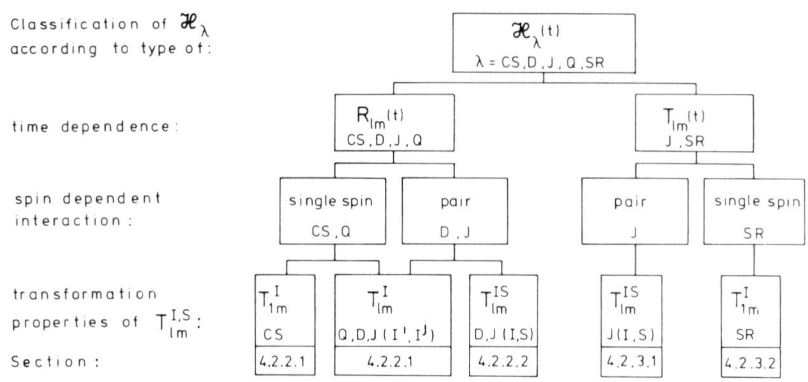

Fig. 4.7. Classification of relaxation mechanisms

i) $R_{lm}^\lambda \to R_{lm}^\lambda(t)$, the parts of the coupling operators that depend on space variables, become time dependent because of molecular motion and this time-dependence determines the relaxation rate. This is the typical case in NMR relaxation, normally encountered for $\lambda = D, Q, CS, J$. According to Eq. (2.29), the R_{lm}^λ are constructed from the principal values ρ_{lm}^λ and Wigner rotation matrices $\mathcal{D}^{(l)}(\Omega^\lambda)$. The case where the orientation of the principal axes system, relative to the laboratory frame is time dependent, because of molecular rotation will concern us most ($\Omega^\lambda = \Omega^\lambda(t)$). Clearly $R_{lm}^\lambda(t)$ can also result from fluctuating principal values $\rho_{lm}^\lambda(t)$ or from a combination of both time dependencies.

ii) $T_{lm}^{\lambda} \rightarrow T_{lm}^{\lambda}(t)$, the spin dependent operators are time dependent in the laboratory frame, and this time dependence determines the relaxation rate. Two cases are important here: spin-rotation interaction, where $T_{lm}^{SR}(t)$ results from fluctuating angular velocity of the molecules in the liquid, and the scalar coupling between two nuclei I and S, if the spin S has a short relaxation time due to a different interaction independent of the spin I (*e.g.*, the quadrupole coupling).

A further distinction and classification results from the nature of the spin dependent interaction. The couplings $\lambda = CS, Q, SR$ describe single spin interactions, whereas, $\lambda = D, J$ correspond to pair interactions.

The final classification takes into account the different transformation properties of the spin dependent operators T_{lm}^{I}. These transformation properties ultimately determine the detailed form of the relaxation rates. If we deal with a single spin species I only, we must distinguish whether or not spin operators with $l = 2$ are present. For pair interactions, we similarly must distinguish whether or not we are dealing with like spins I^i, I^j or unlike spins I, S. The last row of Fig. 4.7 finally refers to the section where the various cases are treated.

4.2.2. Time Dependent $R_{lm}^{\lambda}(t)$

The derivation of the relaxation rates for the different interactions $\lambda = CS, D, J, Q$ is quite similar in this case since we always are dealing with second rank tensors in the spin independent part of the Hamiltonian. As a consequence, the information about the time dependent processes in the liquid that can be gathered by exploiting these mechanisms is also quite similar. The different Hamiltonians $\mathcal{H}_{\lambda}(t)$ differ significantly in the spin dependent parts T_{lm}^{λ}, however, and this leads to different results for the relaxation rates. In particular, the T_{lm}^{λ} for $\lambda = CS, Q, D,$ and J for like spins I^i, I^j depend upon spin operators for a single nuclear species only, contrary to dipole-dipole or J-coupling between unlike spins I, S. To be sure, if one works under high resolution conditions, the spins of a given isotope can be unlike spins if their relative chemical shift is sufficiently large and selective pulses are used [*32*]. The Zeeman interaction is different in these cases:

$$\mathcal{H}_Z = -\gamma_I I_0 \mathbf{B}_0 = \omega_I I_0,$$

and (4.28)

$$\mathcal{H}_Z = -(\gamma_I I_0 + \gamma_S S_0) B_0 = \omega_I I_0 + \omega_S S_0.$$

Likewise, the transformation to the interaction representation, according to $e^{i\mathcal{H}_Z t}\, T_{lm}^{\lambda}\, e^{-i\mathcal{H}_Z t}$, is different and the two cases will be treated separately here.

4.2.2.1. Single Spin Species I

For the transformation $e^{i\mathcal{H}_Z t}\, T_{lm}^{\lambda}\, e^{-i\mathcal{H}_Z t}$, we recall the following relations, (cf. [*2*], p. 23):

$$e^{i\omega t I_0} I_0 e^{-i\omega t I_0} = I_0$$

$$e^{i\omega t I_0} I_{\pm 1} e^{-i\omega t I_0} = I_{\pm 1} e^{\pm i\omega t}. \tag{4.29}$$

Since we are dealing with a single spin species only ($\omega_I = \omega$), we get:

$$e^{i\omega t I_0} T^I_{lm} e^{-i\omega t I_0} = T^I_{lm} e^{im\omega t}. \tag{4.30}$$

This means the T^I_{lm} operators, being defined in the laboratory frame, maintain their properties as irreducible tensor operators of the same rank l if we transform to the rotating frame. Nevertheless, we have to make another distinction here: For $\lambda = D$, J, Q the T^λ_{lm} are built out of spin operators only; therefore, $T^\lambda_{lm} = T^I_{lm}$, $l = 0, 1, 2$. For $\lambda = CS$, however, the spin dependent operator $T^{CS}_{lm} = T^I_{lm}$ is built out of spin operators, on the one hand, and components of the static field $\mathbf{B}_0$. Consequently, in *spin space* for $\lambda = CS$, we have T^I_{1m} operators only. This has important consequences for the relaxation rates; the most striking one being the possibility of $T_2 > T_1$.

$\lambda = D, J, Q$

We will derive the relaxation rate for the general case $l = 1, 2$, although $l = 1$ is possible for $\lambda = J$ only. This will enable us, however, to use this general treatment for other relaxation mechanisms also.

Inserting Eq. (4.27) into the master equation (4.25), together with Eq. (4.30) yields for the equation of motion of the density matrix:

$$\frac{d\rho^*}{dt} = -C^2 \sum_{l=1}^{2} \sum_{m', m=-l}^{l} (-1)^{m+m'} e^{i(m+m')\omega t} [T_{lm}, [T_{lm'}, \rho^*]]$$

$$\cdot \int_0^\infty \langle R_{l-m}(t) R_{l-m'}(t-\tau)\rangle e^{-im'\omega\tau} d\tau, \tag{4.31}$$

where we have dropped the indices λ and I in order to keep the notation simple. For the calculation of the relaxation rates, we are interested in the slow change of the elements of the density matrix only; not in rapid oscillations with frequencies of the order of the Larmor frequency ω. Therefore, we can restrict ourselves to secular terms with $m' = -m$:

$$\frac{d\rho^*}{dt} = -C^2 \sum_{l=1}^{2} \sum_{m=-l}^{l} [T_{lm}, [T_{l-m}, \rho^*]]$$

$$\cdot \int_0^\infty \langle R_{l-m}(t) R_{lm}(t-\tau)\rangle e^{im\omega\tau} d\tau. \tag{4.32}$$

Of the correlation functions $\langle R_{l_1 m_1}(t) R_{l_2 m_2}(t-\tau)\rangle$ possible in principle we have only considered those with $l_1 = l_2$ in Eq. (4.31) to begin with. "Mixed" correlation functions with $l_1 \neq l_2$ vanish because of the different transformation properties of $R_{l_1 m_1}$ and $R_{l_2 m_2}$ under rotation. The orthogonality properties of the Wigner rotation matrices do not call for only $l_1 = l_2$ but also for $m_1 = -m_2$ (cf. Section 4.2.4). The arguments for the simplifications incorporated in Eq. (4.32) are, therefore, stronger than it might seem at this point.

We will denote the autocorrelation functions encountered here by $f_{lm}(\tau)$:

$$f_{lm}(\tau) \equiv (-1)^{l+m} \langle R_{l-m}(t)\, R_{lm}(t-\tau)\rangle, \tag{4.33}$$

and their half-sided Fourier transforms by $g_{lm}(\omega)$:

$$g_{lm}(\omega) \equiv \int_0^\infty f_{lm}(\tau)\, e^{i\omega\tau} d\tau. \tag{4.34}$$

In order to solve the equation of motion of the density matrix, one can solve Eq. (4.32) for the individual matrix elements (Redfield theory [*33*], for a detailed elementary description see also Slichter [*34*]). Since we are interested in the relaxation rates only, which can be obtained from the equation of motion of the expectation values of the spin operators, rather than from the complete equation of motion of the density matrix, we can solve Eq. (4.32) directly in operator form [*2*]. By multiplying with I_M ($M = 0, \pm 1$) and taking the trace, we directly obtain the equation of motion for the expectation value of the spin operator I_M

$$\langle I_M \rangle = Tr\,\{\rho^* \cdot I_M\}. \tag{4.35}$$

Now we have

$$\begin{aligned} Tr\{[T_{lm}, [T_{l-m}, \rho^*]]\, I_M\} &= Tr\{[T_{l-m}, [T_{lm}, I_M]]\, \rho^*\} \\ &\equiv \langle [\, T_{l-m}, [T_{lm}, I_M]]\rangle \end{aligned} \tag{4.36}$$

(cf. [*2*], Chapter VIII, Eq. (44).

The double commutators encountered in the master equation and rewritten in Eq. (4.36) can now be reduced to single commutators by using the commutation relations of the irreducible tensor operators with angular momentum operators (cf. monographs on group theory cited in Chapter 2, Refs. [*5–9*].

$$\begin{aligned} [I_{\pm 1}, T_{lm}] &= \mp \frac{1}{\sqrt{2}} \sqrt{l(l+1) - m(m \pm 1)}\; T_{lm \pm 1} \\ [I_0, T_{lm}] &= m\, T_{lm}. \end{aligned} \tag{4.37}$$

Remembering that $[T_{l-m}, [T_{lm}, I_M]] = [[I_M, T_{lm}], T_{l-m}]$, we obtain for the expectation values of the spin operators

$$\left\langle \frac{dI_0}{dt} \right\rangle = -C^2 \sum_{l=1}^{2} \sum_{m=-l}^{l} (-1)^{l+m} \langle [T_{lm}, T_{l-m}]\rangle m g_{lm}(m\omega)$$

$$\left\langle \frac{dI_{\pm 1}}{dt} \right\rangle = -C^2 \sum_{l=1}^{2} \sum_{m=-l}^{l} (-1)^{l+m} \langle [T_{lm\pm 1}, T_{l-m}]\rangle$$

$$\left(\mp \frac{1}{\sqrt{2}}\right) \sqrt{l(l+1) - m(m \pm 1)}\; g_{lm}(m\omega). \tag{4.38}$$

The single commutators can be read simply from the tables of Clebsch-Gordan coefficients since multiplication of two irreducible tensor operators (cf. [*35*], Eq. (5.1.9)) yields

$$T_{l_1 m_1} \cdot T_{l_2 m_2} = \sum_{L=|l_1-l_2|}^{l_1+l_2} \sum_{M=-L}^{L} T_{LM}(l_1 l_2 LM \,|\, l_1 m_1 l_2 m_2). \tag{4.39}$$

For the commutators, therefore, we obtain ($l_1 = l_2$)

$$[T_{lm_1}, T_{lm_2}] = \sum_{L=0}^{2l} \sum_{M=-L}^{L} T_{LM} \{(llLM \,|\, lm_1 lm_2) - (llLM \,|\, lm_2 lm_1)\}. \tag{4.40}$$

The symmetry properties of the Clebsch-Gordan coefficients:

$$(l_1 l_2 LM | l_1 m_1 l_2 m_2) = (-1)^{l_1+l_2-L} (l_2 l_1 LM \,|\, l_2 m_2 l_1 m_1) \tag{4.41}$$

(cf. [*35*], Eq. (3.5.14)) leave terms with odd L only in Eq. (4.40). The Clebsch-Gordan coefficients needed are tabulated in convenient form, *e.g.*, in the appendix to Heine's book [*36*]. With these, one readily derives the commutators collected in Table 4.1.

Now we can give the equation of motion of the expectation values of the spin operators[15]:

$$\begin{aligned}\left\langle \frac{dI_0}{dt} \right\rangle = -C^2 \{ & \sqrt{2}\, [g_{11}(\omega) + g_{1-1}(-\omega)] \langle T_{10} \rangle \\ & + \frac{2}{\sqrt{10}} [4\, g_{22}(2\,\omega) + 4\, g_{2-2}(-2\,\omega) + g_{21}(\omega) + g_{2-1}(-\omega)] \langle T_{10} \rangle \\ & + \frac{4}{\sqrt{10}} [g_{22}(2\,\omega) + g_{2-2}(-2\,\omega) - g_{21}(\omega) - g_{2-1}(-\omega)] \langle T_{30} \rangle \},\end{aligned} \tag{4.42}$$

$$\begin{aligned}\left\langle \frac{dI_{\pm 1}}{dt} \right\rangle = -C^2 \{ & \sqrt{2}\, [g_{1\mp 1}(\mp\omega) + g_{10}(0)] \langle T_{1\pm 1} \rangle \\ & + \frac{2}{\sqrt{10}} [2\, g_{2\mp 2}(\mp 2\,\omega) + 3\, g_{2\mp 1}(\mp\omega) + 2\, g_{2\pm 1}(\pm\omega) + 3\, g_{20}(0)] \langle T_{1\pm 1} \rangle \\ & + \frac{2\sqrt{6}}{\sqrt{10}} [g_{2\mp 2}(\mp 2\,\omega) - g_{2\mp 1}(\mp\omega) + g_{2\pm 1}(\pm\omega) - g_{20}(0)] \langle T_{3\pm 1} \rangle \}.\end{aligned} \tag{4.43}$$

---

[15] Actually, here and in subsequent equations, the T_{1M} should have another label indicating whether they originate from commutators $[T_{1m}, T_{1m'}]$ or from $[T_{2m'}\ T_{2m'}]$. Since this label is provided by the index l in the g_{lm}, we can do without it.

Table 4.1. Commutators of irreducible tensor operators

$l = 1$	$[T_{11}, T_{1-1}] = \sqrt{2}\, T_{10}$	for $\langle I_0 \rangle$
	$[T_{1\pm1}, T_{10}] = \pm\sqrt{2}\, T_{1\pm1}$	for $\langle I_{\pm1} \rangle$
$l = 2$	$[T_{22}, T_{2-2}] = \frac{2}{\sqrt{10}}(T_{30} + 2\, T_{10})$	for $\langle I_0 \rangle$
	$[T_{21}, T_{2-1}] = \frac{2}{\sqrt{10}}(2\, T_{30} - T_{10})$	
	$[T_{2\pm2}, T_{2\pm1}] = \pm\frac{2}{\sqrt{10}}(\sqrt{3}\, T_{3\pm1} + \sqrt{2}\, T_{1\pm1})$	for $\langle I_{\pm1} \rangle$
	$[T_{2\pm1}, T_{20}] = \pm\frac{2}{\sqrt{10}}(\sqrt{2}\, T_{3\pm1} - \sqrt{3}\, T_{1\pm1})$	

The expectation values $\langle T_{1M} \rangle$ and $\langle T_{3M} \rangle$ are proportional, respectively, to the expectation values $\langle I_M \rangle$ and $\langle T^I_{3M} \rangle$ constructed from the nuclear spin operators according to the Wigner-Eckart theorem [cf. Chapter 2, Eqs. (2.14), (2.15)]. The factors of proportionality will be calculated below.

Conditions for the Validity of Bloch Equations

The expectation values of the spin operators are defined as deviations from the equilibrium values [cf. discussion of Eq. (4.25)]. Therefore the Bloch equations have the form

$$\left\langle \frac{dI_0}{dt} \right\rangle = -\frac{1}{T_0}\langle I_0 \rangle, \quad \left\langle \frac{dI_{\pm1}}{dt} \right\rangle = -\frac{1}{T_{\pm1}}\langle I_{\pm1} \rangle. \tag{4.44}$$

The relation to the relaxation times T_1 and T_2 will be given below. The Eqs. (4.42) and (4.43), however, contain terms $\langle T_{30} \rangle$ and $\langle T_{3\pm1} \rangle$, meaning that the time development of $\langle I_M \rangle$ can, in general, not be described by Bloch equations. In practice, however, the terms with $l = 3$ vanish in many cases of importance and we will discuss the conditions for which this is the case here. The arguments concerning I_0 and $I_{\pm1}$ are similar; therefore, we will concentrate on I_0:

i) *Total Spin* $I = 1$ (*i.e.*, dipole-dipole or J-coupling of two spins $I = 1/2$ or quadrupole coupling for $I = 1$. In these cases, $\langle T_{3M} \rangle \equiv 0$ since, according to the Wigner-Eckhart theorem, T_{3M} can have no matrix elements different from 0 (cf. Section 2.2).

ii) *Extreme Narrowing.* In the limit of rapid motions, for which $\omega^2\tau_l^2 \ll 1$ (τ_l: correlation time, see below), the g_{lm} are independent of the frequency and $g_{lm} \sim \tau_l$ independent of m (cf. below Section 4.2.4). Then the sums over g_{2m}, by which the $\langle T_{3M} \rangle$ are multiplied in Eqs. (4.42) and (4.43), vanish. This condition generally is fulfilled for nuclear spin relaxation in nonviscous liquids.

iii) Existence of a *spin temperature,* meaning that the spin system during the whole relaxation process can be described by a—sufficiently high—spin temperature. In this case, the expectation value $\langle T_{30} \rangle$ vanishes even though T_{30} has matrix elements.

Table 4.2. Commutators of T_{lm}^{λ} operators

λ	$[T_{11}, T_{1-1}]$	$[T_{22}, T_{2-2}]$	$[T_{21}, T_{2-1}]$
D	–	$\left\{\frac{1}{2}[I(I+1)-I_0^i I_0^j]\cdot\right.$	$\left\{-\frac{1}{2}(I_{+1}^i I_{-1}^j + I_{-1}^i I_{+1}^j + I_0^i I_0^j)\cdot\right.$
J	$\frac{1}{2}\mathbf{I}^i\mathbf{I}^j(I_0^i + I_0^j(I_0^i + I_0^j)$	$\left.\cdot(I_0^i + I_0^j)\right\}$	$\left.\cdot(I_0^i + I_0^j)\right\}$
Q	–	$[2I(I+1)-1-2I_0^2]I_0$	$[2I(I+1)-\frac{1}{2}-4I_0^2]I_0$

It is convenient, for this purpose, to express T_{30} in terms of commutators. From Table 4.2, we deduce:

$$T_{30} = \frac{1}{\sqrt{10}}\{[T_{22}, T_{2-2}] + 2[T_{21}, T_{2-1}]\}. \tag{4.45}$$

These commutators are calculated below (cf. Table 4.2). For $\lambda = D, J$ we obtain:

$$T_{30} \sim -\frac{3}{2}(I_0^i I_0^{j^2} + I_0^{i^2} I_0^j) + [\frac{1}{2}I(I+1) - (I_{+1}^i I_{-1}^j + I_{-1}^i I_{+1}^j)] \cdot (I_0^i + I_0^j). \tag{4.46}$$

If a spin temperature exists, it is permissible to average I_0^i and I_0^j separately and this gives for the expectation value $\langle T_{30}\rangle$

$$\langle T_{30}\rangle \sim -\frac{3}{2}(\langle I_0^i\rangle\langle I_0^{j^2}\rangle + \langle I_0^{i^2}\rangle\langle I_0^j\rangle)$$

$$+ [\frac{1}{2}I(I+1) - \langle I_{+1}^i I_{-1}^j + I_{-1}^i I_{+1}^j\rangle](\langle I_0^i\rangle + \langle I_0^j\rangle). \tag{4.46a}$$

For high temperatures, in addition, we have

$$\langle I_0^{i^2}\rangle = \langle I_0^{j^2}\rangle = \frac{1}{3}I(I+1)$$

$$\langle I_{+1}^i I_{-1}^j + I_{-1}^i I_{+1}^j\rangle = 0, \tag{4.47}$$

(cf. [*2*], pp. 290 and 291) and this then gives

$$\langle T_{30}\rangle \sim [-\frac{1}{2}I(I+1) + \frac{1}{2}I(I+1)](\langle I_0^i\rangle + \langle I_0^j\rangle) = 0. \tag{4.48}$$

For $\lambda = Q$, we obtain similarly:

$$T_{30} \sim -10\, I_0^3 + 2[3\, I(I+1) - 1]\, I_0, \tag{4.49}$$

and, therefore, if a spin temperature is defined:

$$\langle T_{30}\rangle \sim -10\,\langle I_0^3\rangle + 2\,[3\,I(I+1)-1]\,\langle I_0\rangle. \tag{4.50}$$

In the high temperature limit, the diagonal elements of the density matrix are given by $\rho^* \sim 1 + \epsilon I_0$. This gives for $\langle T_{30}\rangle = Tr(\rho^* T_{30})$:

$$\begin{aligned}\langle T_{30}\rangle &\sim -\,5Tr\,(I_0^4) + [3\,I(I+1)-1]\,Tr\,(I_0^2)\\ &\sim -\frac{5}{15}\,I(I+1)\,(2\,I+1)\,[3\,I(I+1)-1]\\ &\quad+\frac{1}{3}\,I(I+1)\,(2\,I+1)\,[3\,I(I+1)-1] = 0.\end{aligned} \tag{4.50a}$$

Of course, typically, the assumption of a spin temperature in the case of quadrupolar coupling will not be justified, contrary to the situation for dipolar relaxation. Our derivation shows, however, that disregarding this for the sake of the argument, the conditions that must be met for the $l = 3$ terms to vanish are equal in both cases. Thus only if the deviations of the density matrix from a state describable by a spin temperature are severe, the $l = 3$ terms are significant leading to nonexponential relaxation. Since we are mainly concerned with liquids of low viscosity here, we will only keep T_{1M} in the following.

In order to simplify the notation, we introduce the following abbreviations:

$$\begin{aligned}G_{10}^{\lambda}(\omega) &= g_{11}^{\lambda}(\omega) + g_{1-1}^{\lambda}(-\omega)\\ G_{1\pm1}^{\lambda}(\omega) &= g_{1\mp1}^{\lambda}(\mp\omega) + g_{10}^{\lambda}(0)\\ G_{20}^{\lambda}(\omega) &= 4\,g_{22}^{\lambda}(2\,\omega) + 4\,g_{2-2}^{\lambda}(-2\,\omega) + g_{21}^{\lambda}(\omega) + g_{2-1}^{\lambda}(-\omega)\\ G_{2\pm1}^{\lambda}(\omega) &= 2\,g_{2\mp2}^{\lambda}(\mp2\,\omega) + 3\,g_{2\mp1}^{\lambda}(\mp\omega) + 2\,g_{2\pm1}^{\lambda}(\pm\omega) + 3\,g_{20}^{\lambda}(0),\end{aligned} \tag{4.51}$$

and Eqs. (4.42) and (4.43) then read

$$\left\langle\frac{dI_0}{dt}\right\rangle = -\,C^{\lambda^2}\,[\sqrt{2}\,G_{10}^{\lambda}(\omega)\,\langle T_{10}^{\lambda}\rangle + \frac{2}{\sqrt{10}}\,G_{20}^{\lambda}(\omega)\,\langle T_{10}^{\lambda}\rangle], \tag{4.42a}$$

$$\left\langle\frac{dI_{\pm1}}{dt}\right\rangle = -\,C^{\lambda^2}[\sqrt{2}\,G_{1\pm1}^{\lambda}(\omega)\,\langle T_{1\pm1}^{\lambda}\rangle + \frac{2}{\sqrt{10}}\,G_{2\pm1}^{\lambda}(\omega)\,\langle T_{1\pm1}^{\lambda}\rangle]. \tag{4.43a}$$

According to the Wigner-Eckart theorem, the expectation values $\langle T_{1M}\rangle$ appearing in the equations of motion of the expectation values of the spin operators, Eqs. (4.42) and (4.43) are proportional to the expectation values of $\langle I_M\rangle$:

$$\langle T_{1m}^{\lambda}\rangle = a_l^{\lambda}\,\langle I_M\rangle. \tag{4.52}$$

The factors of proportionality are independent of M. They are different, however, for different couplings λ and for different l (cf. Table 4.1 and Footnote [15]). By inserting Eq. (4.52) into Eqs. (4.42a) and (4.43a) we obtain Bloch equations:

$$\left\langle \frac{dI_0}{dt} \right\rangle = -C^{\lambda^2} [\sqrt{2}\, G_{10}^{\lambda}(\omega)\, a_1^{\lambda} + \frac{2}{\sqrt{10}}\, G_{20}^{\lambda}(\omega)\, a_2^{\lambda}]\, \langle I_0 \rangle, \tag{4.42b}$$

$$\left\langle \frac{dI_{\pm 1}}{dt} \right\rangle = -C^{\lambda^2} [\sqrt{2}\, G_{1\pm 1}^{\lambda}(\omega)\, a_1^{\lambda} + \frac{2}{\sqrt{10}}\, G_{2\pm 1}^{\lambda}(\omega) a_2^{\lambda}] \langle I_{\pm 1} \rangle. \tag{4.43b}$$

In order to calculate the factors of proportionality a_l, one has to calculate selected commutators explicitly for the different interactions. From Table 4.1 we get:

$$l = 1 : [T_{11}, T_{1-1}] = \sqrt{2}\, T_{10} = \sqrt{2}\, a_1 I_0$$

$$l = 2 : 2\, [T_{22}, T_{2-2}] - [T_{21}, T_{2-1}] = \sqrt{10}\, T_{10} = \sqrt{10}\, a_2 I_0 . \tag{4.53}$$

After inserting the T_{lm} operators from Table 2.4, one obtains (after straightforward but lengthy calculations) the commutators collected in Table 4.2.

We then obtain for the quadrupole coupling

$$a_1^Q = 0, a_2^Q = \frac{1}{2\sqrt{10}} [4\, I(I+1) - 3] = \frac{1}{2\sqrt{10}} (2\,I - 1)\,(2\,I + 3). \tag{4.54}$$

For the dipole-dipole or J-coupling, the commutators contain terms I^i and I^j, which depend on the total spin F formed out of I^i and I^j : $F = I^i + I^j$, $0 \leqslant F \leqslant 2\,I$. By forming F^2, we obtain:

$$\mathbf{I}^i\, \mathbf{I}^j = \frac{1}{2} [F(F+1) - 2\,I(I+1)]. \tag{4.55}$$

Consequently, the factors of proportionality also depend on F:

$$a_1^D = 0, a_{1,F}^J = \frac{1}{4\sqrt{2}} [F(F+1) - 2\,I(I+1)]$$

$$a_{2,F}^{D,J} = \frac{1}{2\sqrt{2}} [3\,I(I+1) - \frac{1}{2} F(F+1)]. \tag{4.56}$$

In our derivation of relaxation rates we are interested only in the expectation values $\langle I_M \rangle$ and, therefore, we do not need the individual $a_{l,F}$ but a weighted average (cf. Appendix D):

$$\overline{a_1^J} = \frac{1}{6\sqrt{2}}\, I(I+1), \quad \overline{a_2^{D,J}} = \frac{\sqrt{10}}{12}\, I(I+1). \tag{4.57}$$

The explicit expressions $\langle dI_M/dt \rangle$ obtained in this way are collected below in Table 4.3.

Table 4.3. Time constants of Bloch equations (4.44); $\frac{1}{T_0}$: longitudinal relaxation rate $\frac{1}{T_1}$; $\mathrm{Re}(\frac{1}{T_{\pm 1}})$: transverse relaxation rate $\frac{1}{T_2}$

λ	$\frac{1}{T_0}$	$\frac{1}{T_{\pm 1}}$
CS	$\gamma_I^2 B_0^2 \frac{1}{2}[G_{10}^{CS}(\omega) + g_{2-1}^{CS}(-\omega) + g_{21}^{CS}(\omega)]$	$\gamma_I^2 B_0^2 \frac{1}{2}[2 g_{00}^{CS}(0) + g_{1\mp 1}^{CS}(\mp\omega) + g_{2\mp 1}^{CS}(\mp\omega) + \frac{4}{3} g_{20}^{CS}(0)]$
D	$\frac{2}{3}\gamma_I^4 \hbar^2 I(I+1)\,[G_{20}^{D}(\omega)]$	$\frac{2}{3}\gamma_I^4 \hbar^2 I(I+1)\,[G_{2\pm 1}^{D}(\omega)]$
J	$\frac{1}{6} I(I+1)\,[G_{10}^{J}(\omega) + G_{20}^{J}(\omega)]$	$\frac{1}{6} I(I+1)\,[G_{1\pm 1}^{J}(\omega) + G_{2\pm 1}^{J}(\omega)]$
Q	$\frac{e^2 Q^2 (2I+3)}{40 I^2 (2I-1)\hbar^2}\,[G_{20}^{Q}(\omega)]$	$\frac{e^2 Q^2 (2I+3)}{40 I^2 (2I-1)\hbar^2}\,[G_{2\pm 1}^{Q}(\omega)]$

$\lambda = CS.$

As already noted in the beginning of this section, the *chemical shift* has to be treated separately because of the different transformation properties of the corresponding spin dependent operator. According to Table 2.4, together with Eq. (2.14) we have

$$T_{lm}^{CS} = B_0 b_{lm} T_{1m} \tag{4.58}$$

with

$$b_{00} = 1, \quad b_{10} = b_{2\pm 2} = 0, \quad b_{20} = \sqrt{\frac{2}{3}},$$

$$b_{11} = -b_{1-1} = b_{2\pm 1} = \frac{1}{\sqrt{2}}.$$

Inserting this in Eq. (4.31) we obtain

$$\frac{d\rho^*}{dt} = -\gamma_I^2 B_0^2 \sum_{l=0}^{2} \sum_{m,m'=-1}^{1} (-1)^{m+m'} e^{i(m+m')\omega t} b_{lm} b_{lm'} [T_{1m}, [T_{1m'}, \rho^*]]$$

$$\int_0^\infty \langle R_{l-m}(t) R_{l-m'}(t-\tau)\rangle e^{-im'\omega\tau} d\tau. \tag{4.31'}$$

The subsequent derivation is the same as described above. Analogous to Eq. (4.38) we have

$$\left\langle \frac{dI_0}{dt} \right\rangle = -\gamma_I^2 B_0^2 \sum_{l=0}^{2} \sum_{m=-1}^{1} (-1)^{l+m} b_{lm} b_{l-m} \langle [T_{1m}, T_{1-m}] \rangle m g_{lm}(m\omega)$$

$$\left\langle \frac{dI_{\pm 1}}{dt} \right\rangle = -\gamma_I^2 B_0^2 \sum_{l=0}^{2} \sum_{m=-1}^{1} (-1)^{l+m} b_{lm} b_{l-m} \langle [T_{1m\pm 1}, T_{1-m}] \rangle$$

$$\cdot \left(\mp \frac{1}{\sqrt{2}}\right) \sqrt{2 - m(m \pm 1)}\, g_{lm}(m\omega). \qquad (4.38')$$

By use of Table 4.1 and after inserting the b_{lm} factors, we arrive at

$$\left\langle \frac{dI_0}{dt} \right\rangle = -\gamma_I^2 B_0^2 \frac{\sqrt{2}}{2} [-g_{1-1}(-\omega) - g_{11}(\omega) - g_{2-1}(-\omega) - g_{21}(\omega)] \langle T_{10} \rangle, \qquad (4.42')$$

$$\left\langle \frac{dI_{\pm 1}}{dt} \right\rangle = -\gamma_I^2 B_0^2 \frac{\sqrt{2}}{2} [-2 g_{00}(0) - g_{1\mp 1}(\mp\omega) - g_{2\mp 1}(\mp\omega) - \frac{4}{3} g_{20}(0)] \langle T_{1\pm 1} \rangle. \qquad (4.43')$$

[cf. Eq. (4.42)].
As only $\langle T_{1M} \rangle$ appear in Eqs. (4.42′) and (4.43′) we don't have problems concerning the validity of Bloch equations. The factor of proportionality [Eq. (4.52)] is calculated according to the procedure described in the previous section to yield:

$$a_1^{CS} = -\frac{1}{\sqrt{2}}. \qquad (4.59)$$

Calculation of $1/T_1$ and $1/T_2$.
By comparison with Eq. (4.44) we are now able to write down Bloch equations for $\langle I_0 \rangle$ and $\langle I_{\pm 1} \rangle$ and obtain general expressions for the time constants $1/T_0$ and $1/T_{\pm 1}$. The results are given in Table 4.3.

For random stationary motions, the correlation times $f_{lm}(\tau)$ are real and, furthermore, since there is symmetry between the past and the future, we have $f_{lm}(\tau) = f_{lm}(-\tau) = f_{l-m}(\tau)$ [cf. Eq. (4.33)]. From Eq. (4.34) we then deduce $g_{l-m}(-\omega) = g_{lm}(\omega)$. Then the expression for $1/T_0$ given in Table 4.3, together with the definitions of the G_{lm} [Eq. (4.51)] are real and can be identified directly with the longitudinal relaxation rate $1/T_1$. For $1/T_{\pm 1}$ we get complex expressions, the real parts of which are equal and can be identified with the transverse relaxation rate $1/T_2$. The imaginary parts corresponds to small frequency shifts, which are smaller in magnetude than $1/T_2$ and, therefore, smaller than the natural line width.

In order to obtain explicit expressions for the relaxation rates $1/T_1$ and $1/T_2$ from Table 4.3 we have to know the autocorrelation functions and their Fourier transforms. The calculation of $f_{lm}(\tau)$ for anisotropic diffusional motion will be dealt with in Section 4.2.5. Here we will assume, for simplicity, that the time dependence of the R_{lm} can be described by such a Markov process that can be characterized by a single correlation time τ_l for a given l. This case is discussed in detail in Abragam's book [2]. In our notation, we then obtain:

$$Re\, g_{lm}(\omega) \equiv g_l(\omega) = \frac{1}{2l+1} \frac{\tau_l}{1 + \omega^2 \tau_l^2} \sum_{m'=-l}^{l} (-1)^{l+m'} \rho_{l-m'} \rho_{lm'}. \qquad (4.60)$$

Inserting the expressions (2.31) for ρ_{lm}, gives:

$$l = 1: \sum_{m'=-1}^{1} (-1)^{1+m'} \rho_{1-m'}\, \rho_{1m'} = 2\,(\rho_{XY}^2 + \rho_{XZ}^2 + \rho_{YZ}^2).$$

$$l = 2: \sum_{m'=-2}^{2} (-1)^{m'} \rho_{2-m'}\, \rho_{2m'} = \frac{3}{2}\delta^2 \left(1 + \frac{\eta^2}{3}\right). \tag{4.61}$$

With this and the conventional coupling parameters (cf. Table 2.3), we obtain from Table 4.3 explicit expressions for the relaxation rates, collected for convenience in Table 4.4.

Table 4.4. Relaxation rates for isotropic motion

CS	$\frac{1}{T_1} = \gamma_I^2 B_0^2 \left\{\frac{2}{3} [(\rho_{XY}^{CS})^2 + (\rho_{XZ}^{CS})^2 + (\rho_{YZ}^{CS})^2] \frac{\tau_1}{1 + \omega^2\tau_1^2} + \frac{2}{15} \Delta\sigma^2 \left(1 + \frac{\eta_{CS}^2}{3}\right) \frac{\tau_2}{1 + \omega^2\tau_2^2}\right\}$
	$\frac{1}{T_2} = \gamma_I^2 B_0^2 \left\{\sigma_{iso}^2\, \tau_0 + \frac{1}{3} [(\rho_{XY}^{CS})^2 + \ldots] \frac{\tau_1}{1+\omega^2\tau_1^2} + \frac{1}{15} \Delta\sigma^2 \left(1 + \frac{\eta_{CS}^2}{3}\right) [\frac{4}{3}\tau_2 + \frac{\tau_2}{1+\omega^2\tau_2^2}]\right\}$
D[a]	$\frac{1}{T_1} = \frac{2}{5} \gamma_I^4 \hbar^2 I(I+1) \overline{r_{ij}^{-6}} \left[\frac{\tau_2}{1 + \omega^2\tau_2^2} + \frac{4\tau_2}{1 + 4\omega^2\tau_2^2}\right]$
	$\frac{1}{T_2} = \frac{1}{5} \gamma_I^4 \hbar^2 I(I+1) \overline{r_{ij}^{-6}} \left[3\tau_2 + \frac{5\tau_2}{1 + \omega^2\tau_2^2} + \frac{2\tau_2}{1 + 4\omega^2\tau_2^2}\right]$
J	$\frac{1}{T_1} = \frac{2}{9} I(I+1) \left\{[(\rho_{XY}^J)^2 + (\rho_{XZ}^J)^2 + (\rho_{YZ}^J)^2] \frac{\tau_1}{1 + \omega^2\tau_1^2} + \frac{1}{5} \Delta J^2 \left(1 + \frac{\eta_Y^2}{3}\right) [\frac{\tau_2}{1 + \omega^2\tau_2^2} + \frac{4\tau_2}{1 + 4\omega^2\tau_2^2}]\right\}$
	$\frac{1}{T_2} = \frac{1}{9} I(I+1) \left\{[(\rho_{XY}^J)^2 + \ldots] [\tau_1 + \frac{\tau_1}{1 + \omega^2\tau_1^2}] + \frac{1}{5} \Delta J^2 \left(1 + \frac{\eta_Y^2}{3}\right) [3\tau_2 + \frac{5\tau_2}{1 + \omega^2\tau_2^2} + \frac{2\tau_2}{1 + 4\omega^2\tau_2^2}]\right\}$
Q	$\frac{1}{T_1} = \frac{3}{200} \left(\frac{e^2 qQ}{\hbar}\right)^2 \left(1 + \frac{\eta_Q^2}{3}\right) \frac{2I+3}{I^2(2I-1)} \left[\frac{\tau_2}{1 + \omega^2\tau_2^2} + \frac{4\tau_2}{1 + 4\omega^2\tau_2^2}\right]$
	$\frac{1}{T_2} = \frac{3}{400} \left(\frac{e^2 qQ}{\hbar}\right)^2 \left(1 + \frac{\eta_Q^2}{3}\right) \frac{2I+3}{I^2(2I-1)} \left[3\tau_2 + \frac{5\tau_2}{1 + \omega^2\tau_2^2} + \frac{2\tau_2}{1 + 4\omega^2\tau_2^2}\right]$

[a] If the molecule contains several equivalent spins, $\overline{r_{ij}^{-6}}$ has to be replaced by $\sum_j \overline{r_{ij}^{-6}}$. If there are n non-equivalent spins of the same species for a nonselective experiment, the average $\frac{1}{n} \sum_{i=1} \overline{r_{ij}^{-6}}$ has to be taken.

A few comments to Tables 4.3 and 4.4 seem to be in order. In the extreme narrowing limit $\omega^2\tau_l^2 \ll 1$, typically valid for nuclear relaxation in nonviscous liquids, the $G_{lm}(\omega)$ are independent of both ω and m. Then for $\lambda = D, J, Q$ the relaxation rates $1/T_1$ and $1/T_2$ are equal. Our derivation of the relaxation rates makes clear that this is a direct consequence of the fact that, in these cases, the spin dependent operators T_{lm}^{λ} are irreducible tensors of the same rank in the spin space of the I spins as they are in the laboratory frame. This is not the case for $\lambda = CS$ and, as a consequence here, the relaxation rates $1/T_m$ are not proportional to the G_{lm} (cf. Table 4.3). Therefore, even in the extreme narrowing limit we have $T_1 \neq T_2$. If, in addition to molecular motion, we have chemical exchange, this leads to a $l = 0$ contribution to $1/T_2$. The expression given in Table 4.4 contains σ_{iso}, given on an absolute scale (*i.e.*, relative to the free nucleus). Accordingly, for exchange between two different chemical sites, σ_{iso} has to be replaced by the *difference* between the two shielding constants.

In the diffusion model one has the following relation between the rotational diffusion coefficient D_s and the correlation times of reorientation τ_l:

$$\tau_l = \frac{D_s}{l(l+1)} \tag{4.62}$$

(cf. [2] and below, Section 4.2.3), and, accordingly, $\tau_1 = 3\,\tau_2$. Then we can give the expressions of Table 4.4 in terms of a single correlation time (see also [27]). Whereas relaxation through anisotropic J-coupling generally seems to be unimportant for light nuclei, relaxation through anisotropic shielding can give appreciable contributions in the high magnetic fields of superconducting magnets, (cf. Chapter 6). As long as we have to deal with only the *symmetric* part of the *shielding tensor,* we have in the absence of exchange phenomena $T_1/T_2 = 7/6$ and the longitudinal relaxation time is longer than the transverse. The contribution of the antisymmetric part of $\overset{\rightrightarrows}{\sigma}$ to the relaxation rate, however, is larger for $1/T_1$ than for $1/T_2$. This is a direct consequence of the fact that $T_{10}^{CS} = 0$ (cf. Table 2.4). Because of $T_{10}^{CS} = 0$, the antisymmetric part of the shielding tensor neither gives a contribution to the solid state spectrum in first order (cf. Section 3.1.1) nor does it contribute to the transverse relaxation at $\omega = 0$.

If the point symmetry of the nuclear site in a molecule is so low that antisymmetric constituents ρ_{1m}^{CS} can be different from zero, analysis of relaxation rates neglecting the contribution of the $l = 1$ terms will not be justified in general. Introducing

$$[(\rho_{XY}^{CS})^2 + (\rho_{XZ}^{CS})^2 + (\rho_{YZ}^{CS})^2] = (\Delta_a\sigma)^2,$$

we obtain, in this case

$$\frac{1}{T_1^{CS}} = \gamma_I^2 B_0^2 [2\,(\Delta_a\sigma)^2 + \frac{2}{15}\,(\Delta\sigma)^2 (1 + \frac{\eta^2}{3})]\,\tau_2,$$

$$\frac{1}{T_2^{CS}} = \gamma_I^2 B_0^2 [(\Delta_a\sigma)^2 + \frac{7}{45}\,(\Delta\sigma)^2 (1 + \frac{\eta^2}{3})]\,\tau_2. \tag{4.63}$$

Therefore, even for $|\Delta_a\sigma| = |\Delta\sigma|/4$, the contributions of the $l = 1$ and $l = 2$ terms to the relaxation rate are almost equal. Even the case $T_2 > T_1$ could be encountered here.

Of course, for molecules of low symmetry, the reorientation in the liquid likewise will be anisotropic in most cases; then one has to use the more general expressions given below, cf. Tables 4.9 and 4.10. The direct application of Eq. (4.63), therefore, is limited. The general observations discussed here remain valid, however.

4.2.2.2. Unlike Spins I and S

The spin dependent parts of the Hamiltonians describing the dipole-dipole and the J-coupling between unlike spins I and S are also given by irreducible tensors T^D_{2m} and T^J_{00}, T^J_{1m}, and T^J_{2m}, respectively (cf. Table 2.4). In the interaction representation defined by

$$\mathcal{H}_Z = \omega_I I_0 + \omega_S S_0 \tag{4.28a}$$

the $T^{D,J}_{lm}$ have different transformation properties. The transformation $e^{i\mathcal{H}_Z t}\, T^\lambda_{lm}\, e^{-i\mathcal{H}_Z t}$ can be interpreted as transformation to two rotating spin frames, one for the I and one for the S spins with angular velocities ω_I and ω_S, respectively. This is only a minor complication, however, since for like spins we have described already the space and spin dependent parts of the coupling Hamiltonians in different coordinate frames. Here, in addition, we have to treat I and S in different frames.

The transformation $e^{i\mathcal{H}_Z t}\, T^\lambda_{lm}\, e^{-i\mathcal{H}_Z t}$ can be performed analogous to Eq. (2.10)

$$e^{i\mathcal{H}_Z t}\, T_{LM}\, e^{-i\mathcal{H}_Z t} = \sum_{m=-1}^{1} (1\,m\,1\,M-m\mid 1\,1\,LM)\, T^I_{1m}\, T^S_{1M-m}\, e^{i[m\,\omega_I + (M-m)\,\omega_S]t} \tag{4.64}$$

where according to Eq. (2.14) $T^I_{1m} = I_m$, $T^S_{1m} = S_m$.

The derivation of the relaxation rates starting from the master Eq. (4.25) is performed completely analogous to the treatment of like spins. Instead of Eq. (4.32) we obtain[16]

$$\frac{d\rho^*}{dt} = -C^2 \sum_{L=0}^{2} \sum_{M=-L}^{L} \sum_{m=-1}^{1} (1\,m\,1\,M-m\mid 1\,1\,LM)\,(1\,-m\,1\,-M+m\mid 1\,1\,L\,-M)$$
$$\cdot\,[I_m S_{M-m}, [I_{-m} S_{-M+m}, \rho^*]] \int_0^\infty \langle R_{L-M}(t)\, R_{LM}(t-\tau)\rangle\, e^{i[m\,\omega_I + (M-m)\,\omega_S]\tau}\, d\tau. \tag{4.32'}$$

The autocorrelation function $f_{LM}(\tau)$ and their Fourier transform $g_{LM}(\omega)$ are still defined by Eqs. (4.33) and (4.43).

In order to determine the equation of motion of the expectation values of the spin operators $\langle I_m\rangle$ and $\langle S_m\rangle$, we multiply by $(I_m + S_m)$ and take the trace [cf. Eq. (4.35)]. The derivation that follows will be described for the I spin only. Of course, the same expressions also can be used for the S spin if I and S are inter-

[16] In the following, we will often have to deal with Clebsch-Gordan coefficients $(1\ m_1\ 1\ m_2 \mid 1\,1\,LM)$. Because of this structure $(l_1 = l_2)$, we can do without extra brackets even if m_1, m_2, or M stand for expressions and not just for integers. We give an example as an illustration: $(1\ -m\ 1\ M-m \mid 1\,1\,LM-2\,m) \equiv (1\ (-1)\ 1\ (M-m) \mid 1\,1\,L\ (M-2\,m))$.

changed. The double commutators again are reduced to single commutators and we obtain instead of Eq. (4.38)

$$\left\langle\frac{dI_0}{dt}\right\rangle = -C^2 \sum_{L=0}^{2} \sum_{M=-L}^{L} \sum_{m=-1}^{1} (1\, m\, 1\, M-m \mid 11\, LM)(1\, -m\, 1\, -M+m \mid 11\, L\, -M)$$

$$(-1)^{L+M} \langle [I_m S_{M-m}, I_{-m} S_{m-M}] \rangle m g_{LM}(\omega_m),$$

$$\left\langle\frac{dI_{\pm 1}}{dt}\right\rangle = -C^2 \sum_{L=0}^{2} \sum_{M=-L}^{L} \sum_{m=-1}^{1} (1\, m\, 1\, M-m \mid 11\, LM)(1\, -m\, 1\, -M+m \mid 11\, L\, -M)$$

$$(-1)^{L+M} \langle [I_{m\pm 1} S_{M-m}, I_{-m} S_{m-M}] \rangle \left(\mp\frac{1}{\sqrt{2}}\right) \sqrt{2-m(m\pm 1)}\, g_{LM}(\omega_m),$$

with

$$\omega_m = m\,\omega_I + (M-m)\,\omega_S, \tag{4.38''}$$

where the argument in $g_{LM}(\omega_m)$ is given by $\omega_m = m\,\omega_I + (M-m)\,\omega_S$. The commutators needed here are collected in Table 4.5. Restricting again to the high temperature approximation (see above) for the equation of motion of the expectation values of the spin operators $\langle I_m \rangle$, we only need consider the simple expressions at the right of the arrows. We then obtain

$$\begin{aligned}
\left\langle\frac{dI_0}{dt}\right\rangle = -C^2 \Big\{ & \frac{1}{6} S(S+1)\,[2\, g_{00}(-\omega_I+\omega_S) + 2\, g_{00}(\omega_I-\omega_S) \\
& + g_{1-1}(-\omega_I) + g_{11}(\omega_I) + g_{10}(-\omega_I+\omega_S) + g_{10}(\omega_I-\omega_S) \\
& + 2\, g_{2-2}(-\omega_I-\omega_S) + 2\, g_{22}(\omega_I+\omega_S) + g_{2-1}(-\omega_I) + g_{21}(\omega_I) \\
& + \frac{1}{3} g_{20}(-\omega_I+\omega_S) + \frac{1}{3} g_{20}(\omega_I-\omega_S)]\,\langle I_0 \rangle \\
& + \frac{1}{6} I(I+1)\,[-2\, g_{00}(-\omega_I+\omega_S) - 2\, g_{00}(\omega_I-\omega_S) \\
& - g_{10}(-\omega_I+\omega_S) - g_{10}(\omega_I-\omega_S) + 2\, g_{2-2}(-\omega_I-\omega_S) \\
& + 2\, g_{22}(\omega_I+\omega_S) - \frac{1}{3} g_{20}(-\omega_I+\omega_S) - \frac{1}{3} g_{20}(\omega_I-\omega_S)]\,\langle S_0 \rangle \Big\},
\end{aligned} \tag{4.42''}$$

$$\begin{aligned}
\left\langle\frac{dI_{\pm 1}}{dt}\right\rangle = -C^2 \frac{1}{6} S(S+1)\, \Big\{ & 2\, g_{00}(\mp\omega_I \pm \omega_S) + 2\, g_{00}(0) \\
& + g_{1\mp 1}(\mp\omega_I) + g_{1\mp 1}(\mp\omega_S) + g_{1\pm 1}(\pm\omega_S) + g_{10}(\mp\omega_I \pm \omega_S) \\
& + 2\, g_{2\mp 2}(\mp\omega_I \mp \omega_S) + g_{2\mp 1}(\mp\omega) + g_{2\mp 1}(\mp\omega_S) + g_{2\pm 1}(\pm\omega_S) \\
& + \frac{1}{3} g_{20}(\mp\omega_I \pm \omega_S) + \frac{4}{3} g_{20}(0) \Big\}\, \langle I_{\pm 1} \rangle.
\end{aligned} \tag{4.43''}$$

Table 4.5. Commutators for the coupling of unlike spins[17]

$[I_{+1}S_0, I_{-1}S_0] = -S_0^2 I_0$	$\rightarrow -\frac{1}{3}S(S+1)\langle I_0\rangle$
$[I_0 S_{-1}, I_0 S_{+1}] = I_0^2 S_0$	$\rightarrow +\frac{1}{3}I(I+1)\langle S_0\rangle$
$[I_{+1}S_{\pm 1}, I_{-1}S_{\mp 1}] = \frac{1}{2}[(\pm I(I+1) \mp I_0^2)S_0 + (S(S+1) - S_0^2)I_0]$	$\rightarrow \pm\frac{1}{3}I(I+1)\langle S_0\rangle + \frac{1}{3}S(S+1)\langle I_0\rangle$
$[I_{\pm 1}S_0, I_0 S_0] = \mp S_0^2 I_{\pm 1}$	$\rightarrow \mp\frac{1}{3}S(S+1)\langle I_{\pm 1}\rangle$
$[I_{\pm 1}S_{\pm 1}, I_0 S_{\mp 1}] = [\frac{1}{2}(\pm S(S+1) \mp S_0^2 + S_0) \mp I_0 S_0]I_{\pm 1}$	$\rightarrow \pm\frac{1}{3}S(S+1)\langle I_{\pm 1}\rangle$
$[I_{\pm 1}S_{\mp 1}, I_0 S_{\pm 1}] = [\frac{1}{2}(\pm S(S+1) \mp S_0^2 - S_0) \pm I_0 S_0]I_{\pm 1}$	$\rightarrow \pm\frac{1}{3}S(S+1)\langle I_{\pm 1}\rangle$

This shows that for dipole-dipole or J-coupling of unlike spins I and S, Bloch equations are expected for $\langle I_{\pm 1}\rangle$ only, even within the high temperature approximation. For isotropic motion, the corresponding relaxation rates $1/T_2$ are given below in Table 4.6.

Table 4.6. Relaxation rates for the I spins due to dipole-dipole or J-coupling to unlike S spins for isotropic motion. Footnote to Table 4.4 applies likewise

D

$$\frac{1}{T_1^{II}} = \frac{2}{5}\gamma_I^2\gamma_S^2\hbar^2 S(S+1)r_{IS}^{-6}\left[\frac{1}{3}\frac{\tau_2}{1+(\omega_I-\omega_S)^2\tau_2^2} + \frac{\tau_2}{1+\omega_I^2\tau_2^2} + \frac{2\tau_2}{1+(\omega_I+\omega_S)^2\tau_2^2}\right]$$

$$\frac{1}{T_2} = \frac{1}{5}\gamma_I^2\gamma_S^2\hbar^2 S(S+1)r_{IS}^{-6}\left[\frac{4}{3}\tau_2 + \frac{1}{3}\frac{\tau_2}{1+(\omega_I-\omega_S)^2\tau_2^2} + \frac{2\tau_2}{1+\omega_S^2\tau_2^2} + \frac{\tau_2}{1+\omega_I^2\tau_2^2} + \frac{2\tau_2}{1+(\omega_I+\omega_S)^2\tau_2^2}\right]$$

J

$$\frac{1}{T_1^{II}} = \frac{2}{9}S(S+1)\Big\{3J_{iso}^2\frac{\tau_0}{1+(\omega_I-\omega_S)^2\tau_0^2} + [(\rho_{XY}^J)^2 + (\rho_{XZ}^J)^2 + (\rho_{YZ}^J)^2]\left[\frac{\tau_1}{1+(\omega_I-\omega_S)^2\tau_1^2} + \frac{\tau_1}{1+\omega_I^2\tau_1^2}\right] + \frac{1}{5}\Delta J^2\left(1+\frac{\eta_Y^2}{3}\right)\left[\frac{1}{3}\frac{\tau_2}{1+(\omega_I-\omega_S)^2\tau_2^2} + \frac{\tau_2}{1+\omega_I^2\tau_2^2} + \frac{2\tau_2}{1+(\omega_I+\omega_S)^2\tau_2^2}\right]\Big\}$$

$$\frac{1}{T_2} = \frac{1}{9}S(S+1)\Big\{3J_{iso}^2\left[\tau_0 + \frac{\tau_0}{1+(\omega_I-\omega_S)^2\tau_0^2}\right] + [(\rho_{XY}^J)^2 + \dots]\left[\frac{\tau_1}{1+\omega_I-\omega_S)^2\tau_1^2} + \frac{\tau_1}{1+\omega_I^2\tau_1^2} + \frac{2\tau_1}{1+\omega_S^2\tau_1^2}\right] + \frac{1}{5}\Delta J^2\left(1+\frac{\eta_Y^2}{3}\right)\left[\frac{4}{3}\tau_2 + \frac{1}{3}\frac{\tau_2}{1+(\omega_I-\omega_S)^2\tau_2^2} + \frac{2\tau_2}{1+\omega_S^2\tau_2^2} + \frac{\tau_2}{1+\omega_I^2\tau_2^2} + \frac{2\tau_2}{1+(\omega_I+\omega_S)^2\tau_2^2}\right]\Big\}$$

[17] According to our introduction of the spin Hamiltonians, we have to replace the Clebsch-Gordan coefficient $(1\,m\,1\,-m\,|\,1\,1\,0\,0)$ by $(-1)^m$ [cf. Eq. (2.11a)].

For $\langle dI_0/dt \rangle$, we obtain two coupled differential equations (cf. [2], p. 295)

$$\left\langle \frac{dI_0}{dt} \right\rangle = -\frac{1}{T_1^{II}} \langle I_0 \rangle - \frac{1}{T_1^{IS}} \langle S_0 \rangle$$

$$\left\langle \frac{dS_0}{dt} \right\rangle = -\frac{1}{T_1^{SI}} \langle I_0 \rangle - \frac{1}{T_1^{SS}} \langle S_0 \rangle, \tag{4.65}$$

where $1/T_1^{II}$ and $1/T_1^{IS}$ can be read directly from Eq. (4.42″). By exchanging I and S, we obtain $1/T_1^{SI}$ and $1/T_1^{SS}$.

An example of extraordinary importance for I, S coupling is $I = {}^{13}C$ and $S = {}^{1}H$. If one works without proton decoupling and determines the relaxation rates of the individual lines in the ^{13}C multiplet the coupled equations must be solved. This problem is treated in detail in the literature [*37–40*]. Furthermore, the correlated motion of the protons and anisotropic motion of the molecules has to be taken into account carefully for carbons in methyl or methylene groups [*41–43*]. Similar effects also occur in proton relaxation, if relaxation rates for the individual lines of a strongly coupled multiplet are determined separately [*44, 45*]. Double resonance methods allowing separate determination of auto- and cross-correlation terms are described in a recent review article by Werbelow and Grant [*46*].

Most frequently, however, the ^{13}C relaxation rates are determined with proton decoupling. In order to understand what is observed under these conditions we remind the reader again that $\langle I_0 \rangle$ and $\langle S_0 \rangle$ in Eq. (4.65) represent the deviations from the corresponding equilibrium values, cf. Eq. (4.44)

$$\langle I_0 \rangle = \langle I_Z \rangle - \langle I_Z \rangle_0,$$

$$\langle S_0 \rangle = \langle S_Z \rangle - \langle S_Z \rangle_0. \tag{4.66}$$

Decoupling of the S spins assures $\langle S_Z \rangle = 0$ and, correspondingly, $\langle S_0 \rangle = -\langle S_Z \rangle_0$ to be constant. Therefore, the two equations at (4.65) are decoupled. Since, in addition, $\langle S_Z \rangle_0$ is proportional to $\langle I_Z \rangle_0$, the only effect of the decoupling of importance in this context is to change the observerable nuclear magnetization of the I spins (nuclear Overhauser effect [*47*] analogous to the electron-nuclear Overhauser effect [*48*]). The time constant of the longitudinal relaxation is given simply by $1/T_1 = 1/T_1^{II}$, however. This quantity is given for isotropic motion in Table 4.6. In the extreme narrowing limit, we again have $T_1 = T_2$ as for dipolar or J-coupling of like spins. It should be noted that, strictly speaking, these arguments hold for spin pairs only. For CH_2 or CH_3 groups, the correlated motion of the protons can cause nonexponential recovery of the $\langle I_0 \rangle$ magnetization after a 180° pulse even under proton decoupling [*45, 46*].

It should be noted that the expressions for the relaxation rates, due to coupling of like spins, cannot be obtained by setting $I = S$ and $\omega_I = \omega_S$. If one would do this, $1/T_2$ from Table 4.6 would, as an example, give a value too small by a factor 2/3 [*2*].

The scalar contribution to the relaxation corresponds to Abragam's „*scalar relaxation of the first kind*“ ([*2*], p. 308), since it has been calculated for fluctuation of the scalar coupling constant. The correlation time τ_0 then can be identified with the exchange time of chemical exchange.

4.2.3 Time Dependent $T^{\lambda}_{lm}(t)$

4.2.3.1 Scalar Relaxation *I, S*

Scalar relaxation of the I spins due to the S spins is possible also if the scalar coupling constant J_{iso} is time independent; the S spins, however, have short relaxation times T_1 and T_2 due to another relaxation mechanism independent of the I, S coupling. Typically, S will be a quadrupolar nucleus relaxing rapidly because of quadrupolar coupling. This case has been labeled by Abragam as „*scalar coupling of the second kind*“ and is treated in detail in [2], p. 309 ff. We also will consider only the case where the relaxation times of the spin S, T_1^S, and T_2^S are much longer than the reorientational correlation times τ_1 and τ_2 of the molecule. Then only the scalar part of the I, S coupling Hamiltonian has to be considered here; for the $l = 1$ and $l = 2$ terms Section 4.2.2 applies.

The coupling Hamiltonian then is given simply by:

$$\mathcal{H}_J = J_{iso}^2 \sum_{m=-1}^{m} I_m S_{-m} \tag{4.67}$$

(cf. Table 2.3 and 2.4). After transformation into the interaction representation with $\mathcal{H}_Z$ given by Eq. (4.28a) we can again use our general derivation given in Section 4.2.2. Neglecting the effect the I, S coupling has for the S spins, we assume that the S spins are in thermal equilibrium during the whole relaxation process of the I spins. We then obtain for ρ^*, the density matrix of the I spins

$$\frac{d\rho^*}{dt} = -J_{iso}^2 \sum_{m=-1}^{1} [I_m, [I_{-m}, \rho^*]].$$

$$\int_0^\infty \langle S_{-m}(t)\, S_m(t-\tau)\rangle\, e^{im(\omega_I - \omega_S)\tau} d\tau. \tag{4.68}$$

The autocorrelation function of the S spins in the rotating frame of angular velocity ω_S is given simply by:

$$\langle S_0(t)\, S_0(t-\tau)\rangle = \frac{1}{3} S(S+1)\, e^{-\tau/T_1^S}$$

$$\langle S_{\pm 1}(t) S_{\mp 1}(t-\tau)\rangle = \frac{1}{3} S(S+1)\, e^{-\tau/T_2^S}. \tag{4.69}$$

The equations of motion of the expectation values of the I spin operators $\langle I_M\rangle$ are obtained from Eq. (4.68) in the usual way, described in Section 4.2.2, by multiplying with I_M and taking the trace. The commutators are evaluated readily as:

$$\left\langle \frac{dI_0}{dt} \right\rangle \sim \langle [I_{-m}, [I_m, I_0]]\rangle = -|m| \langle I_0\rangle, \tag{4.70a}$$

$$\left\langle \frac{dI_{\pm 1}}{dt} \right\rangle \sim \langle [I_{-m}, [I_m, I_{\pm 1}]]\rangle = (-1)^m \langle I_{\pm 1}\rangle, m = 0, \mp 1. \tag{4.70b}$$

The "selection rule" for m in Eq. (4.70b) expresses the fact that the commutators $[I_{\pm 1}, I_{\pm 1}]$ vanish, of course.

From Eqs. (4.69) and (4.70a) and (4.70b) the equations of motion for $\langle I_m \rangle$ can directly be obtained in the form of Bloch equations with

$$\frac{1}{T_1^I} = \frac{2}{3} S(S+1) J_{\text{iso}}^2 \frac{T_2^S}{1 + (\omega_I - \omega_S)^2 T_2^{S^2}}$$

$$\frac{1}{T_2^I} = \frac{1}{3} S(S+1) J_{\text{iso}}^2 \left(T_1^S + \frac{T_2^S}{1 + (\omega_I - \omega_S)^2 T_2^{S^2}} \right). \tag{4.71}$$

4.2.3.2. Spin-Rotation Interaction

In all but one of the cases treated so far, the spin dependent operators T_{lm}^{λ} were time independent in the laboratory frame. Only for the scalar coupling between I and S treated in the preceeding section this was not the case. Of the coupling tensor J, however, only the $l = 0$ term had to be considered, which is a scalar, independent of the coordinate frame used; therefore, we always formulated our spin Hamiltonians in the laboratory frame so far. In the case of spin-rotation interaction, on the other hand, T_{lm}^{SR} is time dependent because the angular momentum of a molecule in a liquid is not well defined but can be considered as a fluctuating and, therefore, a time dependent quantity. In most liquids, the spin-rotational relaxation is determined by this time dependence. Therefore the coordinate system most suited here for the formulation of $\mathcal{H}_{SR}$ is the principal axes system of the moment of inertia tensor $\overrightarrow{\overrightarrow{\theta}}$ of the molecule. In this frame the components of the angular momentum are given simply by $J_i = \omega_i \theta_i$, where here ω is the angular velocity of the molecule. We could derive the relaxation rate using our general derivation of Section 4.2.2 and $\mathcal{H}_{SR}$ in the general form, Eq. (2.33). This, however, would be a rather tedious approach since we have formulated $\mathcal{H}_{SR}$ in a molecule fixed system. The R_{lm}^{SR} are time independent and, therefore, their transformation properties under rotations as described by Eq. (2.33) are not of interest here. As a consequence, there are no "selection rules" $l_1 = l_2$, $m_1 = -m_2$ for the autocorrelation functions as discussed in connection with Eqs. (4.31) and (4.32). We, therefore, will find it much more convenient to express $\mathcal{H}_{SR}$ directly as the coupling of the nuclear spin to the magnetic field caused by the rotation of the molecule:

$$\mathcal{H}_{SR} = \sum_{m=-1}^{1} (-1)^m I_m A_{-m}, \tag{4.72}$$

where we construct the irreducible tensor elements of first rank A_{-m} from the cartesian vector components A_i, according to our general recipe, Eq. (2.8);

$$A_i = \sum_{j=x, y, z} c_{ij} J_j, \quad i = x, y, z. \tag{4.73}$$

Here c_{ij} are the cartesian components of the complete spin-rotation tensor, including antisymmetric constituents. The A_i are proportional to the magnetic field components B_i^J at the site of nucleus I due to the rotation of the molecule: $A_i = -\gamma_I B_i^J$.

In order to evaluate the relaxation rates, we need the spin operators in the laboratory system or in the rotating frame, respectively. Let us label the spin operators in the rotating frame with $I_{m'}$, then the spin operators in the principal axes system of $\vec{\vec{\theta}}$, I_m, are given by:

$$I_m = \sum_{m'=-1}^{1} I_{m'}\, e^{im'\omega_0 t}\, \mathfrak{D}^{(1)}_{m'm}\underbrace{(-\gamma, -\beta, -\alpha)}_{-\Omega}. \tag{4.74}$$

With this the master equation, Eqs. (4.25) and (4.32), reads

$$\frac{d\rho^*}{dt} = -\sum_{m_1=-1}^{1}\sum_{m_2=-1}^{1}(-1)^{m_1+m_2}\sum_{m'=-1}^{1}[I_{m'}, [I_{-m'}\,\rho^*]]$$
$$\cdot\int_0^\infty \langle \mathfrak{D}^{(1)}_{m'm_1}[-\Omega(t)]\mathfrak{D}^{(1)}_{-m'm_2}[-\Omega(t-\tau)]\,A_{-m_1}(t)\,A_{-m_2}(t-\tau)\rangle\, e^{im'\omega_0\tau}d\tau, \tag{4.75}$$

where we have restricted ourselves already to secular parts. Again we can obtain the equation of motion for $\langle I_M\rangle$ in the usual way

$$\left\langle\frac{dI_0}{dt}\right\rangle \sim \langle[I_{-m'}, [I_{m'}, I_0]]\rangle = -|m'|\,\langle I_0\rangle \tag{4.70a$'$}$$

$$\left\langle\frac{dI_{\pm1}}{dt}\right\rangle \sim \langle[I_{-m'},[I_{m'}, I_{\pm1}]]\rangle = (-1)^{m'}\langle I_{\pm1}\rangle, m' = 0, \mp 1 \tag{4.70b$'$}$$

[cf. discussion of Eq. (4.70b)]. The factors of proportionality will now be calculated from Eq. (4.75). In this general equation, we encounter autocorrelation functions

$$f(\tau) = \langle \mathfrak{D}^{(1)}_{m'm_1}[-\Omega(t)]\,\mathfrak{D}^{(1)}_{-m'm_2}[-\Omega(t-\tau)]\,A_{-m_1}(t)\,A_{-m_2}(t-\tau)\rangle. \tag{4.76}$$

These mixed correlation functions depend on both the autocorrelation of the orientation of the molecule relative to the laboratory frame and the autocorrelation of the angular momentum of the molecule relative to the molecular system [cf. definition of A_m, Eqs. (4.72) and (4.73)]. Such mixed correlation functions can be calculated strictly only within a certain model for the reorientation of the molecule (see, e.g., [*49*]). For isotropic liquids, we can assume, however, that the orientation of the molecule relative to the laboratory frame and the angular momentum of the molecule are not correlated. Since the laboratory frame is distinguished from other directions of space by the magnetic field $\mathbf{B}_0$ only, this assumption seems natural in view of the weakness of the magnetic interactions. The correlation function, Eq. (4.76), can then be *factorized:*

$$f(\tau) = \langle \mathfrak{D}^{(1)}_{m'm_1}[-\Omega(t)]\,\mathfrak{D}^{(1)}_{-m'm_2}[-\Omega(t-\tau)]\rangle\,\langle A_{-m_1}(t)\,A_{-m_2}(t-\tau)\rangle. \tag{4.76a}$$

The correlation function for the reorientation will be calculated in Section 4.2.4. In most cases of practical interest the angular momentum loses its correlation much more rapidly than the orientation and we will restrict ourselves to this case here. This means that for the relevant values of τ in Eq. (4.75) the orientation at time t and at time $t - \tau$ are almost the same and, therefore, we are allowed to approximate:

$$\langle \mathfrak{D}^{(1)}_{m'm_1}[-\Omega(t)]\,\mathfrak{D}^{(1)}_{-m'm_2}[-\Omega(t-\tau)]\rangle$$

$$\approx \langle \mathfrak{D}^{(1)}_{m'm_1}[-\Omega(t)]\,\mathfrak{D}^{(1)}_{-m'm_2}[-\Omega(t)]\rangle = (-1)^{m'+m_1}\,\frac{1}{3}\,\delta_{-m_1\,m_2}. \tag{4.77}$$

Finally, the contribution of spin-rotation interaction to the relaxation rate is of importance for rapid rotations only for which the extreme narrowing condition is fulfilled; therefore $e^{im'\omega_0\tau} \cong 1$. We then have to deal with simple time integrals rather than Fourier transforms of correlation functions. We call these time integrals a_m in order to distinguish them from the g_{lm}, Eq. (4.34)

$$a_m = (-1)^m \int_0^\infty \langle A_{-m}(t)\,A_m(t-\tau)\rangle\,d\tau. \tag{4.78}$$

Then the terms in Eq. (4.75) for the different m' values are alike for $\langle I_0\rangle$ and $\langle I_{\pm 1}\rangle$ and we obtain:

$$\left\langle \frac{dI_M}{dt} \right\rangle = -\frac{2}{3} \sum_{m=-1}^{1} a_m \langle I_M\rangle \tag{4.79}$$

with $M = 0, \pm 1$. The expectation values of the spin operators, therefore, obey Bloch equations with $T_1 = T_2$. The expression for the relaxation rates become especially simple if we assume that rotations about different axes of the moment of inertia tensor are not correlated. This is not strictly true for an asymmetric rotor (see, e.g. [*23*], Appendix B) but, as long as two principal moments of inertia are not too different, one may neglect this correlation. Then we have

$$a_0 = \sum_j c_{zj}^2 \int_0^\infty \langle J_z(t) J_z(t-\tau)\rangle\,d\tau$$

$$a_{\pm 1} = \frac{1}{2} \sum_j c_{xj}^2 \int_0^\infty \langle J_x(t)\,J_x(t-\tau)\rangle\,d\tau + c_{yj}^2 \int_0^\infty \langle J_y(t)\,J_y(t-\tau)\rangle\,d\tau, \tag{4.80}$$

and accordingly

$$\frac{1}{T_1} = \frac{1}{T_2} = \frac{2}{3} \sum_{i,j=x,y,z} \sum c_{ij}^2 \int_0^\infty \langle J_j(t)\,J_j(t-\tau)\rangle\,d\tau. \tag{4.81}$$

The integrals over correlation functions of the components of the angular momentum are proportional to the corresponding correlation times τ_{J_j} ($j = x, y, z$)

$$\tau_{J_j} = \int_0^\infty \frac{\langle J_j(t)\,J_j(t-\tau)\rangle}{\langle J_j(t)\,J_j(t)\rangle}\,d\tau. \tag{4.82}$$

In order to calculate $\langle J_j(t)\, J_j(t)\rangle = \langle J_j^2 \rangle$, one normally uses the mean angular velocity of the free rotor at the same temperature

$$\langle J_j{}^2 \rangle = \frac{1}{\hbar^2}\, \theta_j^2 \,\langle \omega_j^2 \rangle = \frac{1}{\hbar^2}\, \theta_j^2 \,\frac{kT}{\theta_j}\,. \tag{4.83}$$

Inserting this in Eq. (4.81) yields as final result:

$$\frac{1}{T_1^{SR}} = \frac{1}{T_2^{SR}} = \frac{2}{3}\frac{kT}{\hbar^2} \sum_{i,j=x,y,z} \sum \theta_j c_{ij}^2 \tau_{J_j}. \tag{4.84}$$

We stress again that c_{ij} are cartesian tensor components of the spin-rotation tensor in the principal axes system of the $\vec{\vec{\theta}}$ tensor, including antisymmetric constituents. Special cases of Eq. (4.84) have been derived in a number of papers [*49–53*]. The expression for the relaxation rates, Eq. (4.84), is useful especially if the principal axes systems of c and $\vec{\vec{\theta}}$ coincide since, for the symmetric part ($l = 2$), we have $c_{ij}^{(2)} = 0$ for $i \neq j$. The simplified expressions for some cases of higher symmetry often met in practice, are collected in Table 4.7. Both the symmetry of the molecule as a whole and the site symmetry of the nuclear position have to be considered. The transformation of the c tensor needed in cases of low symmetry is given, in general form, in Appendix C.

4.7. Relaxation rates through spin-rotation interaction

	$1/T_1^{SR} = 1/T_2^{SR}$
Asymmetric rotor, $\mathfrak{c}$ may be asymmetric	$\frac{2}{3}\frac{kT}{\hbar^2} \sum_{i,j=x,y,z} \theta_j\, c_{ij}^2\, \tau_{J_j}$
$\vec{\vec{\theta}}$ and $\mathfrak{c}$ have common principal axes system, $\mathfrak{c}$ symmetric	$\frac{2}{3}\frac{kT}{\hbar^2} (\theta_x\, c_{xx}^2\, \tau_{J_x} + \theta_y\, c_{yy}^2\, \tau_{J_y} + \theta_z\, c_{zz}^2\, \tau_{J_z})$
As above but moreover symmetric top	$\frac{2}{3}\frac{kT}{\hbar^2} (2\,\theta_\perp\, c_\perp^2\, \tau_{J_\perp} + \theta_\parallel\, c_\parallel^2\, \tau_{J_\parallel})$
Spherical top	$\frac{2}{3}\frac{kT}{\hbar^2} (2\,\theta_\perp\, c_\perp^2 + \theta_\parallel\, c_\parallel^2)\, \tau_J$
Linear molecule	$\frac{4}{3}\frac{kT}{\hbar^2}\, \theta\, c^2\, \tau_J$

It should be noted that in cases of low symmetry antisymmetric constituents of the spin-rotation tensors $\mathfrak{c}$ enter the expression for the relaxation rate with equal weight as the symmetric ones. Therefore, special care also must be taken here when analyzing relaxation rates. As we will show in Chapter 5.2, $\mathfrak{c}$ is not symmetric, in general, if we have off-diagonal elements $c_{ij} \neq 0$ in the principal axes system of $\vec{\vec{\theta}}$ and $\theta_x \neq \theta_y \neq \theta_z$.

4.2.4. Time Correlation Functions and Spectral Densities

In order to make use of the general expressions for the relaxation rates [see Table 4.3, Eq. (4.65) together with Eqs. (4.42″) and (4.43″)], we have to know the autocorrelation functions $f_{lm}(\tau)$ and their Fourier transforms $g_{lm}(\omega)$ [see Eqs. (4.33) and (4.34)]. The correlation functions depend on the rotation of the molecules for intramolecular couplings. For intermolecular interactions ($\lambda = D$) they depend also on relative translational motions of the molecules.

By selective experiments (e.g., by partial deuteriation), detailed information about the translational and rotational motion of molecules relative to each other can be gathered. Here we must restrict ourselves, however, to refer to the literature on this interesting subject; in particular to the review article by Hertz [*54*] and some more recent papers [*55–56*]. Molecular pair correlation functions, including angular correlations, have recently been treated in general form by Zeidler [*57*]. This paper also contains a large number of references on intermolecular dipole-dipole relaxation. Most recently, dilution experiments by deuteration have also been applied to polymers [*58*].

In this review we are mainly concerned about molecular rotations which render the coupling parameters R_{lm}^{λ} time dependent. We have in mind ^{13}C and ^{15}N nuclei in organic molecules in particular where the coupling tensors ρ_{lm}^{λ} are dominated by intramolecular interactions. Restricting to *intramolecular* contributions generally is justified for all couplings, except the dipole-dipole coupling. For $\lambda = D$, on the other hand, intermolecular contributions can be comparable with or even larger than the intramolecular ones, especially in proton relaxation.

4.2.4.1 Anisotropic Rotations Within the Diffusion Model

As mentioned above, anisotropic rotational motions have been treated in context with NMR relaxation in a number of papers [*20–25*]. All of these papers use Debye's diffusion model [*13*] to describe the hindered rotation of a molecule in a liquid. Despite its limited range of applicability we will also use it here since, within that model, we can derive explicit expressions for $g_{lm}(\omega)$ for the general case of a completely asymmetric molecule. Extensions of the classic diffusion model are treated in the following section.

The subject matter of the following paragraphs has been treated by Versmold [*24*] who, in addition, considered internal rotations, and in Huntress' review [*23*], based on the work of Favro [*59*]. It should be noted, however, that Perrin [*60*] treated anisotropic rotational diffusion with regard to dielectric relaxation more than 40 years ago. Nevertheless, we repeat part of the derivations in [*23, 24, 59*] here to get a treatment consistent with our notation.

We want to calculate the autocorrelation function $f_{lm}(\tau)$ [Eq. (4.33)]. Since we are dealing with a stationary process, we have:

$$
\begin{aligned}
f_{lm}(\tau) &= (-1)^{l+m}\langle R_{l-m}(t)\, R_{lm}(t-\tau)\rangle \\
&= (-1)^{l+m}\langle R_{l-m}(t+\tau)\, R_{lm}(t)\rangle
\end{aligned}
$$

$$= \langle R_{lm}(t)\, R^{*}_{lm}(t+\tau)\rangle, \tag{4.85}$$

where we have made use of $R^{*}_{lm} = (-1)^{l+m}\ R_{l-m}$.

In order to calculate $f_{lm}(\tau)$, we need to know the R_{lm} in the laboratory frame. We could express R_{lm} in terms of the principal values ρ_{lm} of the corresponding coupling λ according to Eq. (2.29). For a molecule with less than cubic symmetry, the principal axes system for the coupling λ and the diffusion tensor D, however, need not necessarily coincide. We will find it more useful, therefore, to express the R_{lm} in terms of the irreducible tensor elements ρ^{D}_{lm} given in the principal axes system of the diffusion tensor D:

$$R_{lm}(t) = \sum_{m'=-l}^{l} \rho^{\mathsf{D}}_{lm'}\ \mathfrak{D}^{(l)}_{m'm}[\Omega(t)]. \tag{4.86}$$

Since the principal axes system of the diffusion tensor has a fixed orientation with respect to the molecule, the ρ^{D}_{lm}, are time independent, if ρ_{lm} is time independent because the $\rho^{\mathsf{D}}_{lm'}$ are obtained from ρ_{lm}, according to Eq. (2.29), with *fixed* eulerian angles (α, β, γ).

This gives

$$f_{lm}(\tau) = \sum_{m'=-l}^{l}\ \sum_{m''=-l}^{l} \rho^{\mathsf{D}}_{lm'}\rho^{\mathsf{D}*}_{lm''}\langle\mathfrak{D}^{(l)}_{m'm}(t)\mathfrak{D}^{(l)*}_{m''m}(t+\tau)\rangle. \tag{4.87}$$

The Wigner rotation matrices contain all the time dependence:

$$\langle\mathfrak{D}^{(l)}_{m'm}(t)\mathfrak{D}^{(l)*}_{m''m}(t+\tau)\rangle =$$

$$\iint\mathfrak{D}^{(l)}_{m'm}(\Omega_0)\, \mathfrak{D}^{(l)*}_{m''m}(\Omega)\, P(\Omega_0)\, P(\Omega_0\,|\,\Omega,\tau)\, d\,\Omega_0 d\,\Omega, \tag{4.88}$$

where Ω_0 and Ω describe the orientation of the molecule characterized by the principal axes system of D, which we will call molecular system henceforth, relative to the laboratory system at times t and $t+\tau$, respectively. It should be noted that integration over Ω means integration over all values possible for all three eulerian angles. $P(\Omega_0)$ is the a priori probability density to find a molecular axes system with the orientation Ω_0 at time t; therefore $P(\Omega_0) = 1/8\ \pi^2$. $P(\Omega_0|\Omega,\tau)d\Omega$ is the conditional probability to find a molecular axes system with an orientation between $\Omega - d\Omega$ and $\Omega + d\Omega$ at time $t+\tau$ when we *know* the same molecular axes system had the orientation Ω_0 at time t. This conditional probability can be calculated as a Green's function $G(\Omega_0|\,\Omega,\tau)$ for the time evolution during the diffusional process [*59*]. Because of the analogy between the diffusion equation and the Schrödinger equation for the free rotor[18], the Green's function G can be evaluated in terms of the eigenfunctions ψ_N and energies E_N of the free rotor; replacing $\hbar^2/2\ \theta_i$ by D_i [*59*]:

$$G(\Omega_0|\,\Omega,\tau) = \sum_N \psi^{*}_N(\Omega_0)\ \psi_N(\Omega)\, e^{-E_N\tau}. \tag{4.89}$$

[18] A detailed discussion of this analogy is given by Huntress [*23*].

For a completely asymmetric rotor, the eigenfunctions ψ_N cannot be given, in general, in closed form. For low values of J, however, exact eigenfunctions ψ_{NM}^J can be obtained by linear combination of the eigenfunctions φ_{KM}^J of the symmetric rotor [59]:

$$\psi_{NM}^J = \sum_K a_K^J(N)\, \varphi_{KM}^J, \tag{4.90}$$

where the coefficients $a_K^J(N)$ are independent of M. As we will show immediately, we only need consider $J = 1, 2$ for nuclear relaxations, for which (Eq. (4.90) is valid exactly. Inserting Eqs. (4.89) and (4.90) in Eq. (4.88) yields

$$\begin{aligned}\langle \mathfrak{D}_{m'm}^{(l)}(t)\, \mathfrak{D}_{m''m}^{(l)*}(t+\tau)\rangle = \frac{1}{8\pi^2} \sum_J \sum_{M,N} e^{-E_N^J \tau} \\ \cdot \int\int [\sum_K a_K^{J*}(N)\, \varphi_{KM}^{J*}(\Omega_0)] \cdot [\sum_{K'} a_{K'}^J(N)\, \varphi_{K'M}^J(\Omega)] \\ \cdot \mathfrak{D}_{m'm}^{(l)}(\Omega_0)\, \mathfrak{D}_{m''m}^{(l)*}(\Omega)\, d\Omega_0 d\Omega.\end{aligned} \tag{4.91}$$

The eigenfunctions φ_{KM}^J of the symmetric top are given directly by the Wigner rotation matrices

$$\varphi_{KM}^J(\Omega) = (-1)^{M-K} \sqrt{\frac{2J+1}{8\pi^2}}\, \mathfrak{D}_{-K-M}^{(J)}(\Omega), \tag{4.92}$$

and this gives

$$\begin{aligned}\langle \mathfrak{D}_{m'm}^{(l)}(t)\, \mathfrak{D}_{m''m}^{(l)*}(t+\tau)\rangle = \frac{1}{64\pi^2} \sum_{J,M,N} (2J+1)\, e^{-E_N^J \tau} \\ \cdot [\sum_K (-1)^{M-K} a_K^J(N) \int \mathfrak{D}_{m'm}^{(l)}(\Omega_0) \mathfrak{D}_{-K-M}^{(J)*}(\Omega_0)\, d\Omega_0] \\ \cdot [\sum_{K'} (-1)^{M-K'} a_{K'}^J(N) \int \mathfrak{D}_{m''m}^{(l)*}(\Omega) \mathfrak{D}_{-K'-M}^{(J)}(\Omega)\, d\Omega].\end{aligned} \tag{4.93}$$

Now we can make use of the orthogonality of the Wigner rotation matrix elements

$$\int \mathfrak{D}_{m_1 m_1'}^{(l_1)*}(\Omega)\, \mathfrak{D}_{m_2 m_2'}^{(l_2)}(\Omega)\, d\Omega = \frac{8\pi^2}{2l_1+1}\, \delta_{l_1 l_2}\, \delta_{m_1 m_2}\, \delta_{m_1' m_2'} \tag{4.94}$$

which leaves the sum over the different values of the energy, E_N, only[19]:

$$\langle \mathfrak{D}_{m'm}^{(l)}(t) \mathfrak{D}_{m''m}^{(l)*}(t+\tau)\rangle = \frac{(-1)^{m'+m''}}{2l+1} \sum_{N=-l}^{l} a_{-m'}^{l*}(N)\, a_{-m''}^{l}(N)\, e^{-E_N^l \tau}, \tag{4.95}$$

[19] We have restricted ourselves to correlation functions of the type of Eq. (4.33) by keeping only the secular terms of Eq. (4.31). It is of interest that, even for more general correlation functions $\langle R_{lm}(t)\, R_{l'm'}(t-\tau)\rangle$, only the terms with $l = l'$ and $m = -m'$ do not vanish within the diffusion model, according to Eq. (4.94).

and the result is independent of m'. This was to be expected since, for an isotropic liquid, the result must be independent of the eulerian angle γ corresponding to a rotation about the direction of the external magnetic field. This gives the general equation for $f_{lm}(\tau)$

$$f_{lm}(\tau) = \sum_{m'=-l}^{l} \sum_{m''=-l}^{l} \rho^{\mathsf{D}}_{lm'} \, \rho^{\mathsf{D}*}_{lm''} \frac{(-1)^{m'+m''}}{2l+1} \sum_{N=-l}^{l} a^{l*}_{-m'}(N) \, a^{l}_{-m''}(N) \, e^{-E^{l}_{N}\tau}. \tag{4.96}$$

The eigenfunctions and energies for the asymmetric rotor are given, by Favro [*59*] and by Huntress [*23*]. For convenience, we again give the coefficients $a^l_m(N)$ and the energies E^l_N here in Table 4.8 where we use the same convention for the diffusion tensor D as defined above for the coupling tensors, cf. Eq. (2.30):

$$\mathsf{D} = \begin{pmatrix} \mathsf{D}_{xx} & 0 & 0 \\ 0 & \mathsf{D}_{yy} & 0 \\ 0 & 0 & \mathsf{D}_{zz} \end{pmatrix} = \mathsf{D}_S \begin{pmatrix} 1 & 0 & 0 \\ 0 & 1 & 0 \\ 0 & 0 & 1 \end{pmatrix} + \mathsf{D}^* \begin{pmatrix} -\frac{1}{2}(1+\eta_{\mathsf{D}}) & 0 & 0 \\ 0 & -\frac{1}{2}(1-\eta_{\mathsf{D}}) & 0 \\ 0 & 0 & 1 \end{pmatrix}$$

with:

$$\mathsf{D}_S = \frac{1}{3}(\mathsf{D}_{xx} + \mathsf{D}_{yy} + \mathsf{D}_{zz}), \; \mathsf{D}^* = \mathsf{D}_{zz} - \mathsf{D}_S, \; \eta_{\mathsf{D}} = \frac{\mathsf{D}_{yy} - \mathsf{D}_{xx}}{\mathsf{D}^*}.$$

The index K takes on all values from $-l$ to $+l$ for each eigenfunction; in Table 4.8, however, only the nonvanishing coefficients $a^l_K(N)$ and the corresponding values of K are given. With the aid of Table 4.8, we now obtain explicit expressions from

Table 4.8. Eigenvalues of energy and coefficients of eigenfunctions for asymmetric top, $A = \sqrt{1 + \frac{1}{3}\eta^2_{\mathsf{D}}}$

l	N	E^l_N	K	$a^l_K(N)$
1	1	$2\,\mathsf{D}_S + \frac{1}{2}\mathsf{D}^*(1-\eta_{\mathsf{D}})$	± 1	$+(\sqrt{2})^{-1}$
	-1	$2\,\mathsf{D}_S + \frac{1}{2}\mathsf{D}^*(1+\eta_{\mathsf{D}})$	± 1	$\pm(\sqrt{2})^{-1}$
	0	$2\,\mathsf{D}_S - \mathsf{D}^*$	0	1
2	2	$6\,\mathsf{D}_S + 3\,\mathsf{D}^*A$	0	$-\eta_{\mathsf{D}}/(\sqrt{6}\cdot\sqrt{A^2+A})$
			± 2	$\pm(1+A)/(2\sqrt{A^2+A})$
	-2	$6\,\mathsf{D}_S + 3\,\mathsf{D}^*$	± 2	$\pm(\sqrt{2})^{-1}$
	1	$6\,\mathsf{D}_S - \frac{3}{2}\mathsf{D}^*(1+\eta_{\mathsf{D}})$	± 1	$+(\sqrt{2})^{-1}$
	-1	$6\,\mathsf{D}_S - \frac{3}{2}\mathsf{D}^*(1-\eta_{\mathsf{D}})$	± 1	$\pm(\sqrt{2})^{-1}$
	0	$6\,\mathsf{D}_S - 3\,\mathsf{D}^*A$	0	$(1+A)/(\sqrt{2}\cdot\sqrt{A^2+A})$
			± 2	$+\eta_{\mathsf{D}}/(2\sqrt{3}\,\sqrt{A^2+A})$

Eq. (4.96). The prefactor belonging to a given value E_N^l is given by a sum over terms of the type $\rho_{lm}^{\mathrm{D}}\ \rho_{lm}^{\mathrm{D}*}; \rho_{lm}^{\mathrm{D}}\ \rho_{l-m}^{\mathrm{D}}$; or $\rho_{20}^{\mathrm{D}}\ \rho_{2\pm2}^{\mathrm{D}*}$ which can be combined to components of ρ^{D} in *cartesian* representation (cf. Appendix B). In this way we get very compact expressions for the correlation functions:

$$f_{1m}(\tau) = \frac{2}{3}[(\rho_{xy}^{(1)\mathrm{D}})^2 e^{-E_0^1\tau} + (\rho_{xz}^{(1)\mathrm{D}})^2 e^{-E_{+1}^1\tau} + (\rho_{yz}^{(1)\mathrm{D}})^2 e^{-E_{-1}^1\tau}], \tag{4.98}$$

$$\begin{aligned} f_{2m}(\tau) = \frac{1}{5}\Bigg\{ & \frac{1}{12\,A(A+1)}[3(1+A)(\rho_{zz}^{(2)\mathrm{D}}) - \eta_{\mathrm{D}}(\rho_{xx}^{(2)\mathrm{D}} - \rho_{yy}^{(2)\mathrm{D}})]^2 e^{-E_0^2\tau} \\ & + 2\,[(\rho_{yz}^{(2)\mathrm{D}})^2 e^{-E_1^2\tau} + (\rho_{xz}^{(2)\mathrm{D}})^2 e^{-E_{-1}^2\tau} + (\rho_{xy}^{(2)\mathrm{D}})^2 e^{-E_{-2}^2\tau}] \\ & + \frac{1}{4\,A(A+1)}[-\eta_{\mathrm{D}}(\rho_{zz}^{(2)\mathrm{D}}) + (1+A)(\rho_{xx}^{(2)\mathrm{D}} - \rho_{yy}^{(2)\mathrm{D}})]^2 e^{-E_{+2}^2\tau}\Bigg\}, \end{aligned} \tag{4.99}$$

with $A = \sqrt{1 + \frac{1}{3}\eta_{\mathrm{D}}^2}$.

The prefactors to the exponentials in Eq. (4.96) are thus obtained by a single orthogonal transformation of the real matrix $\vec{\vec{\rho}}$ from the principal axes system of the interaction λ in the principal axes system of the diffusion tensor. The eulerian angles for that transformation are defined in Appendix C. In this way, we can give the correlation functions $f_{lm}(\tau)$ and their Fourier transforms $g_{lm}(\omega)$ in closed form.

$$f_{lm}(\tau) = \sum_{k=-l}^{l} c_{lk}\, e^{-\tau/\tau_{lk}}, \tag{4.100}$$

$$Re\, g_{lm}(\omega) = g_l(\omega) = \sum_{k=-l}^{l} \frac{c_{lk}\,\tau_{lk}}{1+\omega^2\tau_{lk}^2}. \tag{4.101}$$

The correlation times τ_{lk} and the coefficients c_{lk} are given in Table 4.9.

The general expressions for the relaxation rates for the different interactions λ are again collected in Table 4.10 so that explicit expressions for the relaxation rates can be obtained easily. Here we have made use of the fact that the $g_l(\omega)$ are independent of m and depend on the absolute value of the frequency only. Here the $g_l(\omega)$ from Eq. (4.101) together with Table 4.9 can directly be inserted; furthermore, we can express the principal values δ_λ in terms of the conventional coupling parameters, cf. Table 2.3. To be sure, the general expressions of Tables 4.9 and 4.10 look as complicated as they have to be, since they give explicit expressions for the relaxation rates for anisotropic rotational diffusion of an asymmetric rotor for all intramolecular couplings, including antisymmetric constituents.

If one analyses experimental relaxation rates, considerably simpler expressions often will be adequate. These are contained as special cases in our general formulae. The following simplifications can often be used:

i) All antisymmetric contributions ($l = 1$) vanish because of symmetry.
ii) The diffusion tensor is axially symmetric; $\eta_{\mathrm{D}} = 0, A = 1$.
iii) The coupling tensor is axially symmetric, $\eta_\lambda = 0$.

Table 4.9. Correlation times τ_{lk} and coefficients c_{lk} for anisotropic motion (cf. Eq. (4.101), $A = \sqrt{1 + \frac{1}{3}\eta_D^2}$

l	k	τ_{lk}	c_{lk}
1	1	$\frac{1}{2D_S}[1 + \frac{D^*}{4D_S}(1 - \eta_D)]^{-1}$	$\frac{2}{3}[-\rho^{\lambda}_{XY}\sin\beta\sin\gamma + \rho^{\lambda}_{XZ}(\cos\alpha\cos\gamma - \sin\alpha\cos\beta\sin\gamma) + \rho^{\lambda}_{YZ}(\sin\alpha\cos\gamma + \cos\alpha\cos\beta\sin\gamma)]^2$
	−1	$\frac{1}{2D_S}[1 + \frac{D^*}{4D_S}(1 + \eta_D)]^{-1}$	$\frac{2}{3}[-\rho^{\lambda}_{XY}\sin\beta\cos\gamma - \rho^{\lambda}_{XZ}(\cos\alpha\sin\gamma + \sin\alpha\cos\beta\cos\gamma) + \rho^{\lambda}_{YZ}(-\sin\alpha\sin\gamma + \cos\alpha\cos\beta\cos\gamma)]^2$
	0	$\frac{1}{2D_S}[1 - \frac{D^*}{2D_S}]^{-1}$	$\frac{2}{3}[\rho^{\lambda}_{XY}\cos\beta - \rho^{\lambda}_{XZ}\sin\alpha\sin\beta + \rho^{\lambda}_{YZ}\cos\alpha\sin\beta]^2$
2	2	$\frac{1}{6D_S}[1 + \frac{D^*A}{2D_S}]^{-1}$	$\frac{\delta_{\lambda}^2}{20A(1+A)}\left\{\frac{-\eta_D}{2}(3\cos^2\beta - 1 - \eta_{\lambda}\sin^2\beta\cos 2\alpha) + \frac{1+A}{2}\left[3\sin^2\beta\cos 2\gamma - \eta_{\lambda}\begin{pmatrix}\cos 2\alpha\cos 2\gamma(\cos^2\beta + 1)\\ -2\sin 2\alpha\sin 2\gamma\cos\beta\end{pmatrix}\right]\right\}^2$
	−2	$\frac{1}{6D_S}[1 + \frac{D^*}{2D_S}]^{-1}$	$\frac{\delta_{\lambda}^2}{40}\left\{3\sin^2\beta\sin 2\gamma - \eta_{\lambda}\begin{bmatrix}\cos 2\alpha\sin 2\gamma(\cos^2\beta + 1)\\ + 2\sin 2\alpha\cos 2\gamma\cos\beta\end{bmatrix}\right\}^2$
	1	$\frac{1}{6D_S}[1 - \frac{D^*}{4D_S}(1 + \eta_D)]^{-1}$	$\frac{\delta_{\lambda}^2}{40}\left\{3\sin 2\beta\sin\gamma + \eta_{\lambda}\begin{bmatrix}\cos 2\alpha\sin 2\beta\sin\gamma\\ + 2\sin 2\alpha\sin\beta\cos\gamma\end{bmatrix}\right\}^2$
	−1	$\frac{1}{6D_S}[1 - \frac{D^*}{4D_S}(1 - \eta_D)]^{-1}$	$\frac{\delta_{\lambda}^2}{40}\left\{3\sin 2\beta\cos\gamma + \eta_{\lambda}\begin{bmatrix}\cos 2\alpha\sin 2\beta\cos\gamma\\ - 2\sin 2\alpha\sin\beta\sin\gamma\end{bmatrix}\right\}^2$
	0	$\frac{1}{6D_S}[1 - \frac{D^*A}{2D_S}]^{-1}$	$\frac{\delta_{\lambda}^2}{60A(1+A)}\left\{\frac{3}{2}(1+A)(3\cos^2\beta - 1 - \eta_{\lambda}\sin^2\beta\cos 2\alpha) - \frac{\eta_D}{2}\left[3\sin^2\beta\cos 2\gamma - \eta_{\lambda}\begin{pmatrix}\cos 2\alpha\cos 2\gamma(\cos^2\beta + 1)\\ - 2\sin 2\alpha\sin 2\gamma\cos\beta\end{pmatrix}\right]\right\}^2$

iv) Eulerian angles α, β, or γ are 0 or $\pi/2$ (e.g., for planar molecules).

v) Extreme narrowing ($\omega^2\tau_{lk}^2 \ll 1$).

The expressions of Table 4.9 are written in such a way that these simplifications can be incorporated easily since entire terms vanish in these cases. An internal check of Table 4.9 is provided by using these general expressions for the special case of isotropic motion ($\eta_D = 0$, $D^* = 0$, $A = 1$). Then the simple expressions for $g_l(\omega)$, Eqs. (4.60) and (4.61) are reproduced independent of α, β, and γ.

The g_l contain the coupling parameters δ_{λ} and, therefore, are *not spectral densities.* Only for isotropic motion or axially symmetric coupling ($\eta_{\lambda} = 0$), the coupling parameters and the spectral densities can easily be separated for $l = 2$ and the usual expressions for the relaxation rates are obtained, cf. [2] and Tables 4.4 and 4.6. It should be noted, however, that even in the general case, for the $l = 2$ terms, the coupling constant δ_{λ} enters as a common factor δ_{λ}^2 for g_2^{λ} only and, therefore, reduced $g_2^{\lambda} = g_2^{\lambda}/\delta_{\lambda}^2$ are independent of the strength of the coupling and only depend on the asymmetry parameter η_{λ}, a number between 0 and 1.

Table 4.10. Relaxation rates for intramolecular couplings expressed through $g_l(\omega)$ [cf. Eq. (4.101)]

CS		$\frac{1}{T_1} = \gamma_I^2 B_0^2\, [g_1^{CS}(\omega) + g_2^{CS}(\omega)]$ with $\omega = \gamma_I B_0$
		$\frac{1}{T_2} = \gamma_I^2 B_0^2 \frac{1}{2} [2\, g_0^{CS}(0) + g_1^{CS}(\omega) + g_2^{CS}(\omega) + \frac{4}{3} g_2^{CS}(0)]$
D	like spins	$\frac{1}{T_1} = \frac{4}{3} \gamma_I^4 \hbar^2 I(I+1)\, [g_2^D(\omega) + 4\, g_2^D(2\,\omega)]$
		$\frac{1}{T_2} = \frac{2}{3} \gamma_I^4 \hbar^2 I(I+1)\, [3\, g_2^D(0) + 5\, g_2^D(\omega) + 2\, g_2^D(2\,\omega)]$
	unlike spins	$\frac{1}{T_1^{II}} = \frac{4}{3} \gamma_I^2 \gamma_S^2 \hbar^2 S(S+1)\, [\frac{1}{3}\, g_2^D(\omega_I - \omega_S) + g_2^D(\omega_I) + 2 g_2^D(\omega_I + \omega_S)]$
		$\frac{1}{T_2^I} = \frac{2}{3} \gamma_I^2 \gamma_S^2 \hbar^2 S(S+1) [\frac{4}{3} g_2^D(0) + \frac{1}{3} g_2^D(\omega_I - \omega_S) + g_2^D(\omega_I) + 2\, g_2^D(\omega_S) + 2\, g_2^D(\omega_I + \omega_S)\}$
J	like spins	$\frac{1}{T_1} = \frac{1}{3} I(I+1)\, [g_1^J(\omega) + g_2^J(\omega) + 4\, g_2^J(2\,\omega)]$
		$\frac{1}{T_2} = \frac{1}{6} I(I+1)\, [g_1^J(0) + g_1^J(\omega) + 3\, g_2^J(0) + 5\, g_2^J(\omega) + 2\, g_2^J(2\,\omega)]$
	unlike spins	$\frac{1}{T_1^{II}} = \frac{1}{3} S(S+1)\, [2\, g_0^J(\omega_I - \omega_S) + g_1^J(\omega_I - \omega_S) + g_1^J(\omega_I) + \frac{1}{3} g_2^J(\omega_I - \omega_S) + g_2^J(\omega_I) + 2\, g_2^J(\omega_I + \omega_S)]$
		$\frac{1}{T_2^I} = \frac{1}{6} S(S+1)\, [2\, g_0^J(0) + 2\, g_0^J(\omega_I - \omega_S) + g_1^J(\omega_I - \omega_S) + g_1^J(\omega_I) + 2\, g_1^J(\omega_S) + \frac{4}{3} g_2^J(0) + \frac{1}{3} g_2^J(\omega_I - \omega_S) + g_2^J(\omega_I) + 2\, g_2^J(\omega_S) + 2\, g_2^J(\omega_I + \omega_S)]$
Q		$\frac{1}{T_1} = \frac{e^2 Q^2 (2\,I+3)}{20\, I^2 (2\,I-1)\hbar^2}\, [g_2^Q(\omega) + 4\, g_2^Q(2\,\omega)]$
		$\frac{1}{T_2} = \frac{e^2 Q^2 (2\,I+3)}{40\, I^2 (2\,I-1)\hbar^2} [3\, g_2^Q(0) + 5\, g_2^Q(\omega) + 2\, g_2^Q(2\,\omega)]$

Of course, the situation is completely analogous for the $l = 1$ terms if one divides, for example, by $(A^{D})^2 = (\rho_{XY}^{D})^2 + (\rho_{XZ}^{D})^2 + (\rho_{YZ}^{D})^2$, being the square of the length of the vector **A** corresponding to the $l = 1$ coupling, cf. discussion of Eqs. (2.30)–(2.32).

The expressions of Table 4.9, in general, can be used only if the orientation of the principal axes system of the interaction λ, relative to the *diffusion* tensor principal axes system, is known. For symmetric tops, the principal axes systems of the moment of intertia tensor $\overset{\rightrightarrows}{\theta}$ and of the diffusion tensor coincide, but this need not be the case for an asymmetric top. If they do coincide, the eulerian angles α, β, γ can easily be determined, if the orientation of the principal axes system of the coupling, relative to a molecular frame, is known. If they do not coincide, these angles must be considered as parameters which have to be determined, e.g., by determining relaxation rates for different λ and for different nuclei in the molecule for which these eulerian angles $\alpha_i^\lambda, \beta_i^\lambda, \gamma_i^\lambda$ are different.

As an example, Fig. 4.8 shows reduced $g_2(\omega)$ in the neighborhood of $\omega \cdot \tau_{lk} = 1$ for an asymmetric top with $D_s/D^* = 1, \eta_D = 1/4$, i.e., $D_{zz} = 2\ D_s$, $D_{yy} = 5/8\ D_s$, $D_{xx} = 3/8\ D_s$. This corresponds to highly asymmetric rotational diffusion. The dotted line (Case A) corresponds to $Z_\lambda \| z_D, X_\lambda \| x_D$; the dashed line (Case B) corresponds to $Z_\lambda \| y_D, Y_\lambda \| x_D, \eta_\lambda = 1/2$ in both cases. The solid line, for comparison, shows $g_2(\omega)$ for isotropic motion. The case we have in mind is a planar molecule (e.g., a benzene derivative) where the rotation about the normal to the molecular plane is faster than about axes in the plane. If we combine relaxation studies exploiting different interactions, where for one interaction we have case A but for another we have case B, anisotropic motion can clearly be elucidated and we will present examples of such studies in Chapter 6.

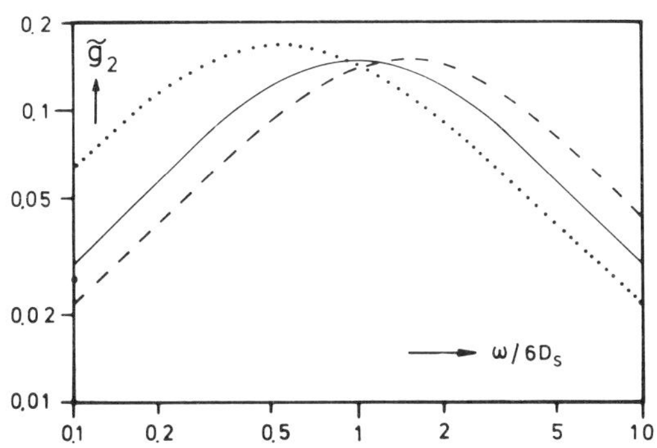

Fig. 4.8. Reduced $\tilde{g}_2(\omega)$ for anisotropic rotational diffusion (- - -, ...) (see text for details). Included for comparison is isotropic motion $\eta_\lambda = 0$) (——)

4.2.4.2 Relation Between the Correlation Times τ_l and τ_J

So far we have treated the correlation times of the components of the angular momentum (τ_J) and the correlation times of angular reorientation (τ_{lk}) independent of each other (cf. Tables 4.7 and 4.9). These correlation times are related closely, however. The relation between τ_J and τ_l is especially simple within the diffusion model since the rotational diffusion coefficient can be expressed directly in terms of the autocorrelation function of the corresponding angular velocity component.

$$D_{ii} = \int_0^\infty \langle \omega_i(0)\ \omega_i(t) \rangle dt \tag{4.102}$$

(cf. [*23, 51*]). The derivation of Eq. (4.102) is completely analogous to the derivation of the corresponding relationship between the translational diffusion coefficient and the velocity autocorrelation function given by Zwanzig [*61*] and is described in detail by Huntress [*23*]. As long as the principal axes system of the diffusion tensor D and that of the moment of inertia tensor $\vec{\vec{\theta}}$ are parallel, we can write $J_i = \omega_i \cdot \theta_i$ and, therefore, we have

$$D_{ii} = \langle \omega_i(0)^2 \rangle \int_0^\infty \frac{\langle \omega_i(0)\ \omega_i(t) \rangle}{\langle \omega_i(0)^2 \rangle} dt = \langle \omega_i(0)^2 \rangle\ \tau_{J_i} \tag{4.103}$$

[cf. Eq. (4.82)]. If we identify $\langle \omega_i(0)^2 \rangle$ with the mean square angular velocity of the free rotor at the same temperature [Eq. (4.83)], we get:

$$\mathsf{D}_{ii} = \frac{kT}{\theta_i} \tau_{J_i}. \tag{4.104}$$

It should be noted that this result was obtained without special assumption about the particular form of the autocorrelation function of the angular velocity. In turn validity of Eq. (4.104) does not imply that this correlation function can be expressed as a single exponential $e^{-|t|/\tau_{J_i}}$ (cf. Kivelson [62]). For *isotropic* motion, we finally have

$$\mathsf{D}_s = \frac{1}{l(l+1)\,\tau_l} \tag{4.105}$$

as given above already, Eq. (4.62), and then from Eq. (4.104)

$$\tau_l \cdot \tau_J = \frac{\theta}{l(l+1)\,kT}. \tag{4.106}$$

This relation, due to Hubbard [50], is often referred to as Hubbard's relationship. Furthermore, from Eq. (4.105) we see that in the diffusion model for isotropic motion, we have

$$\tau_1 = 3\,\tau_2. \tag{4.105a}$$

In Chapter 6 we will discuss experimental examples verifying Eq. (4.106) quantitatively.

4.2.4.3 Extended Diffusion Model

For simple nonassociated liquids, deviations from the diffusion model are expected for higher temperatures. In this region, an extension of the diffusion model, due to Gordon [63], has been quite successful for spherical top molecules and linear molecules. The extended diffusion model is a "gas-like" model of the liquid. We follow the reorientation of a given molecule starting at time t_0. We assume the molecule can rotate freely with a certain angular velocity ω_0 during a time τ_0. At time $t_1 = t_0 + \tau_0$ the molecule undergoes a strong collision with a neighbouring molecule, the collision time being so short that the orientation of the molecule does not change appreciably during the collision. After that, the molecule reorients freely again with angular velocity ω_1 until, at time $t_2 = t_1 + \tau_1$, the next collision happens. Three cases have been treated:

i) In the *J-diffusion* model, one assumes that at each collision the angular velocity $\vec{\omega}$ is randomized, both with respect to its magnitude and its orientation [63].

ii) In the *M-diffusion* model, the magnitude of $\vec{\omega}$ remains constant and only its orientation is randomized at each collision [63].

iii) For *internal rotation,* the rotation axis and, therefore, the orientation of $\vec{\omega}$ for the internal rotation (e.g., of a methyl group) remains constant for a large number of collisions; its absolute value, however, is randomized at each collision. This can be considered as a special case of the J-diffusion model for the one-dimensional rotor [*64*].

The classic Debye diffusion model is contained in the J-diffusion model as the limit for small angular steps between collisions. The correlation time τ_J then can be visualized as the mean time between two consecutive collisions. The J-diffusion model takes in account the effect of large angular steps between collisions and allows the smooth transition from Debye diffusion for liquids to the perturbed free rotor limit for gases. In the latter model, one assumes a molecule undergoes many complete rotations between consecutive collisions. Thus for anisotropic and, therefore, orientation dependent couplings with the quasi-static nuclear spin, an average over the different M states has to be taken. This average is dominated by the $M = 0$ term and largely independent of the speed of rotation. In dilute gases, therefore, both τ_l and τ_J are a measure of the mean time between two subsequent collisions and, contrary to the diffusion limit for which τ_l and τ_J are inversely proportional to each other [cf. Eq. (4.106)] in the perturbed free rotor limit we have direct proportionality:

$$\tau_l = \frac{\tau_J}{2l+1}. \tag{4.107}$$

The simple idea of the extended diffusion model can be worked out classically (see Fixman and Rider [*65*], McClung [*49*], and Maryott *et al.* [*66*]). Contrary to Eq. (4.100) which gives a single exponential for isotropic motion, the autocorrelation functions obtained here are nonexponential. The extended diffusion model, however, is of interest for rapid rotation, in particular, for which, in NMR, the extreme narrowing condition is fulfilled. For the slow time scale of NMR, therefore, only the time integral over the correlation function is needed [the correlation time $\tau_l \sim g_l(0)$]. For the spherical rotor, the J-diffusion model gives [*49*]:

$$\tau_l = \frac{X}{1-X}\,\tau_J$$

$$\text{with } X = \sqrt{\frac{2}{\pi}}\,\frac{1}{2l+1}\,\frac{1}{(\sqrt{\overline{\omega^2}})^3}\sum_{a=-l}^{l}\int_0^\infty \frac{\omega^2 e^{-\frac{\omega^2}{2\overline{\omega^2}}}}{1+a^2\omega^2\tau_J^2}\,d\omega \quad \left(\overline{\omega^2} = \frac{kT}{\theta}\right). \tag{4.108}$$

For the numerical calculation, it is convenient to express X in terms of the complementary error function, erfc (β), which may be evaluated readily [*66–67*]:

$$X = \frac{1}{2l+1}\left\{1 + 4\sum_{a=1}^{l}\beta^2[1 - F(\beta)]\right\}$$

$$F(\beta) = \sqrt{\pi}\,\beta\,e^{\beta^2}\,\text{erfc}\,(\beta)$$

$$\beta = \frac{1}{\sqrt{2}}\,\sqrt{\overline{\omega^2}}\,\tau_J\,a. \tag{4.109}$$

For practical applications, it is useful, furthermore, to consider reduced correlation times:

$$\tau_2^* = \sqrt{\frac{kT}{\theta}}\,\tau_2, \qquad \tau_J^* = \sqrt{\frac{kT}{\theta}}\,\tau_J. \tag{4.110}$$

The relation between τ_2^* and τ_J^* can now be discussed independently of a special system and experimental data can be compared directly with the predictions of the extended diffusion model. For convenience, τ_2^* is plotted vs. τ_J^* in Fig. 4.9 for the *J*- and *M*-diffusion models and for the one dimensional rotor (internal rotation) [*49, 64*]. In spherical and linear molecules, the *J*-diffusion model, indeed, has been shown to predict the deviations from Debye diffusion observed at higher temperatures properly. This is shown in Fig. 4.10, where experimental data for $^{13}CS_2$ [*68*], ClO_3F [*69*], and C_6D_5F [*70*], all obtained for a significant range of τ_2^* and τ_J^*, are compared with the predictions of the *J*-diffusion model.

The application of the *J*-diffusion model to anisotropic rotation of symmetric tops is much less satisfactory. If the model described above is fully maintained for that case, one needs only a single correlation time τ_J even for anisotropic motion since each collision randomizes $\vec{\omega}$. As a consequence in the limit of small angular steps, one gets [*71*]

$$D_\perp = \frac{kT}{\theta_\perp}\,\tau_J \qquad D_\parallel = \frac{kT}{\theta_\parallel}\,\tau_J,$$

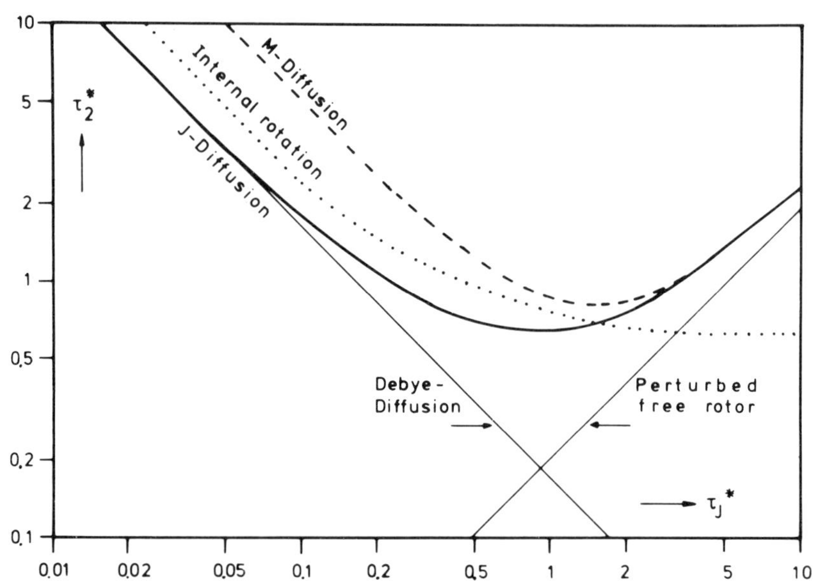

Fig. 4.9. Reduced correlation times τ_2^* and τ_J^* according to extended diffusion model for spherical top (McClung [*49*]) and for internal rotor [*63*]. Also included are limiting slopes of Debye diffusion and pertubed free rotor model

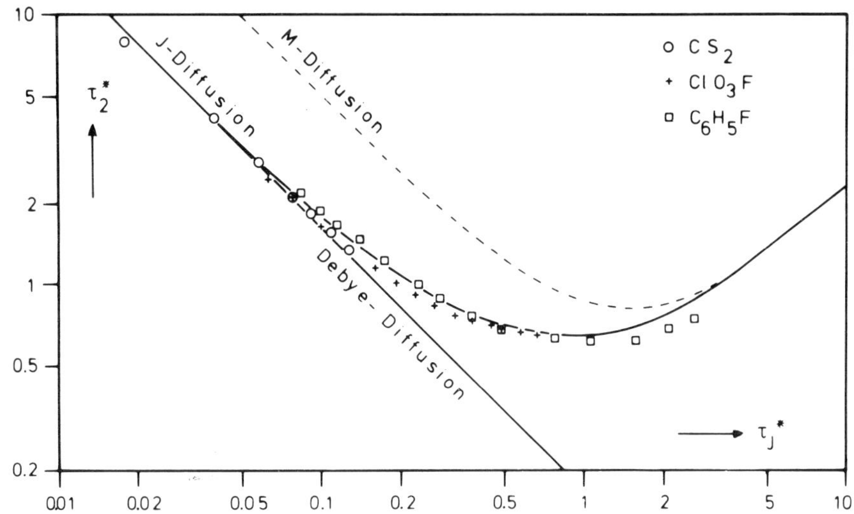

Fig. 4.10. Comparison of experimental data with J-diffusion model, CS_2 [68], ClO_3F (Maryott *et al.* [69]), C_6D_5F (DeZwaan *et al.* [70])

and, therefore

$$\frac{D_\perp}{D_\parallel} = \frac{\theta_\parallel}{\theta_\perp}. \tag{4.111}$$

This would mean that anisotropic rotational diffusion would be due only to the differences in the moments of inertia for rotation about different axes, which is not verified experimentally (cf. Chapter 6). Similarly, treating internal rotation within that model, using a single correlation time τ_J, for the molecule as a whole and for the internal motion [72] does not seem to be very satisfactory. This shows the limitation of this simple model which does not take in account the torques that act on a molecule or a group of a molecule rotating in a liquid.

4.2.4.4 Other Models for Rotation in Liquids

We have put considerable emphasis on the classic diffusion model and its extensions, mainly in order to get explicit expressions for the relaxation rates. In nuclear relaxation studies of liquids, the extreme narrowing condition $\omega_I^2 \cdot \tau_l^2 \ll 1$ will often be fulfilled and one obtains effective correlation times τ_l only. The expressions of Table 4.10 are kept general, however, so that the $g_l(\omega)$ calculated within different models can be inserted directly. A detailed discussion of the models recently proposed to describe the rotation of molecules in liquids cannot be given here and we will restrict ourselves to giving a number of references pertinent to this area. Steele recently published a review article on rotation of molecules [73] which gives a survey on different experimental techniques as well as a comparison of the different theoretical approaches. Autocorrelation functions of angular velocity severely deviating from simple exponentials have been discussed by Kivelson [62]. Generalized Einstein relations $D(\omega) = kT/\theta\beta(\omega)$ have been proposed by Hwang and Freed [74], where $D(\omega)$ and $\beta(\omega)$ are frequency dependent diffusion and damping coefficients, respectively. In the extended stochastic model for the reorientation introduced by Lindenberg and Cukier [75], the extended diffusion model is contained as a special case. Finally, one should not forget the molecular dynamics calculations introduced by Alder and Wainwright [76]. In their classic paper, Rahman and Stillinger [77] obtained remarkable agreement between observed and calculated correlation times τ_1 and τ_2 for water using a simple point charge model. This little survey is by no means exhaustive. In the papers mentioned here, however, one can find a large number of further references on this interesting subject.

5. Coupling Tensors

For the analysis of relaxation data according to Tables 4.9 and 4.10, two different procedures can be used.

i) We can determine correlation times from frequency dependent relaxation studies, according to Eq. (4.101). This technique, called nuclear relaxation spectroscopy, already has been reviewed in this series by Noack [1]. It is used mainly to study relatively slow motions in viscous liquids and polymer solutions and melts. For slow motions only, we find $\omega\tau_l > 1$ and, accordingly, g_l to be frequency dependent.

ii) For liquids of low viscosity, on the other hand, we typically have $\omega^2\tau_l^2 \ll 1$ and g_l is independent of the NMR frequency. If we want to determine the correlation times τ_l and τ_J from relaxation rates in these cases, we have to know the coupling parameters δ_λ and η_λ. This holds, in particular, for the spin-rotation interaction which offers the possibility to determine τ_J.

We will deal, therefore, with the coupling tensors in this chapter. From Table 2.3, we see that only for the intramolecular dipole-dipole interaction the coupling constant can be calculated easily from molecular geometry. In all other cases the coupling constants have to be determined experimentally. Therefore, in Section 5.1 we will give a brief survey of the experimental methods used to measure the coupling parameters. In view of the development of solid state NMR in recent years, the determination of shielding tensors is described in some detail since the combination of relaxation studies in liquids with solid state NMR has proved especially successful.

In Section 5.2 we will deal with the theoretical explanation of the coupling tensors. Besides discussing the difficulties encountered in calculating the coupling parameters, we will concentrate on the relationship between the shielding tensor $\overleftrightarrow{\sigma}$ and the spin rotation tensor c. This relationship is of considerable importance for relaxation studies because it allows reliable calculations of elements of the spin-rotation tensor, hard to get otherwise, from elements of the shielding tensor $\overleftrightarrow{\sigma}$, which have become accessible by solid state NMR in recent years.

5.1. Experimental Determination of the Coupling Tensors

5.1.1. Survey of Experimental Methods

Quite a number of experimental methods are available for determining the coupling parameters. The most reliable ones, naturally, are measurements on either dilute gases or on single crystals. For experimental reasons, however, one will often be restricted to measurements on powders or on molecules dissolved in liquid crystals. For all of the experimental methods to be discussed, excellent books or review articles are available which we can refer to here.

Microwave and molecular beam resonance spectroscopy, applying external electric and magnetic fields, in principle, allow the determination of all internal coupling tensors for essentially isolated molecules in dilute *gases.* Besides the classic books of Ramsey [*2*] and Townes and Schawlow [*3*], the more recent book of Gordy and Cook [*4*] and Flygare's review article [*5*] should be consulted. In our context, the possibility to determine spin-rotation constants is especially important. Furthermore, the direct measurement of coupling tensors (e.g., q or $\vec{\vec{\sigma}}$ in gases) offers the possibility to compare these values with the corresponding ones observed in solids. This is of special interest in systems with strong intermolecular interactions in condensed phases, in particular hydrogen bonds. A classic example is water itself, where the quadrupole coupling constant e^2qQ/h for both ^{2}D and ^{17}O is smaller by about 30% in ice than it is in the free molecule (see, e.g., [*6*]). Unfortunately, these methods are restricted to relatively simple molecules [*2–5*].

From NMR studies of molecules dissolved in *liquid crystals,* coupling parameters even for relatively complex molecules can be determined. Two previous volumes in this series deal with liquid crystals [*7, 8*]. With regard to shielding tensors in particular, the recent review article of Appleman and Dailey [*9*] should be consulted. One always has to keep in mind, however, that in this method one normally obtains a single linear combination of tensor elements only [*7–9*] and the anisotropic parts of the tensors cannot be fully determined in many cases. On the other hand, the surroundings of a molecule dissolved in a liquid crystal will be much more like those in a liquid or in a solution than like those the same molecule will experience in a solid or in the gas phase. Therefore, liquid crystal studies, in principle, offer a possibility to detect changes of the coupling parameters from their values in solids or gases. This could be of considerable interest for relaxation studies. The analysis of NMR data on liquid crystals will be much more reliable if the orientation of the principal axes system of the coupling tensor is known (e.g., from solid state NMR). Most recently, NMR studies on smectic phases have proved quite successful since the orientation of the dissolved molecules, relative to the magnetic field, can be changed [*10–12*]. Finally, it should be noted the first survey of experimental values for the anisotropy of the shielding of spin $I = 1/2$ nuclei was obtained by such liquid crystal studies. Likewise, experimental values for anisotropic *J*-coupling have been determined almost exclusively by that method. Especially clear-cut examples are the determination of the ^{19}F–^{19}F coupling tensor in 1,2 difluoroethane [*13*] and of the anisotropy of the ^{13}C–^{199}Hg coupling in dimethyl mercury [*14*].

In *solids,* in addition to the NMR methods to be described in the following section, we have a number of techniques allowing the determination of quadrupole coupling parameters. First of all there is Mössbauer spectroscopy (for more recent monographs see, e.g., [*15–17*]). Besides its importance for studying magnetic materials, this method has provided important contributions elucidating the bonding in organometallic compounds. Perturbed-angular correlation studies allow the measurement of quadrupole coupling constants for nuclei like ^{19}F with vanishing quadrupole moments in the ground state [*18*]. Here the review article of Shirley and Haas [*19*] seems to be particularly well suited for NMR spectroscopists. Most recently, the quadrupole moment of the ^{20}F nucleus has been determined from the asymmetry of the β-decay of polarized targets [*20*] so that this method also can be used to determine quadrupole coupling constants [*21*].

5.1.2. Solid State NMR

Solid state NMR, especially of spin $I = 1/2$ nuclei, has been developed enormously during the last few years, mainly due to J. Waugh and his co-workers at MIT who succeeded in designing pulse techniques that drastically enhance the *resolution* of solid state spectra and, in addition, significantly increase the *sensitivity* for nuclei with low natural abundance (^{13}C in particular). By now, one is used to calling this field "high resolution NMR in solids", as expressed by the titles of the two books on this subject that have been published recently [*22–23*]. There the pulse techniques are discussed in detail and, therefore, we will give only a short survey of the different techniques without even specifying the pulse sequences as such. It is hoped that such a survey, covering a few pages only, may be useful for readers not interested in the details of pulsed NMR but interested in the basic ideas of these experiments.

In all cases, the dipole-dipole interaction of the nuclear spins, typically dominating the solid state spectra, is eliminated. Before the development of these pulse techniques, the resolution of solid state NMR spectra already has been increased by spinning the sample about an axis at "the magic angle" ϑ_m to the applied field, $\vartheta_m = 54°44'$; therefore, $3\cos^2\vartheta_m - 1 = 0$ (for a review see Andrew [*24*]). In this method, the anisotropic coupling tensors R^{λ}_{2m} are eliminated simultanously for all interactions. The information one can get from ordinary sample spinning experiments, therefore, is restricted to the isotropic parts of the coupling tensors and thus is very similar to that obtained from NMR of the melt or of a solution of that compound. If pulse techniques are used, one can retain complete information about the anisotropic parts of the shielding tensor or even about the dipolar coupling itself (see below).

Likewise, the sensitivity and resolution in nuclear quadrupole resonance also has been increased considerably by double resonance techniques in recent years [*25, 26*]. This technique, originally due to Hahn [*25*], has been developed as a standard method by Edmonds and his co-workers and we can refer to his most recent review article [*27*]. This article, besides a detailed description of the apparatus, also contains large tables of up-dated experimental values of quadrupole coupling parameters for ^{14}N, ^{17}O, and ^{2}D.

5.1.2.1 Measurement of $\overleftrightarrow{\sigma}$ Tensors

NMR in High Magnetic Fields

The magnetic shielding, described by the $\overleftrightarrow{\sigma}$ tensor, causes line shifts and splittings in the solid state spectrum that are proportional to the strength of the applied magnetic field (cf. Table 2.4). All other couplings are field independent, at least to a high degree of approximation.[20] The simplest method, in principle, to determine $\overleftrightarrow{\sigma}$, despite strong dipolar coupling, is provided by the use of the high magnetic fields of superconducting magnets.

As an illustrative example, Fig. 5.1 shows the powder spectrum of solid white phosphorus (P_4) at 25 K and a frequency of 92 MHz corresponding to a field of 5.3 T [*29*]. Clearly, this method works even for nuclei with high magnetic moments (as is ^{31}P) if the shielding anisotropy is sufficiently large. Here the frequency splitting ($\nu_{\|} - \nu_{\perp}$ = 37 kHz corresponding to $\Delta\sigma$ = –405 ppm) by far exceeds the dipolar line width of 4.5 kHz and we obtain the powder spectrum characteristic for an axially symmetric $\overleftrightarrow{\sigma}$ tensor (cf. Fig. 3.2a).

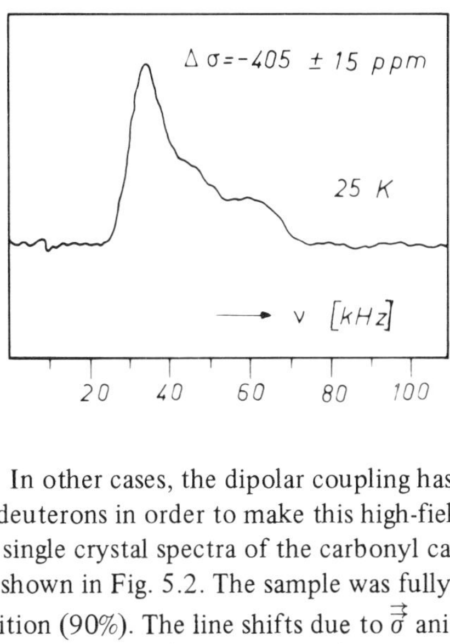

Fig. 5.1. ^{31}P spectrum of solid white phosphorus P_4 at 92MHz and 25 K [*29*]

In other cases, the dipolar coupling has to be reduced by replacing the protons by deuterons in order to make this high-field method successful. As an example, ^{13}C single crystal spectra of the carbonyl carbon of benzophenone at 61 MHz [*30*] are shown in Fig. 5.2. The sample was fully deuterated and enriched in ^{13}C in that position (90%). The line shifts due to $\overleftrightarrow{\sigma}$ anisotropy then are significantly larger than the dipolar width ($\Delta\sigma$ = 155 ppm).

Further examples are given in Chapter 6. Although this technique cannot compete with the pulse methods described in the following sections, it is important to realize that for nuclei like ^{13}C, ^{15}N, ^{19}F, ^{31}P, where $\Delta\sigma$ reaches values of more than a hundred ppm, application of high magnetic fields alone often is sufficient to determine the $\overleftrightarrow{\sigma}$ tensor. Therefore, the pulse methods now are often combined with working in high fields of 4–8.5 T.

---

[20] The possibility of $\overleftrightarrow{\sigma}$ being field-dependent has been discussed by Ramsey [*28*].

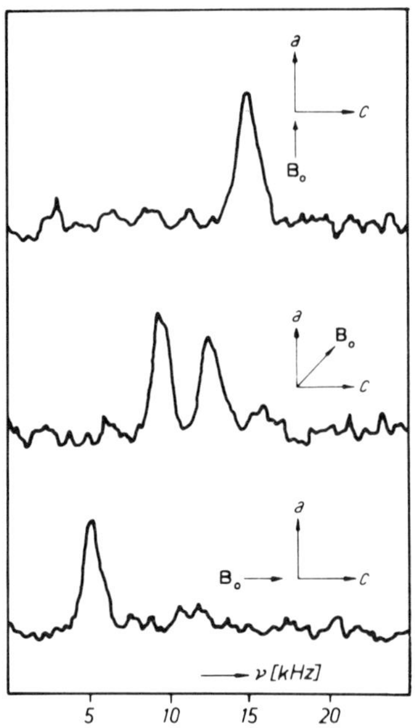

Fig. 5.2. ^{13}C spectra of the carbonyl carbon (90% enriched) in a single crystal of perdeuterated benzophenone; rotation about crystallographic b axis [*30*]. [From Kempf, J., *et al.*: Chem. Phys. Letters **17**, 39 (1972)]

Heteronuclear Decoupling and Cross-Polarization

Applying high magnetic fields alone will be successful for determining the shielding tensor for the spins I only if either the shielding anisotropy is especially large or the I spins are magnetically dilute. These conditions are not fulfilled, for example, for ^{13}C in CH groups of organic compounds. In benzene, the $^{13}C-^{1}H$ distance is 1.09Å only, corresponding to a dipolar splitting in the ^{13}C spectrum of 40 kHz maximum (for $C-H \parallel \mathbf{B}_0$). Therefore, one has to eliminate the $I-S$ coupling ($I = {}^{13}C$, $S = {}^{1}H$) by irradiating the sample at the resonance frequency of the S spin (ω_S). This is completely analogous to proton decoupling in high resolution ^{13}C spectroscopy in liquids, the important difference being that there the isotropic J coupling $^{13}C-^{1}H$ of the order of a few hundred Hz has to be eliminated, whereas here, the dipolar coupling of the order of 20 kHz has to be wiped out, with corresponding requirements on the spectrometer.

In addition to achieving decoupling, the *rf*-field of frequency ω_S is used for another important purpose. As shown first by Hartmann and Hahn [*31*], a transfer of polarization of the S spins to the I spins can be achieved; thereby, increasing the sensitivity for the rare I spins. The direct method where, after transfer of polarization, the I spins are detected directly while simultaneously irradiating the S spins, has been developed by Pines, Gibby, and Waugh [*32, 33*]. This "proton enhanced nuclear induction spectroscopy" has been developed rapidly to a standard NMR method and, as an illustrative example, Fig. 5.3 shows the ^{13}C spectrum of solid benzene [*33*]. The shielding tensor is axially symmetric ($\Delta\sigma$ = 180 ppm) since, at the temperature of 223 K (where the spectrum has been taken), the benzene molecules rotate rapidly about their sixfold axes. The increase in resolution compared with simply applying high magnetic fields, is obvious.

Fig. 5.3. Proton enhanced and decoupled ^{13}C spectrum of solid benzene at 25 MHz and 223 K (Pines *et al.* [*33*]). [From Pines, A., *et al.*: J. Chem. Phys. **59**, 569 (1973)].

The indirect method, where the rare I spins are detected via the S spins [*34–36*] did not reach practical importance although its sensitivity, in principle, should be higher than that of the direct method [*36*]. Most recently, heteronuclear decoupling has been achieved even for ^2D, despite the presence of the quadrupole coupling, by double quantum transitions [*37*]. In many cases, with solid polymers in particular, the ^{13}C spectra do not show much structure even under proton decoupling, the reason being that the spectra result from a superposition of many powder patterns for different nuclear sites. By rotating the sample at a moderate speed about the magic angle axis (see above), the effect of the shielding anisotropy can be eliminated and spectra that can be interpreted more readily [*38, 39*] are obtained.

Homonuclear Decoupling

If we want to determine shielding tensors of abundant nuclei, protons, or ^{19}F, in particular, the homonuclear dipole-dipole interaction between I^i and I^j must be eliminated. This was first achieved by Waugh, Huber, and Haeberlen [*40*] applying a sequence of pulse "cycles" consisting of four $\pi/2$ pulses of different phases. These strong *rf* pulses impose a *common* motion on the I spins in such a way that, when averaged over such a cycle, the mean value of the dipolar Hamiltonian $\overline{\mathcal{H}_D}$, vanishes. Besides the early discussion of this coherent averaging [*41*] and the relaxation phenomena of interest for such multiple pulse experiments [*42*], the reader again is referred to Haeberlen's book [*22*] which gives a clear and comprehensive treatment of selective averaging.

Contrary to the sample spinning method where the transformation properties of the space dependent operators R^{λ}_{lm} are exploited to achieve line-narrowing, in multiple pulse experiments, the transformation properties of the spin dependent operators T^{λ}_{lm} in spin space are exploited to achieve decoupling. From Table 3.1 we know that the homonuclear dipole-dipole, the J-, and the quadrupole coupling, provided $\mathcal{H}_Q \ll \mathcal{H}_Z$, are all described by irreducible tensors T^{λ}_{20} in the rotating frame. The spin dependent operator T^{CS}_{20}, however, is proportional to $T^I_{10} = I_0$. The spin dependent operators for $\lambda = D, J, Q$, on the one hand, and for $\lambda = CS$, on the other hand, are irreducible tensors of different rank in spin space and accordingly have different transformation properties; therefore, one can find pulse sequences for which $\overline{\mathcal{H}}_{D,J,Q} = 0$, whereas, $\overline{\mathcal{H}}_{CS} \neq 0$. The information about the shielding tensor $\vec{\vec{\sigma}}$, therefore, is maintained, apart from a scaling factor. At this point, we wish to remind the reader of the derivation of the relaxation rates (see Section 4.2.2.1). Because of the different transformation properties of the T_{lm}^{λ} operators in spin space, we also had to treat the case $\lambda = CS$ separately with considerable consequences for the relaxation rates.

Let us also give an illustrative example here. Fig. 5.4 shows ^{1}H multiple pulse spectra of a spherical single crystal of ferrocene at room temperature [*43, 44*]. For the same reason as in benzene, (cf. Fig. 5.3), one observes a single axially symmetric shielding tensor for each of the two molecules in the unit cell only, because the ferrocene molecules rotate rapidly about their five-fold axes. The absolute value of the ^{1}H shielding anisotropy $\Delta\sigma = -6.5 \pm 0.1$ ppm, however, is much smaller than that of ^{13}C in benzene. The higher resolution of the lower spectrum (Fig. 5.4b [*44*]) compared with the spectrum of the original paper (Fig. 5.4a [*43*]) demonstrates recent progress in multiple pulse NMR.

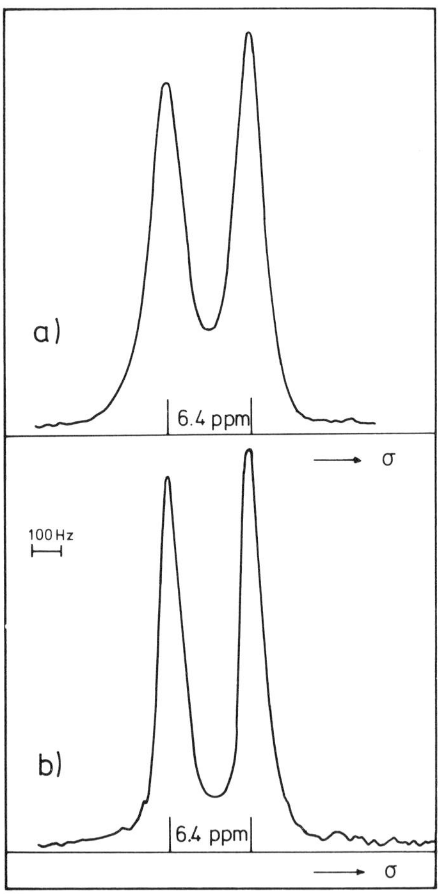

Fig. 5.4a and b. ^{1}H multiple pulse spectra of a single crystal sphere of ferrocene at 90 MHz and room temperature [*43, 44*]. The orientation of the crystal is such that $\mathbf{B}_0 \parallel C_5$ for site A and $\mathbf{B}_0 \perp C_5$ for site B in the unit cell. Time basis of multiple pulse cycle $\tau = 2.6\ \mu s$ (a) and $\tau = 2.0\ \mu s$ (b) (Courtesy of U. Haeberlen)

If one is interested in shielding tensors of protons in solids, containing, in addition, other nuclei with high magnetic moments (e.g., ^{19}F), one has to achieve both homo- and heteronuclear decoupling. This is done by irradiating the ^{19}F spins by π pulses properly timed with respect to the WAHUHA cycle (for details see Haeberlen [*22*], Mehring [*23*], or van Hecke *et al.* [*45*]).

5.1.2.2 Structural Information from Dipole-Dipole Coupling

Until now we have considered the dipole-dipole coupling as an "enemy" who has to be wiped out. On the other hand, this coupling allows a precise determination of internuclear vectors through the r_{IS}^{-3} dependence of its strength and the angular dependence described by an axially symmetric second rank tensor. Solid state spectra dominated by the dipole-dipole coupling have, in fact, been used since the pioneer work of Pake [*46*] mainly to study solid hydrates. Such "broad line" NMR investigations, aimed to obtain structural information, have experienced a revival recently since, by Fourier transform NMR, samples with relaxation times T_1 of the order or hours can be investigated (cf. [*48, 49*]).

As an example, Fig. 5.5 shows the ^1H broad line spectrum of a single crystal of KHF_2 [*48*]. In this solid, we have F–H–F hydrogen bonds with the extremely short F–F distance of 2.3 Å only. The orientation of the crystal corresponding to Fig. 5.5 was chosen in such a way that $\mathbf{B}_0$ was parallel to the F–F vector for one site in the unit cell and perpendicular to that vector for the other site. The frequency splitting in that spectrum is more than 300 kHz. The slight asymmetry of the spectrum results from the anisotropy of the shielding [*48*].

The dipole-dipole coupling, however, is a relatively far-reaching interaction and, therefore, is rather unselective. Such broad line methods, therefore, are restricted to simple structures. Through the decoupling experiments described in the preceding sections, this far reaching unselective dipolar interaction can be eliminated while simultaneously maintaining selected dipolar coupling, which there can be used to

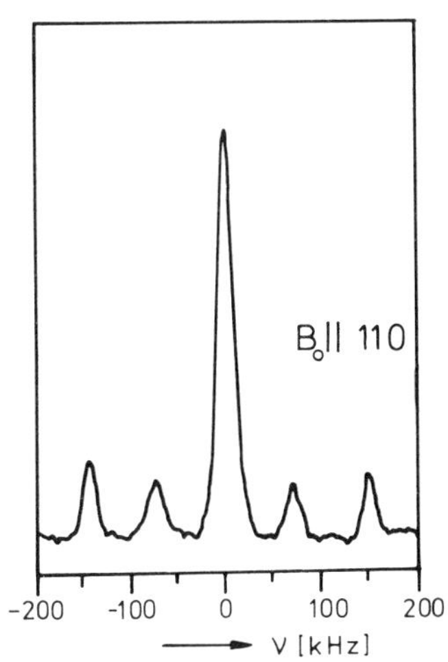

Fig. 5.5. ^1H spectrum of a single crystal of KHF_2 at 90 MHz [*48*]. The orientation of the crystal is such that $\mathbf{B}_0 \parallel$ *F-H-F* for site A and $\mathbf{B}_0 \perp$ *F-H-F* for site B in the unit cell

obtain structural information. Corresponding pulse methods have been developed recently, mainly by various modifications of the pulse sequences described above.

Example: $^{13}C-^{14}N$ *Coupling*

In organic compoinds containing nitrogen, the $^{13}C-^{1}H$ coupling can be eliminated. The $^{13}C-^{14}N$ coupling then remains, resulting in simple spectra, from which C–N vectors can be deduced. As an example Fig. 5.6 shows a proton decoupled ^{13}C spectrum of a single crystal of glycine [*50*]. The orientation of the crystal was chosen in such a way that the methylene carbons of the different molecules in the unit cell are magnetically equivalent and the spectrum shows the characteristic triplet due to the $^{13}C-^{14}N$ dipolar coupling. The two lines at left correspond to the carboxylic carbons. The line width in this spectrum is determined by residual $^{13}C-^{14}N$ coupling to adjacent molecules [*50*]. For the analysis of the spectra, one has to remember that for ^{14}N in organic solids, the condition $\mathcal{H}_Q \ll \mathcal{H}_Z$ often is not fulfilled and the expression of Section 3.2.2 have to be used (cf. [*51*]).

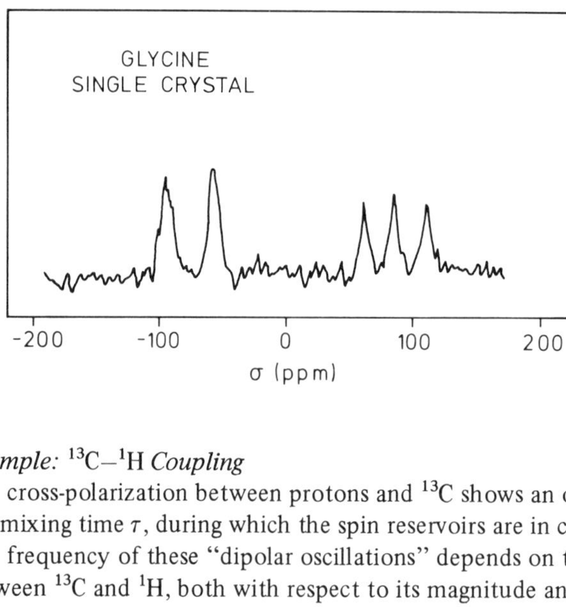

Fig. 5.6 Proton enhanced and decoupled ^{13}C spectrum of a single crystal of glycine at 25 MHz (Griffin *et al,* [*50*]). The lines at the left are due to the carbonyl carbons and those at the right to the methylene carbons split by adjacent ^{14}N nuclei. [From Griffin, R. G., *et al.*: J. Chem. Phys. **63**, 3676 (1976]

Example: $^{13}C-^{1}H$ *Coupling*

The cross-polarization between protons and ^{13}C shows an oscillatory behaviour when the mixing time τ, during which the spin reservoirs are in contact, is varied [*31, 52*]. The frequency of these "dipolar oscillations" depends on the dipolar interaction between ^{13}C and ^{1}H, both with respect to its magnitude and its angular variation. This can be exploited to obtain information about the dipolar coupling of specific spin pairs if the mixing time τ is varied and, moreover, the I^iI^j coupling is eliminated by irradiating with a multiple pulse sequence [*53–56*]. As an early example Fig. 5.7 shows ^{13}C spectra of solid benzene at 200 K for different values of τ [*54b*]. Characteristic distortions corresponding to the different dipolar oscillation frequencies across the powder spectrum are evident. For analysis of the data double Fourier transformation is required (cf. also [*57*]).

These methods of exploiting the structural information available through the dipolar interaction also allow direct and unique determination of the principal axes system of the $\overset{\rightrightarrows}{\sigma}$ tensor relative to *molecular axes.* This offers considerable advantages

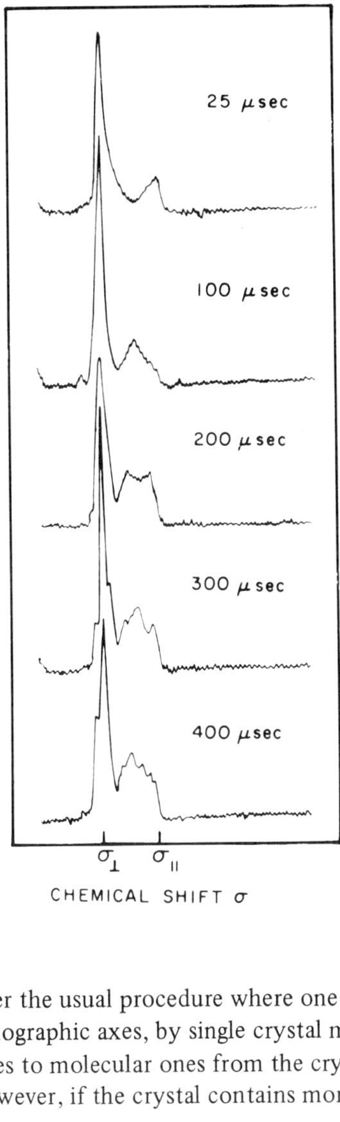

Fig. 5.7. ^{13}C spectra of solid benzene at 200 K (Hester *et al.* [*54*]) for different mixing times τ indicated at the right. [From Hester, R. K., *et al.*: J. Chem. Phys. **63**, 3606 (1976)]

over the usual procedure where one determines the principal axes relative to crystallographic axes, by single crystal measurements, and then assigns these principal axes to molecular ones from the crystal structure. This assignment is not unique, however, if the crystal contains more than one magnetic site.

5.2 Theory of Coupling Tensors

Since a complete description of the theory of the coupling tensors is far beyond the scope of this book, we had to restrict ourselves severely and the following sections have been included, having in mind the following aspects:

i) The coupling tensors which have been introduced in a rather formal way in Chapter 2 are traced back to well-known interactions between nuclei and electrons.

ii) The inherent difficulties in calculating these coupling parameters are emphasized. Although theoretical chemists, in recent years, have made considerable progress to overcome these difficulties, a detailed discussion of their work will not be given here. It is relevant to those who wish to obtain information about the electronic structure of molecules through the coupling tensors. If we want to extract dynamic information from relaxation or line shape data, we can get the coupling tensors with sufficient accuracy from experiment only; not from calculation.

iii) The spin-rotation interaction calls for special attention since, even for only moderately complicated molecules, the spin-rotation tensor $\mathfrak{c}$, at present, cannot be determined experimentally, either by microwave or molecular beam resonance spectroscopy, nor can it be calculated reliably from first principles. Therefore, we will stress the close relationship between the spin-rotation tensor $\mathfrak{c}$ and the shielding tensor $\vec{\vec{\sigma}}$. Since the latter now can be determined experimentally, this relationship offers an indirect way to obtain "experimental" values for the $\mathfrak{c}$ tensor.

iv) Finally, the theoretical expressions given here, in particular for the $\vec{\vec{\sigma}}$ tensor, will enable us to understand the characteristics of the shielding tensors presented in Chapter 6.

5.2.1. Comparison of Couplings

In Chapter 2, we already made a distinction between direct and indirect couplings:
Direct couplings $\lambda = Z, D, Q$;
Indirect couplings $\lambda = CS, J, SR$.

Accordingly, the theoretical calculation of the coupling parameters will be treated separately for these two cases.

5.2.1.1 Direct Couplings

The *Zeeman* interaction, although typically the strongest interaction for spin $I = 1/2$ nuclei, need not be discussed here since it depends only on γ_I (a nuclear parameter) but is independent of the compound to which the nucleus belongs.

Likewise, the *dipole-dipole* coupling between nuclei can be calculated classically from the molecular geometry within the Born-Oppenheimer approximation. In solids, one often will be interested in the second moment, based on the truncated dipolar Hamiltonian (cf. Table 3.1) only. The corresponding formulae are worked out, in detail, in Abragam's book [*58*].

Finally, the calculation of the *quadrupole* coupling is relatively simple, at least in principle. Here we have to deal with the electric coupling of the nuclear quadrupole moment with the electric charge distribution of the molecule or the crystal, respectively. Therefore, calculation of the cartesian tensor component q_{ab} of the field gradient tensor $\mathfrak{q}$ involves calculation of the expectation values of $\mathfrak{q}_{Op}$ over the ground state of the molecule only:

$$q_{ab} = \langle 0 | \sum_i \frac{3\, a_i b_i - r_i^2 \delta_{ab}}{r_i^5} | 0 \rangle - \sum_K{}' \frac{Z_K}{r_K^5} (3\, a_K b_K - r_K^2 \delta_{ab}). \qquad (5.1)$$

Here $a, b = x, y, z$. The position of the electron i is described by the vector $\mathbf{r}_i$ with components a_i and b_i, the origin being the nucleus under investigation. In addition, we have a corresponding contribution with opposite sign from the other nuclei of the molecule; Σ' indicates that the sum excludes the nucleus under investigation). Although simple in principle, the calculation of the expectation value [Eq. (5.1)] is by no means trivial. In fact, comparison of calculated quadrupole coupling parameters with observed ones provides a sensitive means to probe the quality of calculated electronic wave functions of a molecule. The interested reader is referred to background material in Lucken's book [*59*] and to a recent review on this subject [*60*]. In passing, it should be mentioned that here also we encounter shielding or antishielding of an external field gradient due to the inner electrons of the atom to which the nucleus of interest belongs (Sternheimer antishielding, cf. [*59*]). The quadrupole coupling, which can be observed experimentally, thus in fact results from both direct and indirect couplings.

5.2.1.2. Indirect Couplings

The calculation of indirect couplings is considerably more involved than the calculation of the field gradient tensor. The theoretical explanation of indirect couplings has been given by Ramsey in a number of fundamental papers 25 years ago (cf. Refs. in [*2*]). In addition to the complete and detailed treatment of these couplings in Ramsey's [*2*] and Abragam's [*58*] books, for more recent developments two review articles are available (cf. Appleman and Dailey [*9*], who treat magnetic shielding and susceptibility effects, and Murrell [*61*] who treats *J*-coupling). Therefore, we can restrict ourselves here to the most important contributions to the couplings and leave out all unnecessary details (cf. also Chapter 8 of Townes and Schawlow [*3*]).

The electrostatic interaction of the electrons among themselves and with the nuclei by far exceeds the strength of the magnetic couplings. Therefore, in standard molecular orbital (MO) theory, the electronic eigenfunctions of a molecule are calculated on the basis of a Hamiltonian neglecting magnetic couplings. The electronic eigenfunction of the ground state is labeled ψ_0^0 and the eigenfunctions of excited states ψ_n^0. The magnetic couplings then are treated by perturbation theory where we usually have to go to second order to find the dominant terms.[21]

The standard expression for the additional energy reads:

$$\Delta E = \Delta E^{(1)} + \Delta E^{(2)},$$

with

$$\Delta E^{(1)} = \langle 0 | \mathcal{H}_P | 0 \rangle$$

$$\Delta E^{(2)} = \sum_{n>0} (E_0 - E_n)^{-1} \langle 0 | \mathcal{H}_P | n \rangle \langle n | \mathcal{H}_P | 0 \rangle, \tag{5.2}$$

where $\langle 0|$ and $\langle n|$ represent the eigenfunctions ψ_0^0 and ψ_n^0, respectively, and n runs over all excited states of the molecule. The perturbation operator $\mathcal{H}_P$ describes all

[21] Contributions to the shielding tensor obtained in third order perturbation theory are discussed in [*62*]. For light nuclei, (*e.g.*, ^{13}C and ^{15}N), these contributions turned out to be unimportant.

magnetic interactions between the electrons and the nuclei and the couplings of the electrons with the external field $\mathbf{B}_0$ or the total angular momentum $\mathbf{J}$. This operator $\mathcal{H}_P$ is written out fully in Abragam's book [*58*]; the most important terms are given below. Clearly, in order to calculate the coupling tensors $\vec{\vec{\sigma}}$, $\mathfrak{c}$, and $\mathfrak{J}$, we need consider only those terms of $E^{(1)}$ and $E^{(2)}$ which are bilinear in $(\mathbf{I}, \mathbf{B}_0)$, $(\mathbf{I}, \mathbf{J})$, or $(\mathbf{I}^i, \mathbf{I}^j)$, respectively. In all three cases, $\vec{\vec{\sigma}}$, $\mathfrak{c}$, and $\mathfrak{J}$, we, in fact, obtain contributions in first order perturbation theory (cf. [*58*]). As we will see shortly, these first order contributions to $\vec{\vec{\sigma}}$ are relatively simple; to $\mathfrak{c}$ they are even trivial to calculate. The first order contributions to the $\mathfrak{J}$-coupling are negligible, in general [*58, 61*]; therefore, we will not deal with these contributions here but will include only the corresponding terms in the final expressions without derivation. The most important contributions to the indirect coupling parameters are due to the interaction of the nuclear dipole moment with the orbital angular momentum and the spin of the electrons $\mathbf{l}_i$ and $\mathbf{s}_i$, respectively, described by the operator $\mathcal{H}_{IE}$ which is well-known from the theory of the hyperfine interaction in atoms:

$$\mathcal{H}_{IE} = \mathcal{H}_{IL} + \mathcal{H}_{IS}^{D} + \mathcal{H}_{IS}^{K}, \tag{5.3}$$

where

$$\mathcal{H}_{IL} = 2\,\beta\,\gamma_I\,\hbar\,\mathbf{I}\sum_i \frac{\mathbf{l}_i}{r_i^3} \tag{5.4a}$$

is the coupling of $\mathbf{I}$ with the magnetic field resulting from the orbital motion of the electrons:

$$\mathcal{H}_{IS}^{D} = 2\,\beta\,\gamma_I\,\hbar\,\mathbf{I}\sum_i \left(\frac{3\,\mathbf{r}_i\,(\mathbf{s}_i\mathbf{r}_i)}{r_i^5} - \frac{\mathbf{s}_i}{r_i^3}\right), \tag{5.4b}$$

the dipolar, and

$$\mathcal{H}_{IS}^{K} = 2\,\beta\,\gamma_I\,\hbar\,\mathbf{I}\sum_i \frac{8\,\pi}{3}\,\mathbf{s}_i\,\delta(r_i), \tag{5.4c}$$

the contact interaction between $\mathbf{I}$ and the spin of the electrons. The delta function $\delta(r_i)$ represents the value of the integrand at $r_i = 0$ in any integration over the coordinates of the electron i. The sum over i runs over all electrons in the molecule, the coordinates $\mathbf{r}_i$ are measured with respect to the nucleus I, and $\beta = 1/2\; e/m\;\hbar$ is the Bohr magneton. Since the coupling described by the Hamiltonians (5.4a)–(5.4c) are not bilinear in $\mathbf{I}$, $\mathbf{B}_0$, $\mathbf{J}$ they cannot contribute to $\vec{\vec{\sigma}}$, $\mathfrak{c}$, or $\mathfrak{J}$ in first order perturbation theory. In second order, however, we clearly can get contributions since $\mathcal{H}_{IE}$ is part of $\mathcal{H}_P$ and, together with terms proportional to $\mathbf{B}_0$, $\mathbf{J}$, or $\mathbf{I}$, can lead to bilinear terms in $\Delta E^{(2)}$. For $\vec{\vec{\sigma}}$ and $\mathfrak{c}$, we get only contributions from $\mathcal{H}_{IL}$, however. In order to understand this, we have to discuss the reasons for the vanishing of the expectation values of the orbital angular momentum operator $\mathbf{L} = \sum_i \mathbf{l}_i$ and of the electron spin operator $\mathbf{S} = \sum_i \mathbf{s}_i$ in diamagnetic compounds.

Let us consider the so-called *quenching of angular momentum* first:

$$\langle \psi_0^0 | L_z | \psi_0^0 \rangle = 0, \tag{5.5}$$

where we have confined the discussion to the z components of **L** for simplicity. It is easy to show that Eq. (5.5) is fulfilled if ψ_0^0 belongs to a nondegenerated state [*58, 63*]. This is a sufficient but not a necessary condition. As discussed at length by Van Vleck [*63*], the underlying physical reason for the quenching of orbital angular momentum is the fact that, in molecules and, even more so, in condensed phases, the electrostatic interaction between the electrons and the nuclei does not have spherical symmetry as it has in free atoms. Consequently, the eigenfunctions of the energy—neglecting magnetic couplings—, in general, are not eigenfunctions of angular momentum operators.[22] Therefore for sufficiently low symmetry, it is meaningless to speak about *the* angular momentum of the *electrons* in a molecule.

Through interaction of the electrons with well defined vectors, a *homogeneous external field* or the *total angular momentum* of the molecule, part of the orbital angular momentum of the electrons can be "unquenched". This can be described by perturbation theory with the perturbation Hamiltonian:

$$\mathcal{H}_{LB} = \beta\, \mathbf{L} \cdot \mathbf{B}_0$$

$$\mathcal{H}_{LJ} = -\hbar \sum_{a=x,y,z} \theta_a J_a L_a, \tag{5.6}$$

where θ_a is the moment of inertia[23] for rotations about principal axis a of the moment of inertia tensor.

The eigenfunction in first order perturbation theory is given by:

$$\psi_0^1 = \sum_n c_n^1\, \psi_n^0, \tag{5.7}$$

for which the expectation value $\langle \psi_0^1 | L_z | \psi_0^1 \rangle$ can be different from zero and where $z \parallel \mathbf{B}_0$ or $\mathbf{J}$, respectively. The expansion coefficients c_n^1 are proportional to B_0 or J and are given according to standard perturbation theory by

$$c_n^1 \sim (E_0 - E_n)^{-1} \langle n | L_z | 0 \rangle, \quad (n > 0). \tag{5.7a}$$

The energy contribution in second order is obtained from $\langle \psi_0^1 | \mathcal{H}_P | \psi_0^1 \rangle$, which represents nothing else but the standard expression (5.2). Clearly, the term $\mathcal{H}_{IL}$ in $\mathcal{H}_P$ then gives contributions to $\overleftrightarrow{\sigma}$ or c, respectively (see below).

For the electron *spin*, such an unquenching does not occur since, in diamagnetic compounds, the ground state ψ_0^0, to a good approximation, is an eigenstate of the total spin operator with vanishing expectation value. Therefore, in first order, the

---

[22] Linear molecules, of course, have cylindrical symmetry and the eigenfunctions of the energy (ψ_n^0) are eigenfunctions of L_z.

[23] To be precise, θ_a is the part of the moment of inertia due to the nuclei alone (cf. [*3*]).

eigenfunction is unchanged, since for the corresponding perturbation Hamiltonian

$$\mathcal{H}_{SB} = 2\,\beta\,\mathbf{S}\cdot\mathbf{B}_0, \tag{5.6a}$$

the matrix elements $\langle n|S_z|0\rangle$ vanish. The electron spin, therefore, does not contribute to the shielding and the spin-rotation interaction even in second order perturbation theory. For the indirect J-coupling, however, $\mathcal{H}_{IS}^{D}$ and $\mathcal{H}_{IS}^{K}$ give rise to the dominant terms when $\Delta E^{(2)}$ is calculated according to Eq. (5.2).

5.2.2. Magnetic Shielding and Spin-Rotation Interaction

The close relationship between magnetic shielding and spin-rotation interaction is understood easily on the basis of the general discussion of the preceeding section since the two perturbation Hamiltonians of Eq. (5.6) contain the same operator acting on the electrons, namely the total electronic angular momentum operator. Therefore both interactions are treated together here.

For the *magnetic shielding*, by standard second order perturbation theory as described above, we obtain:

$$\begin{aligned}\sigma_{ab} &= \frac{2\,m\,\beta^2}{\hbar^2}\,\langle 0|\sum_i \frac{\delta_{ab}\,r_i^2 - a_i b_i}{r_i^3}|0\rangle \\ &+ \frac{2\,\beta^2}{\hbar^2}\sum_{n>0}(E_0-E_n)^{-1}\left\{\langle 0|\sum_i \frac{l_{ia}}{r_i^3}|n\rangle\langle n|L_b|0\rangle + c.c.\right\} \\ &\equiv \sigma_{ab}^{d} + \sigma_{ab}^{p},\end{aligned} \tag{5.8}$$

as first derived by Ramsey [*2*]. Notation is described in connection with Eqs. (5.1) and (5.3). In addition to the contribution in second order, normally called the *paramagnetic* part σ_{ab}^{p}, we have the *diamagnetic* part σ_{ab}^{d} obtained in first order already. It corresponds to the precession of the electrons of the molecule in the external magnetic field *about the nucleus* for which we want to calculate the shielding because all space variables are measured from that nucleus with our choice of a coordinate system [*2*].

The theoretical expression (5.8) gives the shielding tensor $\overleftrightarrow{\sigma}$, which can be determined experimentally, as a sum of two terms with opposite signs partially cancelling each other. In numerical calculations, therefore, difficulties are encountered if one tries to evaluate the shielding according to Eq. (5.8). Likewise, the sum over all excited states in $\overleftrightarrow{\sigma}^{p}$ is unsatisfactory. This sum is obtained if the expansion coefficients in Eq. (5.7a) are calculated in the standard way. In more recent *ab initio* calculations, one no longer calculates these coefficients according to the standard formula, but within a perturbed Hartree-Fock treatment of the molecule in the magnetic field. Furthermore, gauge invariant orbitals are used which contain the vector potential of the magnetic field explicitly. This assures that physically irrelevant terms in $\overleftrightarrow{\sigma}^{d}$ and $\overleftrightarrow{\sigma}^{p}$, depending on the choice of the coordinate system and largely cancelling each other, are not calculated to begin with. The two procedures commonly used, which have been developed by the groups of Lipscomb [*64*], Pople [*65*], and Ditchfield [*66*], are described and compared by Appleman and Dailey [*9*].

We nevertheless give the standard expression here for the following reasons:

i) $\vec{\vec{\sigma}}^p$ and the sum over excited states can be expressed by the spin-rotation tensor $\mathfrak{c}$, a quantity accessible to experiment.

ii) $\vec{\vec{\sigma}}^d$ can be approximated easily with remarkable success for ^{13}C and ^{15}N.

iii) Based upon Eq. (5.8) the characteristics of the $\vec{\vec{\sigma}}$ tensors for ^{13}C and ^{15}N found experimentally, can be rationalized (cf. Chapter 6) since here $\vec{\vec{\sigma}}^p$ is predominant.

The expression for the *spin-rotation tensor* also was derived first by Ramsey [*2*]. A detailed more recent derivation also has been given by Flygare [*5, 67*]. The spin-rotation tensor, of course, can be given in any cartesian coordinate system but we will find the principal axes system of the moment of inertia tensor to be the most convenient. Then, standard second order perturbation theory yields:

$$\begin{aligned} c_{ab} &= \frac{2\,m\,\beta\,\gamma_I}{\theta_b} \sum_k{}' \frac{Z_K}{r_K^3} (\delta_{ab}\, r_K^2 - a_K b_K) \\ &+ \frac{2\,\beta\,\gamma_I}{\theta_b} \sum_{n>0} (E_0 - E_n)^{-1} \left\{ \langle 0 | \sum_i \frac{l_{ia}}{r_i^3} | n \rangle \langle n | L_b | 0 \rangle + c.c. \right\}. \end{aligned} \tag{5.9}$$

Here, Z_K is the charge of nucleus K in the molecule and Σ' indicates again that the sum over the nuclei excludes the nucleus for which we want to calculate the spin-rotation tensor.

A comment on Eq. (5.9) has to be given here. The spin-rotation tensor which can be determined in microwave spectroscopy[24] describes the coupling of the nucleus with the total angular momentum of the molecule. Therefore, the expression for the matrix element c_{ab} contains the factor θ_b^{-1}. This means that the spin-rotation tensor "automatically" contains antisymmetric constituents, provided it has off-diagonal elements at all, in the principal axes system of $\vec{\vec{\theta}}$ and $\theta_a \neq \theta_b \neq \theta_c$. We will return to this at the end of this section.

The spin rotation tensor $\mathfrak{c}$, likewise, is obtained here as a sum of two terms partially cancelling each other, the contribution of the nuclei and that of the electrons, respectively. Comparison with Eq. (5.8) shows that the contribution to $\vec{\vec{\sigma}}$ and $\mathfrak{c}$, respectively, obtained in second order, are proportional to each other as expected from Eq. (5.3). In the principal axes system of the moment of inertia tensor, therefore, we have a simple relation of the tensor elements of $\vec{\vec{\sigma}}$ and $\mathfrak{c}$:

$$\sigma_{ab} = \frac{2\,m\,\beta^2}{\hbar^2} \left\{ \langle 0 | \sum_i \frac{\sigma_{ab}\, r_i^2 - a_i b_i}{r_i^3} | 0 \rangle - \sum_K{}' \frac{Z_K}{r_K^3} (\sigma_{ab}\, r_K^2 - a_K b_K) + \frac{\theta_b c_{ab}}{2\,m \beta \gamma_I} \right\}. \tag{5.10}$$

---

[24] It should be noted that the conventional definitions of $\mathfrak{c}$ in NMR, where the angular velocity ω is of interest, and in microwave spectroscopy, where the frequency ν is measured, differ by a factor of $2\,\pi$.

This relation allows the reliable calculation of elements of the spin-rotation tensor from the elements of the shielding tensor which, at present, can be determined experimentally more easily. In order to make use of Eq. (5.10) we need to know the diamagnetic part $\vec{\vec{\sigma}}^d$, however. This still calls for a molecular orbital calculation, although only knowledge of the ground state is required.

In practice, a semi-empirical calculation of $\vec{\vec{\sigma}}^d$ has proved extraordinarily successful. If, in the spirit of a population analysis, we assign to each atom of the molecule the number of electrons according to its nuclear charge[25], we can treat $\vec{\vec{\sigma}}^d$ by a moment expansion. We then replace the coordinates of the electrons by those of the corresponding nuclei. This atom-dipole approximation has been developed by Flygare and coworkers [*68, 69, 5*]. The agreement of the values for $\vec{\vec{\sigma}}$, calculated in this way with results from *ab initio* calculations is remarkable [*5*]. The different terms of the expansion also are discussed in detail by Haeberlen [*22*], since this approximation offers a quite satisfactory explanation of the shielding anisotropy for protons in hydrogen bonds (cf. Chapter 6). For ^{13}C and ^{15}N, it is usually sufficient to consider the first two terms in the expansion only:

$$\sigma_{ab}^d = \sigma^d \text{ (free atom)} + \frac{2\, m\, \beta^2}{\hbar^2} \sum_K{}' \frac{Z_K}{r_K^3} (\delta_{ab} r_K^2 - a_K b_K). \qquad (5.11)$$

Then the contribution of the *electrons* "belonging" to a given atom in the molecule, just cancel the *nuclear* contribution of this atom [cf. Eq. (5.10)] and we obtain the simple relation:

$$\sigma_{ab} = \frac{2\, m\, \beta^2}{\hbar^2} \left\{ \sigma^d \text{ (free atom)} + \frac{\theta_b c_{ab}}{2\, m\, \beta\, \gamma_I} \right\}, \qquad (5.12)$$

which is of considerable importance for the analysis of relaxation rates as will be shown in Chapter 6.

It should be noted that in Eq. (5.12) the physically irrelevant terms in Eqs. (5.8) and (5.9) mentioned above have compensated each other. Therefore, corrections to Eq. (5.12) can be incorporated easily. Such corrections can be obtained, for example, from the higher terms in Flygare's expansion [*69, 5*] or from *ab initio* calculation of $\vec{\vec{\sigma}}^d$. In the latter case, we simply have to subtract the term $\sum_K{}' \ldots$ of Eq. (5.11) involving knowledge of the molecular geometry only, to calculate the correction to Eq. (5.12).

Finally, we want to emphasize that we need not know the parameter σ^d (free atom) in order to make use of Eq. (5.12). This parameter is contained in this expression only because we have given $\vec{\vec{\sigma}}$ on an absolute scale. By using a reference compound with known spin-rotation constant, we can eliminate σ^d (free atom) if the relative chemical shift is known (see, e.g. [*70*]). One of the reasons why use of Eq. (5.12) together with a reference compound, yields excellent results most likely

[25] Excess charges can, of course, also be incorporated.

stems from the fact that, for the compound of interest and the reference, the same approximations are involved and we are left with differences of correction terms only. Similarly, the expression for calculating the *anisotropy* of the shielding is independent of σ^d (free atom).

Let us give ^{13}CO as a numerical example. For such a linear molecule, $c_{\parallel} = 0$ since its eigenfunctions are eigenfunctions of L_z [cf. Eq. (5.9)]. From the experimental values of the spin-rotation constant $c = 2\pi \cdot (-32.56)$ kHz [*71*], we calculate, from Eq. (5.12) $\Delta\sigma = \sigma_{\parallel} - \sigma_{\perp} = 384$ ppm. With Huo's value for σ^d [*72*], one gets $\Delta\sigma =$ 401 ppm [*72*]; the error is 4% only.

Let us come back to the antisymmetric constituents of the $\mathfrak{c}$ tensor in connection with relaxation rates. Equations (5.9) and (5.12) show that $\mathfrak{c}$ has antisymmetric constituents, even if $\vec{\vec{\sigma}}$ is symmetric for the same nucleus, but both tensors have off-diagonal elements in the principal axes system of $\vec{\vec{\theta}}$ and $\theta_a \neq \theta_b \neq \theta_c$. These antisymmetric parts can be quite significant if all the principal moments of inertia are significantly different from each other. These terms are quasi-trivial, however, as shown by the following consideration. Had we started the derivation for the relaxation rates from a Hamiltonian for the coupling of the nuclear spin to the angular *velocity* of the molecule rather than to the angular *momentum* (see, e.g., [*73*], the elements of the coupling tensor would have been $c'_{ab} = \theta_b c_{ab}$. The $\mathfrak{c}'$ tensor is antisymmetric only if the same holds for $\vec{\vec{\sigma}}$ tensor (cf. Eq. (5.12). The expression for the relaxation rate then contains $\theta_b^{-1}\, c'^2_{ab}$, which leads to the same result as expressed in Eq. (4.4) since $\theta_b^{-1}\, c'^2_{ab} = \theta_b c^2_{ab}$.

5.2.3 Anisotropic J-Coupling

The anisotropy of the *J*-coupling is only of minor importance for relaxation studies; therefore, we will not give details here but refer to the literature instead [*61, 74*]. In order to obtain expressions independent of the magnetic moment of the nuclei dealt with, one often uses reduced coupling tensors $\mathfrak{K}$ introduced by Pople and Santry [*75*]:

$$\mathfrak{K} = \frac{2\pi}{\hbar\gamma_{I1}\gamma_{I2}}\,\mathfrak{J}^{I^1 I^2}. \qquad (5.13)$$

By standard second order perturbation theory with the perturbation operators $\mathcal{H}_{IL}$, $\mathcal{H}^D_{IS}$, and $\mathcal{H}^K_{IS}$ [cf. Eqs. (5.4a–5.4c)], we obtain terms involving the same perturbation operator in both matrix elements, but with the space coordinate and angular momentum operators being centred about different nuclei I^i and I^j. There is only one mixed term $\sum_{n>0}(E_0 - E_n)^{-1}\langle 0|\mathcal{H}^D_{IS}|n\rangle\langle n|\mathcal{H}^K_{IS}|0\rangle$. In addition, we have a small contribution in first order due to the coupling of the two nuclei I^i and I^j with the same electron. The explicit expressions are available in [*61*] and [*74*]. Numerical calculations are at least as difficult as for $\vec{\vec{\sigma}}$ and $\mathfrak{c}$.

6. Experimental Examples

6.1. Shielding Tensors $\overleftrightarrow{\sigma}$

As already emphasized in Chapter 1, we would like to demonstrate in this review that, besides the standard relaxation mechanisms, dipole-dipole and quadrupole coupling, other mechanisms, anisotropic shielding and spin-rotation interaction in particular, offer new possibilities to get dynamic information. In this chapter we have collected a few representative examples where dynamic information is obtained either through spin-lattice relaxation rates or through line shape analysis. First, however, we shall give a brief survey of the current knowledge of $\overleftrightarrow{\sigma}$ tensors for several reasons:

i) Shielding tensors can be used as probes for dynamic information only if we know the tensor elements.

ii) Shielding tensors can be used to calculate the spin-rotation tensors, (cf. Section 5.2.2). With these "experimental" spin-rotation tensors, we then can determine the correlation time τ_J from spin-rotational relaxation rates.

iii) We can obtain jump frequencies from line shape analysis only if the shielding tensor is known.

When we were able to measure relaxation through anisotropic shielding in liquids for the first time (in liquid $^{13}CS_2$ [*1*]), very little was known about $\overleftrightarrow{\sigma}$ tensors of spin $I = 1/2$ nuclei. In fact, we succeeded in determining the anisotropy of the shielding $\Delta\sigma$ from the relaxation study alone, (see below) and the value we found was confirmed by solid state NMR [*2*] shortly afterwards. Since then, a large amount of experimental material about shielding tensors for the spin $I = 1/2$ nuclei of interest has been gathered by solid state NMR. The experimental values for shielding tensors of ^1H, ^{13}C, ^{15}N, ^{19}F, ^{31}P, and other less fashionable nuclei have been collected in large tables by Mehring [*3*] in his recent review. Therefore, we will not repeat the numbers here. It might be useful, however, to present a number of representative shielding tensors in diagrams. Presenting the data in this way eases the discussion of the question often encountered in practice; namely, for which compounds relaxation through anisotropic shielding possibly can compete with dipolar relaxation. In order to make the comparison of $\overleftrightarrow{\sigma}$ tensors for different molecules as convenient as possible, we have chosen the labels of the different principal axes in these figures such that $\sigma_{XX} \leqslant \sigma_{YY} \leqslant \sigma_{ZZ}$. Thus the Z direction always is the most shielded one. This convention of labeling the principal axes is consistent with our earlier convention, Eq. (2.27), based on the traceless part of $\overleftrightarrow{\sigma}$ alone, in all but a few cases.

6.1.1. ^1H Shielding Tensors

Proton shielding tensors have been obtained mainly by multiple pulse NMR in the last 5 years and have been reviewed extensively by Haeberlen [*4*]. Since shielding effects are rather small for protons, the anisotropy of the shielding $\Delta\sigma$, likewise, is small; typically $\Delta\sigma \leqslant 20$ ppm. Therefore, anisotropic shielding, in general, will be a negligible relaxation mechanism for protons but proton shielding tensors offer a means to study ultraslow motions in the kHz range through line shape analysis (cf. Section 6.2).

Results from single crystal measurements are collected in Fig. 6.1. The anisotropy of the shielding $\Delta\sigma$ is especially large for protons in hydrogen bonds. For KHF_2 with the extremely short H–F distance of 1.15 Å only, it reaches the value of $\Delta\sigma = 47$ ppm. For O···H···O hydrogen bonds, $\Delta\sigma$ typically is about 20 ppm but in ice it is significantly larger ($\Delta\sigma = 34$ ppm). The shielding tensor for protons in hydrogen bonds is approximately axially symmetric; the unique direction being approximately parallel to the O···O direction (cf. Fig. 6.2). In many cases, hydrogen bonds lead to planar structures, at least locally, as shown schematically in Fig. 6.2. In the single crystal studies, the direction of least shielding (*X* axis) always was found to be perpendicular to that

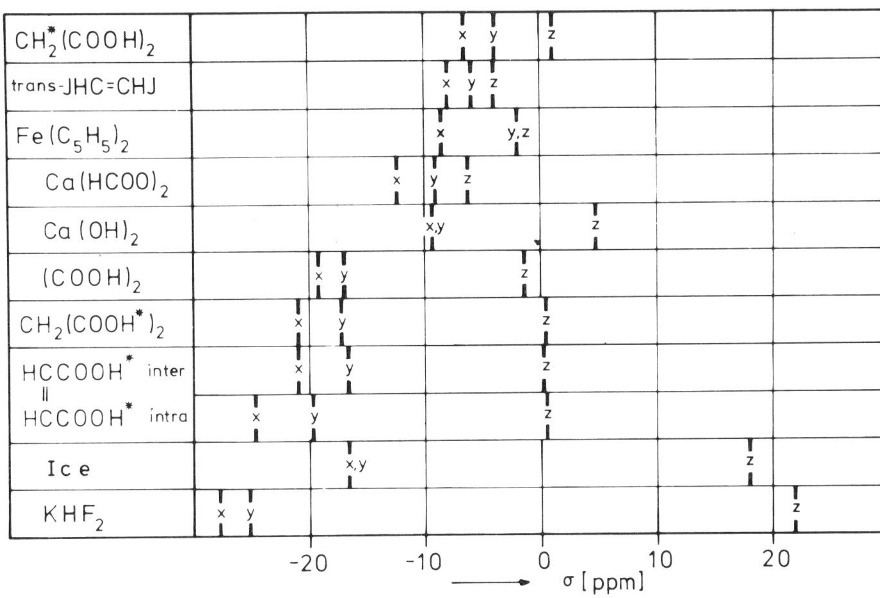

Fig. 6.1. [1]H shielding tensors relative to a spherical smaple of TMS. The proton position studied is indicated by * [*5–13*]

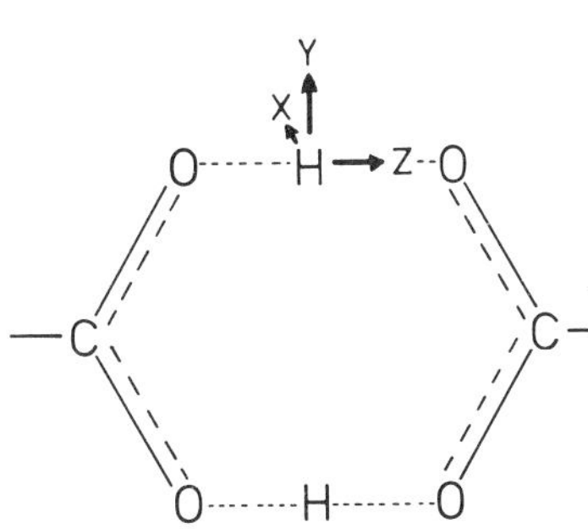

Fig. 6.2. Orientation of principal axes system of $\vec{\vec{\sigma}}$ for ^{1}H in hydrogen bonded carboxylic acids, schematic (cf. [*5, 10, 11*])

plane. These characteristics of the $\overleftrightarrow{\sigma}$ tensors of protons in hydrogen bonds can be explained adequately by Gierke and Flygare's atom-dipole approximation [*14, 4*]. In addition to the terms of Eq. (5.12), however, we have to take in account another term:

$$\sigma_{aa}^{IV} = \frac{2\,m\,\beta^2}{\hbar^2} \sum_K{}' \frac{\langle \frac{1}{3}\,\rho^2 \rangle_K}{r_K^3} \left[2 - \frac{3\,(b_K^2 + c_K^2)}{r_K^2}\right]. \tag{6.1}$$

This "quadrupole term", given here in the notation of Haeberlen [*4*], takes in account the spatial extension of the electrons. In order to calculate $\overleftrightarrow{\sigma}^{IV}$, one approximates the electron distribution by an infinitely thin spherical shell of corresponding negative charge at a radius $\sqrt{\langle \rho^2 \rangle_K / Z_K}$ from the nucleus K. Numerical valus for $\langle \rho^2 \rangle_K / Z_K$ have been given by Malli and Fraga [*15*]. Analysis of the $\overleftrightarrow{\sigma}$ tensors for protons in hydrogen bonds shows that $\overleftrightarrow{\sigma}^{IV}$ predominates the *anisotropy* of the shielding. On the other hand $\overleftrightarrow{\sigma}^{IV}$ does *not* contribute at all to the isotropic shift since it is a traceless tensor. For further details and a clear physical interpretation of $\overleftrightarrow{\sigma}^{IV}$, the reader is referred to Haeberlen [*4*].

For covalently bound protons, the anisotropy of the shielding is considerably smaller in magnitude. In planar structures, for protons adjacent to a $C = C$ or $C = 0$ double bond, one normally finds the least shielded direction to be perpendicular to the molecular plane. Apparently the $\overleftrightarrow{\sigma}$ tensor, typically, is highly asymmetric ($\eta_{CS} \approx 1$) in these cases (cf. Fig. 6.1). Recent *MO* calculations (e.g., of ethylene [*16*]) give a satisfactory description of the anisotropy of the shielding.

6.1.1.1 Susceptibility Correlation

When determining ^{1}H shielding tensors in solids, one carefully has to take in account the contribution to the observed line shifts due to the susceptibility of the sample. This has been demonstrated recently by single crystal work on ferrocene [*7*]. It has been known, of course, for a long time that the magnetic field inside a sample and, consequently, the NMR frequency depends on the *shape* of the sample because of its susceptibility [*17*]. Moreover, the magnetic field inside a sample is homogeneous only if the external field is homogeneous over the sample and the sample is an ellipsoid. In high resolution NMR in liquids, the ellipsoid is approximated by a long cylinder, the sample tube.

In high resolution NMR in solids, not only the external field $\mathbf{B}_0$ but also the *rf* field $\mathbf{B}_1$ has to be homogeneous over the sample. Therefore, if one wants to determine $\overleftrightarrow{\sigma}$ tensors of protons, one not only has to work with single crystals but these single crystals have to be ground into spheres. The deviations from spherical shape should not exceed a few percent [*18, 7*]. This can cause severe problems if the crystals have strongly preferred cleavage planes, (e.g., ferrocene). As an illustration, Fig. 6.3 shows rotation patterns for ferrocene. The open circles were obtained for an "ordinary crystal" as grown by sublimation. The closed circles show the "correct" angular dependence for a spherical crystal. Clearly, the contribution of the suscep-

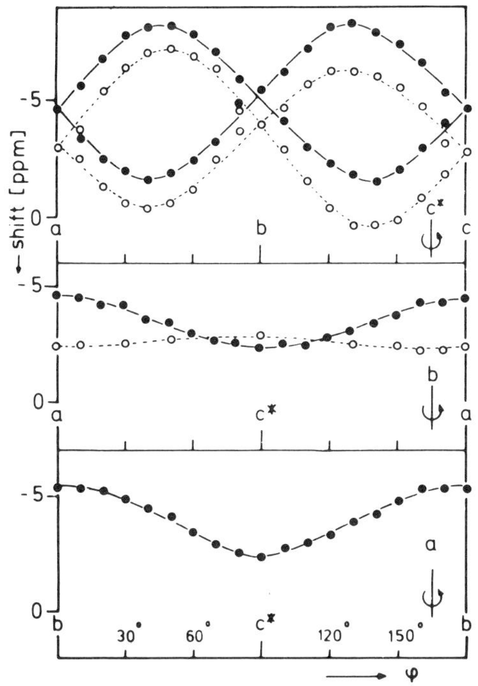

Fig. 6.3. Rotation patterns from multiple pulse spectra of ferrocene [7], • sphere, ○ "natural" crystal. [From Spieß, H. W., *et al.*: Chem. Phys. **12**, 123 (1976)]

tibility of the crystal to the line shifts is of the order of 2–3 ppm and cannot be neglected compared with $\Delta\sigma = -6.5$ ppm.

Use of spherical crystals removes all contributions from the bulk susceptibility [*17*]. However, the apparent shielding tensor, then detected, still can contain significant contributions from neighbouring molecules [*19*]. This is demonstrated again by our data on ferrocene [*7*]. If the shielding tensor were due to *intra*molecular contributions only, one should find it to be axially symmetric at room temperature due to the rapid reorientation of the ferrocene molecules about their C_5 axes. Instead, we found the apparent shielding tensor to be nonaxially symmetric [*7*]; the difference in X and Y components being 0.8 ppm due to *inter*molecular contributions. In reference [*7*] we mistakingly identified these *inter*molecular contributions with the anisotropy of the bulk susceptibility $\Delta\chi$. The fair agreement between $\Delta\chi$ and the intermolecular contribution noted there is, therefore, almost certainly fortuitous [*19*]. Susceptibility effects are especially important in proton NMR since the shielding effects are so small. But even for nuclei like ^{13}C or ^{15}N, the accuracy of the shielding components will be limited to about ±2 ppm if susceptibility effects are not taken in account carefully.

6.1.2. [13]C and [15]N Shielding Tensors

6.1.2.1 Experimental Values for [13]C

Representative examples of ^{13}C shielding tensors are collected in Figs. 6.4 and 6.5. In the upper part of Fig. 6.4 for aliphatic groups, CH_3 in particular; below, for aromatic

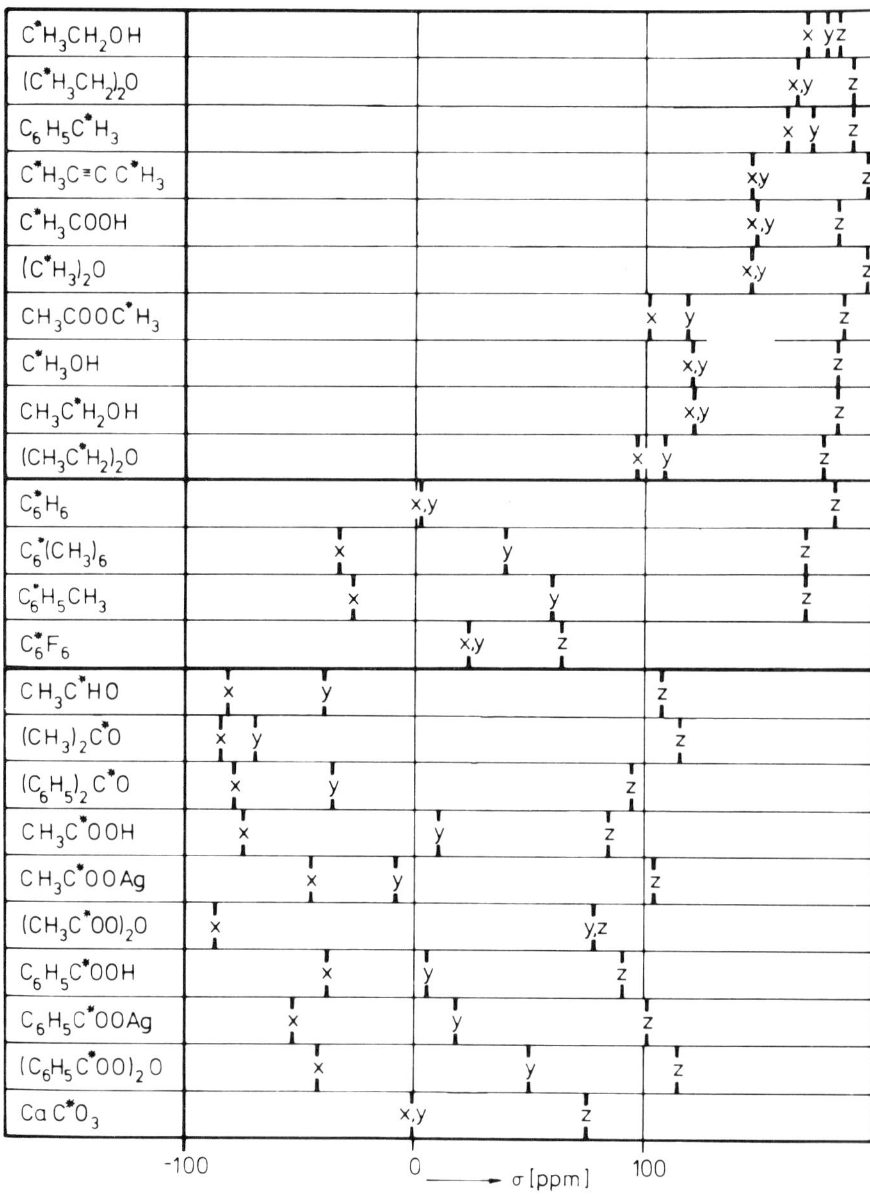

Fig. 6.4. ^{13}C shielding tensors in organic compounds relative to CS_2. The carbon position studied is indicated by * [*20–23*]

and carbonyl groups. In Fig. 6.5, shielding tensors for linear molecules and $C = S$ groups are depicted. The largest absolute values of the shielding anisotropy are encountered here (note the different horizontal scale). From these examples, some general "rules" can be formulated.

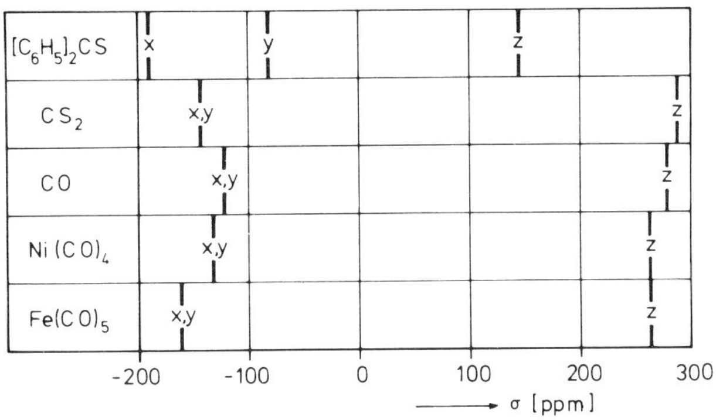

Fig. 6.5. ^{13}C shielding tensors for a $C = S$ group and for linear molecules relative to CS_2 [*21, 23, 24*]

i) Aliphatic Groups

For sp^3 carbons, the shielding anisotropy is relatively small ($\Delta\sigma \leqslant 30$ ppm). A remarkable feature these data display is the constancy of σ_{ZZ}; the most shielded component. As revealed by single crystal measurements [*25*], the Z axes for methyl groups are parallel to their threefold axes. The X and Y components show distinct shifts to lower shielding on substitution by more electronegative groups (note the series $H_3C\text{-}CH_2$, $H_3C\text{-}CO$, $H_3C\text{-}OH$).

ii) Aromatic Rings

For aromatic hydrocarbons, $\Delta\sigma$ is much larger ($\Delta\sigma \approx 180$ ppm). Interestingly, σ_{ZZ} almost is unchanged from the values found for methyl groups. From single crystal measurements [*25*], we know the Z axis is perpendicular to the molecular plane. Another direct proof of this is provided by benzene itself where the shielding tensor at the temperature of the experiment is axially symmetric because of rapid rotation of the molecules about their sixfold axes and the unique axis–perpendicular to the molecular plane–is the most shielded one (cf. Fig. 5.3). If the rotation of the molecules is frozen out, as, for instance, in hexamethylbenzene [*20*] one finds σ_{XX} and σ_{YY} differ significantly. From the single crystal study of durene [*25*], we know the direction of the corresponding principal axes, (cf. Fig. 6.6). The corresponding principal values are very similar for different aromatic rings, suggesting this is the representative orientation of the principal axes system.

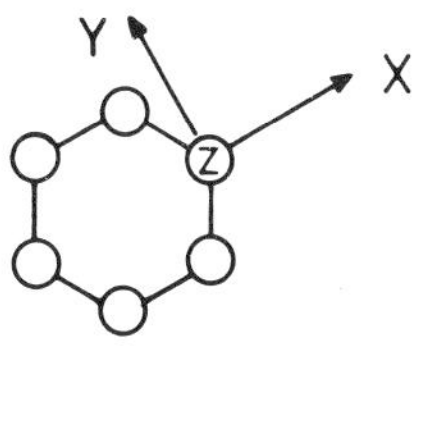

Fig. 6.6. Orientation of the principal axes system of $\vec{\vec{\sigma}}$ for ^{13}C in aromatic hydrocarbons, schematic (cf. Pausak *et al.* [*25*])

iii) Carbonyl Groups

The anisotropy of the shielding $\Delta\sigma$ here also is in the range of 150–180 ppm. At first sight, there seems to be much less systematics. A closer look at the data (cf. Fig. 6.4) reveals, however, that both σ_{ZZ} and σ_{XX} cover small ranges of characteristic values, respectively. The third component, σ_{YY}, on the other hand, can take on essentially every value between −80 ppm and +80 ppm relative to CS_2. The isotropic shifts for the carboxylic acids and their derivatives lie within a range of approximately 25 ppm only. This shows that single elements of the shielding tensor can be much more sensitive against the electronic structure than the isotropic shifts one can determine in liquids. The orientation of the principal axes system of the $\overset{\rightrightarrows}{\sigma}$ tensor, relative to the molecular frame, is depicted in Fig. 6.7 for benzophenone and benzoic acid, as determined from single crystal studies [*26, 21*]. The most shielded direction is found to be perpendicular to the sp^2 plane. In benzophenone, the Y axis is along the $C = 0$ bond and, in benzoic acid, it is as close to the $C = 0$ bond direction as possible for two equivalent CO bonds. These findings have since been confirmed in a number of single crystal studies of carboxyl groups [*27, 28*]. If the two CO bonds are not equivalent (e.g., because of asymmetric hydrogen bonding) the Y axis typically is found to be closer to the $C = 0$ bond.

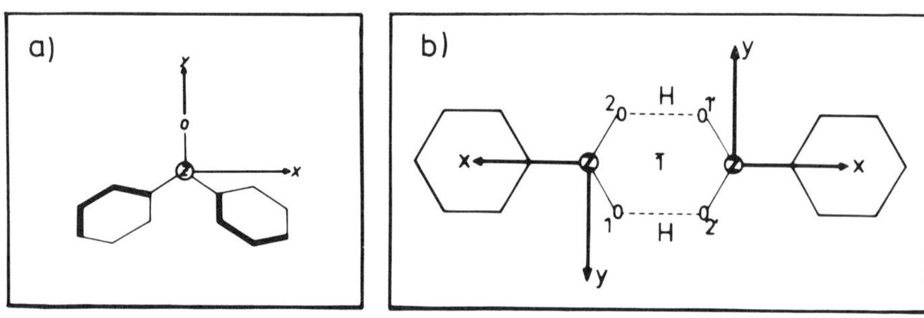

Fig. 6.7a and b. Orientation of principal axes system of $\overset{\rightrightarrows}{\sigma}$ for ^{13}C in carbonyl groups, (a) benzophenone [*26*], [From Kempf, J., *et al.*: Chem. Phys. Letters **17**, 39 (1972)], (b) benzoic acid [*21*]. [From Kempf, J., *et al.*: Chem. Phys. **4**, 269 (1974)]

For planar molecules or planar molecular groups (sp^2 carbons, therefore, we generally find the most shielded direction to be perpendicular to that plane. This makes the shielding tensor a very useful tool to study anisotropic rotation of planar molecules by spin-lattice relaxation (cf. Section 6.3).

iv) Linear Molecules

In linear molecules like CO or CS_2 (cf. Fig. 6.5), the paramagnetic contribution to the shielding vanishes for $\mathbf{B}_0$ parallel to the molecular axis and according to Eq. (5.12) $\sigma_\parallel \approx \sigma^d$ (free atom). In fact, $\sigma_\parallel = \sigma_{ZZ}$ is almost identical in both cases and does not even change appreciably if the CO molecule is bound in a metal carbonyl.

6.1.2.2 Experimental Values for ^{15}N

For ^{15}N, the experimental material is much smaller because of experimental difficulties. The magnetogyric ratio and its natural abundance are smaller than the corresponding values for ^{13}C by factors 0.4 and 0.33, respectively. Comparison of the data for ^{15}N (cf. Fig. 6.8) with those for ^{13}C shows the shielding tensors for both nuclei display similar features; the values for $\Delta\sigma$ being drastically larger in magnitude for ^{15}N, however. In order to illustrate this even more, a number of $\Delta\sigma$ values of ^{13}C and ^{15}N for isoelectronic pairs are collected in Table 6.1.

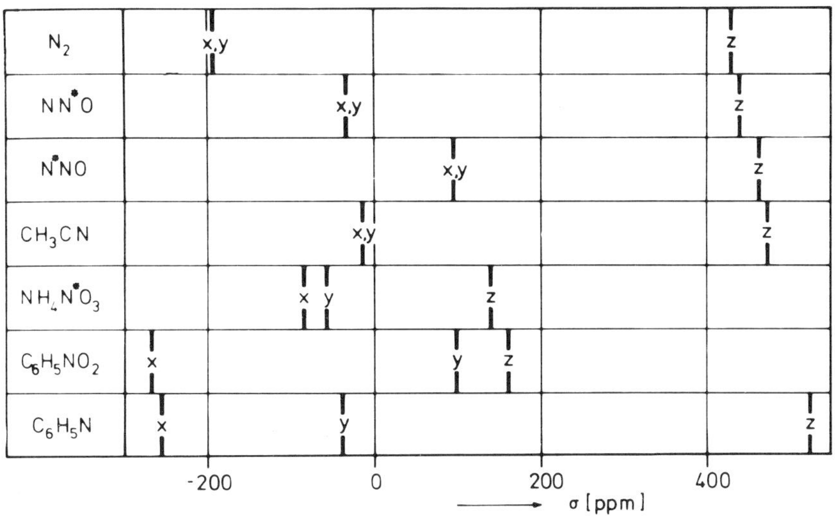

Fig. 6.8. ^{15}N shielding tensors in organic compounds relative to NO_3^-. The nitrogen position studied is indicated by * [*29–34*]

Table 6.1. Anisotropy of the shielding for ^{13}C and ^{15}N in isoelectronic systems

Compound	Nucleus	$\Delta\sigma$ [ppm]	$\frac{(\Delta\sigma)_{^{15}N}}{(\Delta\sigma)_{^{13}C}}$	Refs.
Pyridine	^{15}N	672	3.7	[*34*]
Benzene	^{13}C	180		[*20*]
Nitrobenzene	^{15}N	−398	3.5	[*33*]
Ag Benzoate	^{13}C	−112		[*21*]
NO_3^-	^{15}N	210	2.8	[*32*]
CO_3^{2-}	^{13}C	75		[*22*]
N_2	^{15}N	635	1.6	[*29*]
CO	^{13}C	401		[*23*]

Because of the enormous size of the shielding anisotropy for ^{15}N, shielding tensors can be determined even in protonated samples without decoupling, simply by applying high magnetic fields. To illustrate this, Fig. 6.9 shows the ^{15}N spectrum of solid pyridine [*34*] at 32 MHz corresponding to a magnetic field of 7.8 *T*. The characteristic powder pattern for a non-axially symmetric shielding tensor (cf. Fig. 3.3) is recognized easily.

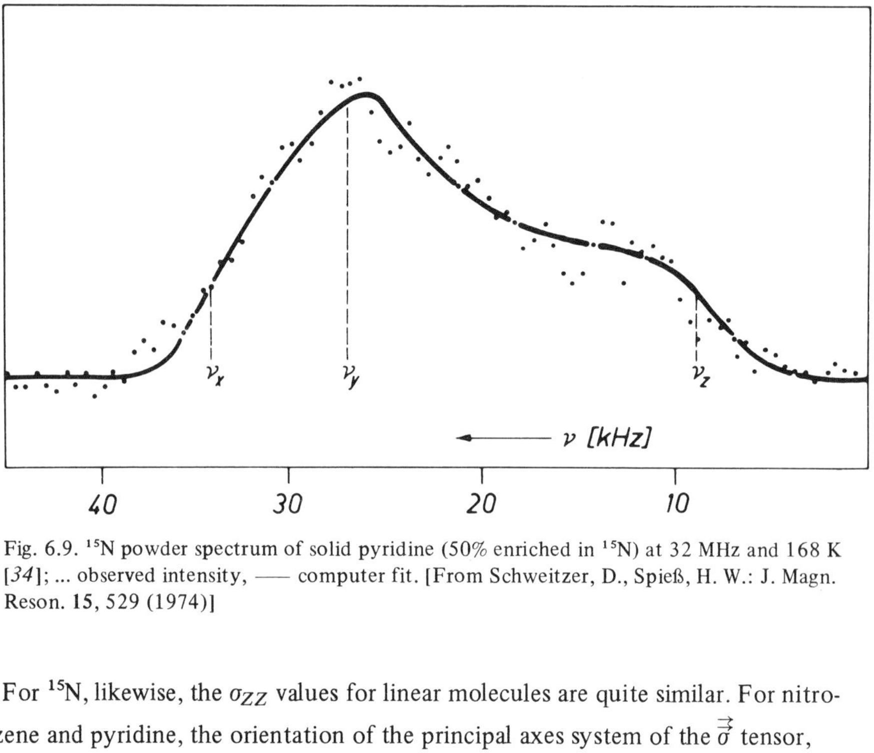

Fig. 6.9. ^{15}N powder spectrum of solid pyridine (50% enriched in ^{15}N) at 32 MHz and 168 K [*34*]; ... observed intensity, —— computer fit. [From Schweitzer, D., Spieß, H. W.: J. Magn. Reson. **15**, 529 (1974)]

For ^{15}N, likewise, the σ_{ZZ} values for linear molecules are quite similar. For nitrobenzene and pyridine, the orientation of the principal axes system of the $\overrightarrow{\overrightarrow{\sigma}}$ tensor, relative to molecular axes, has not been determined experimentally but we can safely assume the axes to be like those found for ^{13}C in benzoic acid and benzene, respectively.

6.1.2.3 Theoretical Explanation

In this section we will try to make plausible the characteristics of the $\overrightarrow{\overrightarrow{\sigma}}$ tensors of ^{13}C and ^{15}N described in the previous sections. We should have some idea, for example, *why* the unique axis of $\overrightarrow{\overrightarrow{\sigma}}$ in planar molecules is perpendicular to the molecular plane. Since we want to use $\overrightarrow{\overrightarrow{\sigma}}$ tensors for studying the dynamic behaviour of molecules, $\overrightarrow{\overrightarrow{\sigma}}$ tensors of ^{13}C and ^{15}N in linear and planar molecules will concern us most since $\Delta\sigma$ has the largest absolute values there. In this context, by planar molecules we do not necessarily mean the molecule as a whole but rather the immediate surroundings of the ^{13}C or ^{15}N nucleus of interest.

The characteristics of the shielding tensors described in the previous sections strongly suggest that the $\vec{\vec{\sigma}}$ tensor is a local property dominated by the immediate surroundings of the nucleus. For electrons "belonging" to remote atoms, the contributions to $\vec{\vec{\sigma}}^d$ and $\vec{\vec{\sigma}}^p$ largely cancel each other [cf. Eqs. (5.10)–(5.12)]. The qualitative features of the $\vec{\vec{\sigma}}$ tensors then are dominated by the paramagnetic part $\vec{\vec{\sigma}}^p$ [cf. Eq. (5.8)] where furthermore, only the valence electrons of those atoms have to be taken in account which are directly bound to the atom to which the nucleus of interest belongs. A detailed discussion of this has been given for carbonyl groups in [*21*]. More recently, such local paramagnetic contributions have been calculated for ^{13}C in benzene, toluene, acetic acid and derivatives [*35*], giving remarkable agreement with the experimental results. In order to ease the discussion, we once again give the theoretical expression for the diagonal element:

$$\sigma_{XX}^p = \frac{2\beta^2}{\hbar^2} \sum_{n>0} (E_0 - E_n)^{-1} \left\{\langle 0 | \sum_i \frac{l_{iX}}{r_i^3} | n\rangle \langle n | L_X | 0\rangle + c.c.\right\}. \tag{6.2}$$

According to Eq. (6.2) σ^p is always negative since $E_0 - E_n < 0$. This means, the *larger* $|\sigma^p|$, the *smaller* the shielding. Let us now consider which excited states give the largest contributions to σ_{XX}^p. The angular momentum operators in Eq. (6.2) are single electron operators; therefore, only excited singlet states, differing from the ground state by the occupation of no more than a single orbital, can contribute. Symmetry further restricts the number of excited states that have to be considered.

The discussion is particularly simple for a linear molecule, like ^{13}CO. where $\sigma_{\|}^p = 0$. From the symmetry of the molecule, we know that only $\sigma \rightarrow \pi^*$ or $\pi \rightarrow \sigma^*$ excitations can contribute to $\sigma_{\perp}^p$. From Huo's calculation [*36*], we have $\Delta\sigma^d$ = –83 ppm and this, together with $\Delta\sigma$ = 401 ppm (cf. Table 6.1), gives $\Delta\sigma^p$ = 484 ppm. This simple example shows that the paramagnetic part for $\mathbf{B}_0$, being perpendicular to the CO bond, indeed can be of the order of several hundred ppm.

In carbonyl groups, the situation is slightly more complicated. A schematic *MO* diagram for a $C = 0$ group is depicted in Fig. 6.10. The single electron excitations dominating the different shielding tensor elements are indicated by arrows. Since

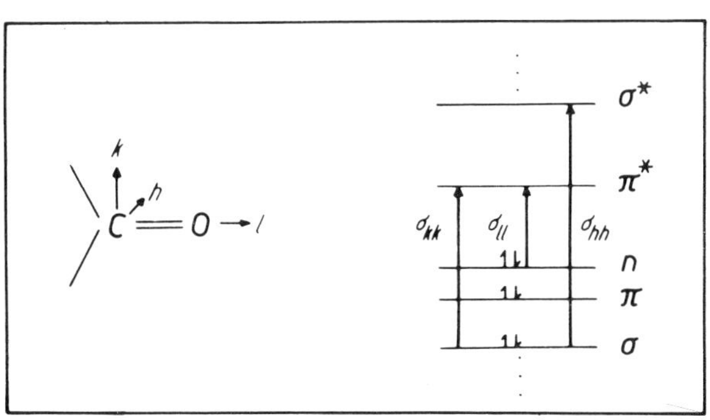

Fig. 6.10. Schematic MO diagram of a carbonyl group

the contributions to $\vec{\vec{\sigma}}^p$ are weighted by $(E_0 - E_n)^{-1}$, only the highest occupied and the lowest empty orbitals have been included. Symmetry determines which one of the excitations can contribute to a given shielding component. If $\mathbf{B}_0$ is perpendicular to the sp^2 plane, only $\sigma \rightarrow \sigma^*$ excitations, with in general high excitation energies, can contribute to the shielding. Consequently, σ^p will be small and the shielding large. Based on the excitation energies only, one would expect the $n \rightarrow \pi^*$ transition to give the largest contribution to σ^p (cf. Fig. 6.10). This is not generally the case, however. When we express the matrix elements of Eq. (6.2) as matrix elements over atomic orbitals [*21*], we see that the expression for σ^p contains the sum of products of the corresponding *MO* coefficients. Because of the r^{-3} dependence of the operator involved, the *MO* coefficient of the carbon 2*p* orbital is most important. This *MO* coefficient is especially small [*21*] for the nonbonding *n* orbital corresponding to the lone pair at the oxygen. Therefore, the contribution of the $\sigma \rightarrow \pi^*$ excitation, although less favored by the energy difference, in general, is larger than that of the $n \rightarrow \pi^*$ excitation and consequently the direction in the sp^2 plane perpendicular to the $C = 0$ bond, is the least shielded one.

We now readily can understand why the intermediate shielding component σ_{YY} shows a much stronger variation than the other ones (cf. Fig. 6.4), if we assume that, in general, it is dominated by the $n \rightarrow \pi^*$ excitation. The excitation energy is the smallest one possible and, consequently, is most sensitive to small changes in the bonding. Moreover, the contribution to σ^p, due to this excitation, is strongly dependent on the *MO* coefficient of the carbon 2*p* orbital in the *n* orbital. Even minor changes of this *MO* coefficient, therefore, can change the corresponding shielding component appreciably (for details see [*21*]).

This consideration also makes plausible that the most shielded direction generally is found to be perpendicular to the sp^2 plane. For this direction, only $\sigma \rightarrow \sigma^*$ excitations are involved having especially high excitation energies, while large paramagnetic contributions, due to the $\sigma \rightarrow \pi^*$, $\pi \rightarrow \sigma^*$, and $n \rightarrow \pi^*$ excitations, can result only if $\mathbf{B}_0$ is in the sp^2 plane.

In aliphatic systems, we are dealing with σ bonds only; therefore, the overall shielding is larger than in planar molecules. There is another interesting point to be noted. In alcohols, the shielding component for $\mathbf{B}_0 \perp$ C-OH shows a marked paramagnetic shift (cf. Fig. 6.4). The value of that shielding component (σ_{XX} in CH_3OH) shows a remarkable agreement with the characteristic value of σ_{ZZ} for carbonyl groups where, likewise, $\mathbf{B}_0 \perp$ C-O and only $\sigma \rightarrow \sigma^*$ excitations are involved.

6.1.3. Shielding Tensors of Other Nuclei

^{19}F shielding tensors, obtained mainly by multiple pulse methods, have been reviewed extensively by Mehring [*3*], Haeberlen [*4*], and by Appleman and Dailey [*37*]. Data for ^{31}P and ^{29}Si also have been given in tabular form by Mehring [*3*]. We, therefore, will not comment here on the experimental values again.

6.2 Rotational Jumps in Solids

In Section 4.1, we have calculated the line shape for an axially symmetric $\vec{\vec{\sigma}}$ tensor when the molecule undergoes a rotational jump motion in the solid. The line shape displayed especially characteristic features if the jump frequency τ^{-1} was of the same order as the width of the powder spectrum in the rigid case $\Delta\omega$ (cf. Figs. 4.4 and 4.6). For this frequency region, line shape analysis offers a rather direct means not only to determine jump frequencies but also to distinguish between different types of motion. There are still not many experimental examples of such line shape analyses at present; therefore, we will discuss a few cases in some detail.

In solid *white phosphorus* (P_4), we did not only find such line shapes but the jump frequencies determined from line shape analysis could be checked independently by T_1 measurements [*24*]. At low temperatures (e.g., 25 K), the characteristic powder spectrum for an axially symmetric $\vec{\vec{\sigma}}$ tensor is observed (cf. Fig. 6.11). At room temperature, on the other hand, the ^{31}P spectrum consists of a sharp single line in the centre of the spectrum showing that the jump process can average out the anisotropic coupling completely. This excludes rotation about a single fixed axis.

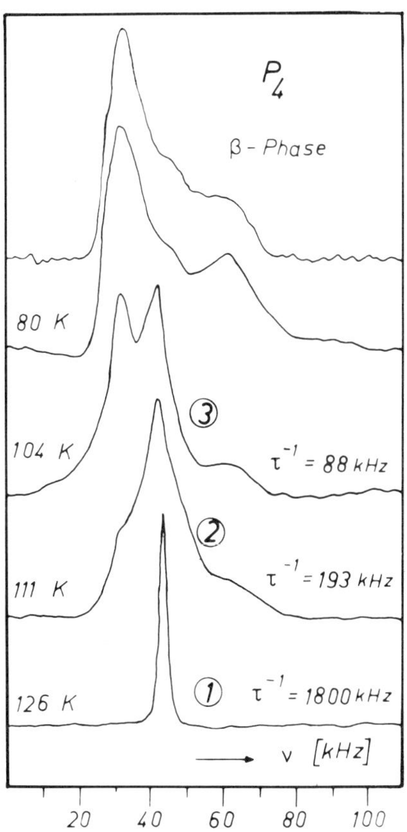

Fig. 6.11. ^{31}P spectra of solid white phosphorus P_4 at 92 MHz and different temperatures [*24*]. The values for the jump frequency τ^{-1} were obtained from T_1. [From Spieß, H. W., *et al.*: Chem. Phys. 6, 226 (1974)]

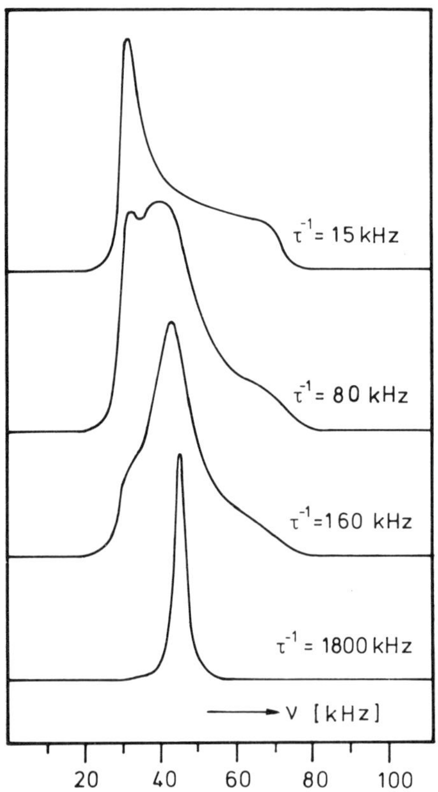

Fig. 6.12. Calculated ^{31}spectra of P_4 according to jump model

Around 100 K, we observe the transition region from fast to slow exchange within a range of about 40 K, as shown in Fig. 6.11. Here, the characteristic line shapes with the shoulder at $\omega_\perp$ (spectrum②), and the double peak spectrum ③ appear. On further lowering the temperature, the experimental spectra do not directly approach the usual powder spectrum. Instead, a second maximum (near $\omega_\parallel$) is observed which vanishes only at very low temperatures.

Theoretical line shapes, calculated as described in Section 4.1, are shown for comparison in Fig. 6.12. The most characteristic features of the spectra are reproduced well by the simple stochastic jump model. The deviations from the powder spectrum observed at lower temperatures (cf. the spectrum at 80 K) and the sharpness of the double peak (cf. spectrum ③) are not reproduced. In the original papers [*38*, *24*], such deviations, in fact, had been calculated. They were due, however, to a numerical artefact resulting from the nonlinear frequency division used there. Although the simple stochastic model does not explain all the details of these spectra it, nevertheless, allows even quantitative analysis of the data. This is shown by the comparison of the values of the jump frequencies obtained by line shape analysis with those determined from spin-lattice relaxation times T_1. The T_1 values are plotted in Fig. 6.13. As an extra complication, we have to deal with several phases in solid white phosphorus.

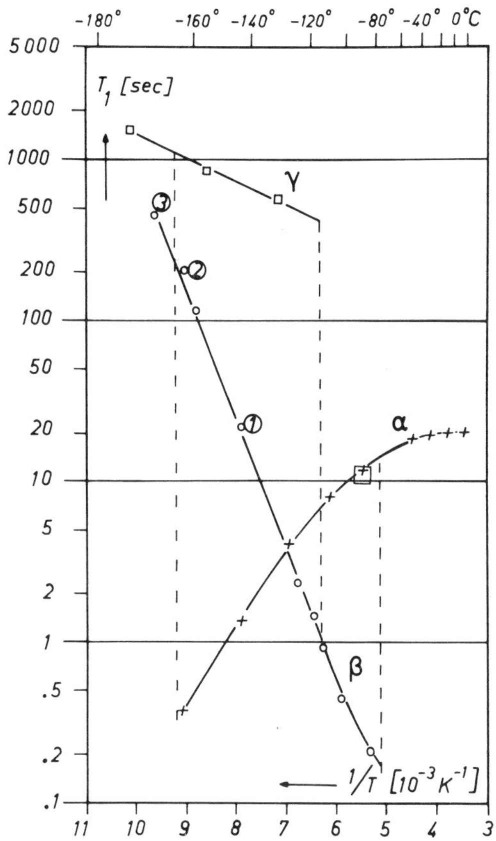

Fig. 6.13. ^{31}P spin-lattice relaxation rates of solid white phosphorus at 92 MHz [*24*]. [From Spieß, H. W., *et al.*: Chem. Phys. **6**, 226 (1974)]

By measuring T_1 and its temperature dependence, however, we can make sure which phase we have at a given instance [*24*]. The points marked ①, ②, ③ correspond to the spectra in Fig. 6.11. Analysis of the T_1 values is especially simple here since, at the frequency of the experiment (92 MHz) and at low temperatures, 97% of the relaxation rate is given by relaxation through anisotropic shielding. For the P_4 tetrahedra, we have a single correlation time τ only and, therefore, from Table 4.4, we obtain:

$$\frac{1}{T_1} = \frac{2}{15} (\Delta\sigma)^2 \omega^2 \frac{\tau}{1 + \omega^2\tau^2} . \tag{6.3}$$

In the β phase of solid P_4, the condition $\omega^2\tau^2 \gg 1$ is fulfilled (cf. Fig. 6.13) and Eq. (6.3) reduced to:

$$\frac{1}{T_1} = \frac{2}{15} (\Delta\sigma)^2 \tau^{-1}. \tag{6.3a}$$

With $\Delta\sigma = -405$ ppm from the low temperature spectrum, we readily obtain from the T_1 data the values for the jump frequencies τ^{-1} which are given in Fig. 6.11. The agreement between the values from the line shape analysis and from the T_1 data demonstrates that one can not only obtain information about the type of motion from line shapes of this kind but one can also measure jump frequencies quantitatively. This provides a more direct way to study slow rotational motions than the measurement of $T_{1\rho}$ [*39*] or T_{1D} [*40*]. (Note that here T_1 is in the range of 20–1000 s.)

From line shape analysis for protons, where the shielding anisotropy is smaller, even slower motions can be studied. A particularly nice example has been given by Pines and coworkers [*41*] who studied deuteron decoupled ^{1}H spectra in *ice* from 163 to 268 K (cf. Fig. 6.14). The spectra show close agreement with the theoretical ones calculated for tetrahedral motion of the protons in the ice lattice and, clearly, jump frequencies below 10 kHz can be measured. Moreover, one gets the clear-cut qualitative result that the reorientational part of the proton motion in ice is a tetrahedral one.

A first example for the direct determination of slow jumps of protons in a single crystal has been provided by Vaughan and coworkers [*42*] by multiple pulse NMR in $CaSO_4 \cdot 2\,H_2O$. A doublet, which is observed for a particular orientation at 90 K, collapses to a singlet at 300 K.

Besides line shapes for cubic symmetry, experimental examples for rotational jump motions about a single axis have been studied (e.g., in hexamethylbenzene [*43*]). The corresponding ^{13}C spectra are reproduced in Mehring's book [*3*]. The ESR spectra of the AsO_4^{4-} radical anoin in irradiated KH_2AsO_4 [*44*] represents examples of $\pi/2$ jumps about a single axis, as shown by the line shape analysis [*44*].

Finally, a particularly simple example for the rotational jump model for highly *viscous liquids* has been established through the ESR spectrum of sulfur dissolved in 60% oleum [*45*], shown in Fig. 6.15. The paramagnetic species probably is a planar S_5^+ ring [*45*] whose ESR spectrum is governed by an axially symmetric g tensor. Clearly, the rotational jump model represented by the dashed lines, gives much better agreement with the experimental spectra than does the brownian motion model, represented by the dotted line in Fig. 6.15. Of course, ESR line shapes of spin lables probably represent the bulk of experimental material about slow rotational motions in viscous systems [*46*] but their discussion is beyond the scope of this review.

Despite the growing number of experimental examples, the effects rotational motion has on the line shapes cannot be considered yet as being fully understood. As an example of how strong the deviations from the powder spectrum in the rigid case can be, Fig. 6.16 shows ^{13}C spectra of solid *ironpentacarbonyl* $Fe(CO)_5$ at various temperatures. At temperatures close to the melting point (252 K) we find a characteristic "two line" spectrum which approaches the usual powder pattern at lower temperatures; the rigid limit being reached below 30 K [*24*]. Special care was taken to avoid possible sources of error. The liquid sample was crystallized by rapid freezing in liquid N_2 in order to avoid partial ordering. Similarly the repetition time between two consecutive pulses was varied between 1 min and 1 h without noticeable effect on the line shape, excluding the possibility of the line shape being caused by angular dependent and, therefore, frequency dependent relaxation rates [*38*]. $Fe(CO)_5$ is a trigonal bipyramid for which an exchange between axial and equatorial carbonyl groups has been postulated in the liquid and the gas phase [*47–48*]. It seems only natural to attribute the ^{13}C line shapes in the solid (Fig. 6.16) to this exchange process but, by

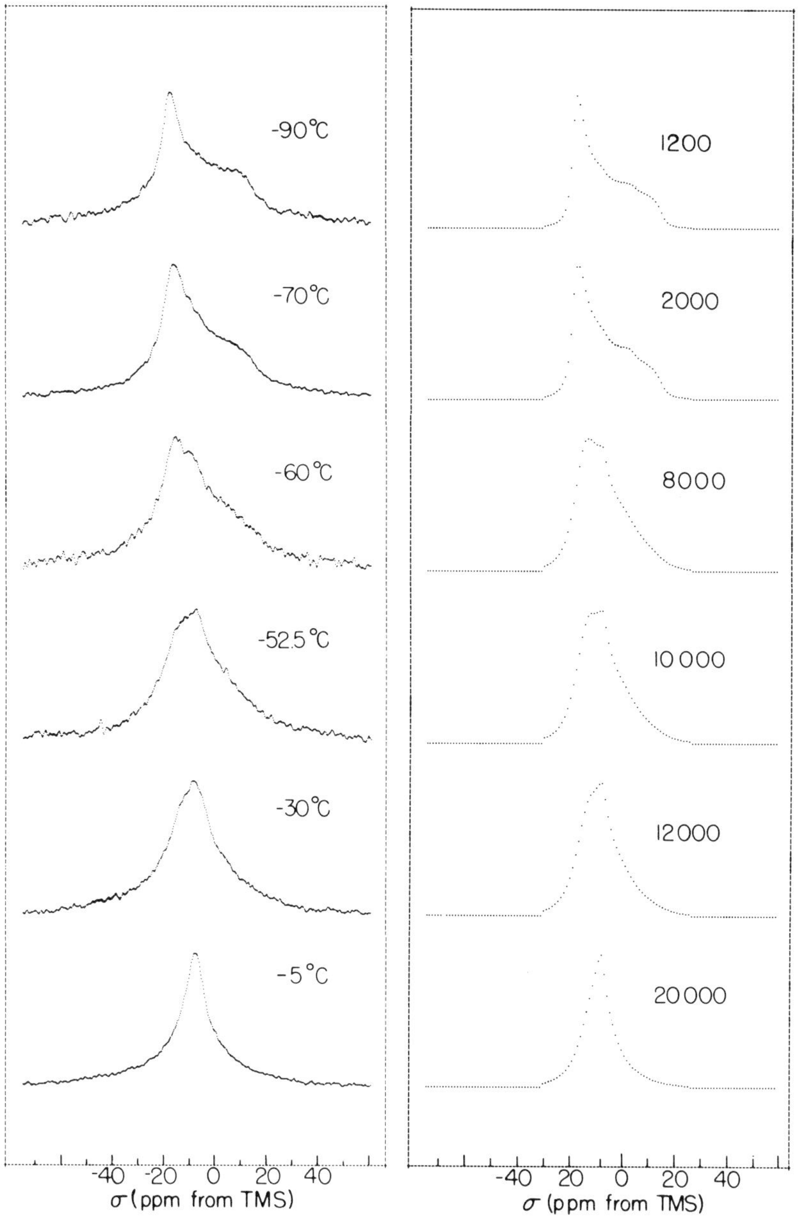

Fig. 6.14. NMR spectra of residual protons in ice with deuterium decoupling (Pines *et al.* [*41*]). The spectra at the right were calculated for tetrahedral jumps; jump rates given in Hz (Courtesy of A. Pines)

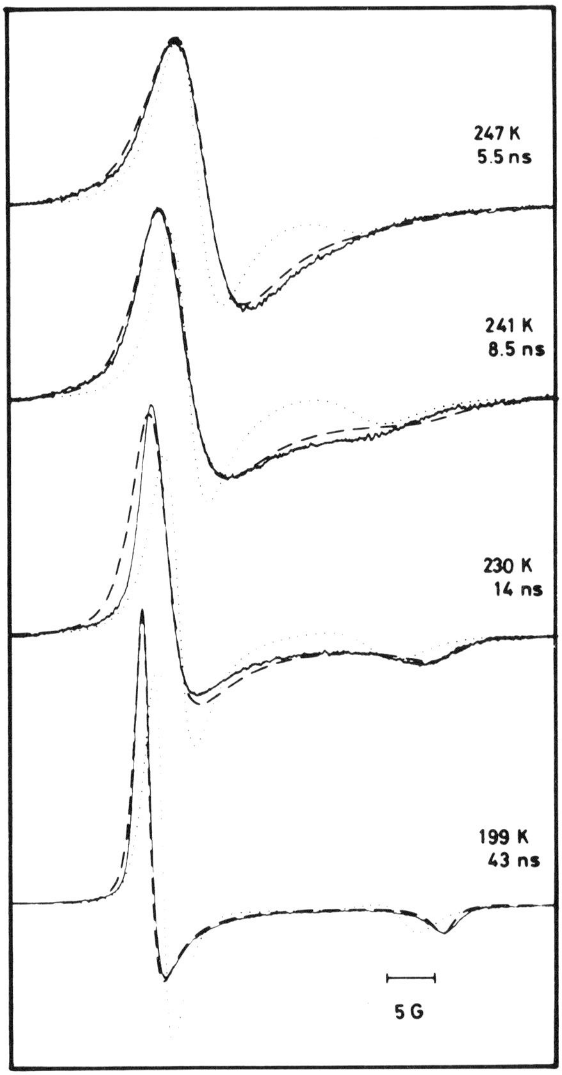

Fig. 6.15. ESR line shapes for slow molecular rotation of S_5^+ ring (Hensen *et al.* [*45*]). —— Experimental spectra, ... rotational brownian motion model, - - - rotational random jump model. The reorientational correlation times are given at the right. [From Hensen, K., *et al.*: J. Chem. Phys. 61, 4365 (1974)]

the simple stochastic model, the deviations from the powder spectrum cannot be explained. From the measurements of T_1 (cf. Fig. 6.16), we obtain $\tau^{-1} = 24$ kHz, 13.4 kHz, and 5.5 kHz for 213 K, 155 K, and 100 K, respectively [*24*], showing that $\tau^{-1} < \Delta\omega = 165$ kHz and that the temperature dependence of τ^{-1} is very small. Similarly, the line shape varies slightly with temperature only, but this does not prove the existence of an exchange. It seems to be quite clear, however, that below 30 K the exchange is frozen out.[26] This was shown independently from the NMR investigations by studying UV spectra of partially oriented $Fe(CO)_5$ in solid CO at 20 K [*49*].

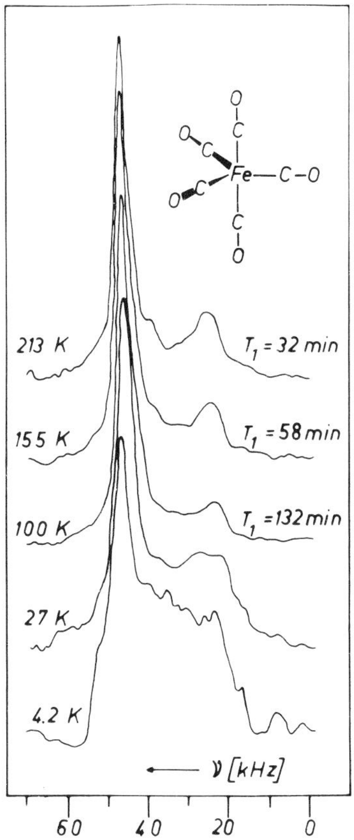

Fig. 6.16. ^{13}C spectra of solid ironpentacarbonyl [$Fe(CO)_5$] at 61 MHz at different temperatures [*24*]. Also included are experimental values of the spin-lattice relaxation time T_1. [From Spieß, H. W., *et al.*: Chem. Phys. **6**, 226 (1974)]

Let us close this section with two examples of ^2D NMR. Figure 6.17 shows single crystal ^2D spectra of $Li_2SO_4 \cdot D_2O$ obtained through wide line NMR by Berglund and Tegenfeld [*50*] over the temperature range corresponding to intermediate exchange. The agreement between experimental and theoretical spectra is excellent allowing accurate determination of jump frequencies for the flip of the D_2O molecules.

In *solid polymers,* the study of the ^1H line shape by wide line NMR methods is one of the standard ways to obtain information about crystallinity and dynamic behaviour (see, e.g., [*51, 52*]). Since the dipolar coupling governing the ^1H wide line spectra is rather unselective, the line shape, in general, cannot be calculated easily and analysis of the data is difficult. The information about reorientational motion of polymer segments and chains can be obtained much more directly through ^2D

[26] At 4.2 *K*, we did not measure T_1. From the magnetization observed, we can deduce T_1 must be of the order of weeks at that temperature. The spectrum at 4.2 *K* was observed by letting the sample built up some magnetization at 25 *K* (for 15 min or so) and then rapidly cooling it down to 4.2 *K*.

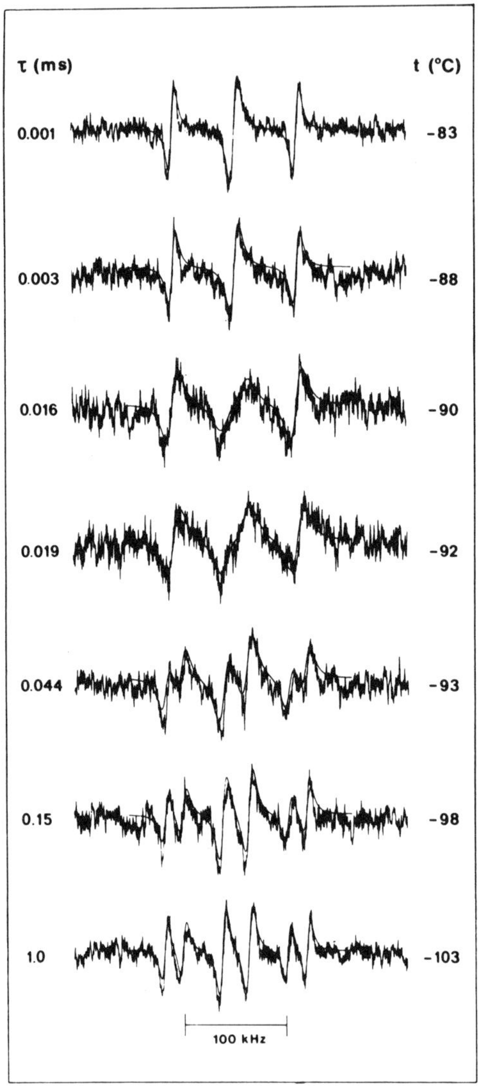

Fig. 6.17. Experimental and calculated ^{2}D wide-line spectra of a single crystal of $Li_2SO_4 \cdot D_2O$ (all taken at the same orientation of the crystal relative to the external magnetic field). The theoretical spectra were calculated for 180° flips of the D_20 molecules (Berglund and Tegenfeld [*50*]). (Courtesy of J. Tegenfeld)

resonance [*53*]. This is shown in Fig. 6.18, where ^{2}D wide line spectra of solid *poly-ethylene–d*$_4$ at 294 K and 353 K are plotted. The central part only of the powder spectrum is shown in the derivative mode. At room temperature, we see, in addition to the rigid powder spectrum, a broad central absorption. The reorientational mobility of the amorphous regions of the sample, therefore, is not rapid enough to completely average out the anisotropic quadrupole coupling. At higher temperatures, however, we also see a sharp central component so that the spectrum at 353 K can be decomposed clearly as arising from three components: the crystalline region giving rise to the rigid powder pattern, the intermediate flexible part of the sample being responsible for the broad central absorption, and the highly flexible amorphous

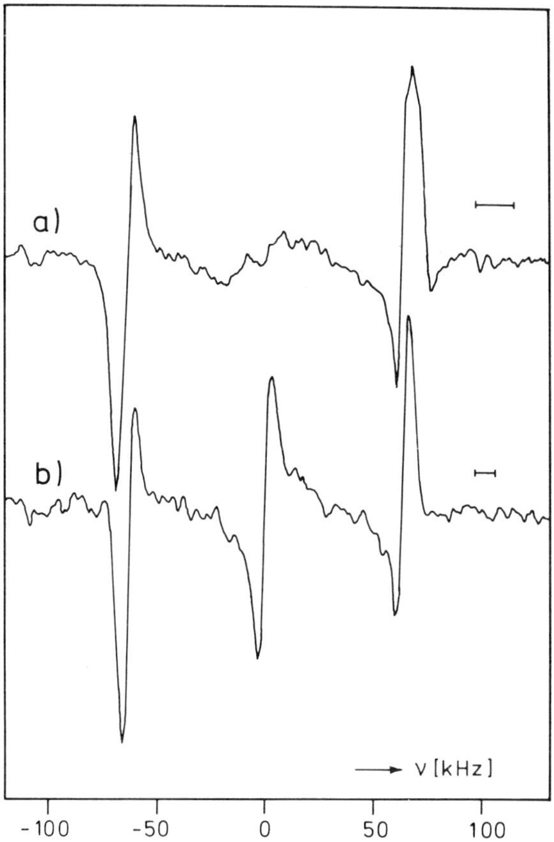

Fig. 6.18a and b. ^{2}D wide-line spectra of solid polyethylene [*53*]; (a) 293 K, (b) 353 K. Note that the modulation amplitude, indicated by the bars, is different for the two spectra (Courtesy of D. Hentschel)

regions giving rise to the sharp central peak. Clearly, close to the melting point, the molecules in the amorphous regions are flexible enough to completely average out the quadrupole splitting corresponding to $e^2qQ/h = 162$ kHz [*53*].

Line shape analyses of this kind provide an additional tool besides the established methods to study slow motions by NMR [*39, 40*]. The frequency range over which the line shape is sensitive against the motion is 1–20 kHz for protons; up to several hundred kHz for other spin $I = 1/2$ nuclei, if superconducting magncts are employed; and up to about 2 MHz for ^{2}D due to the quadrupole coupling.

6.3 Spin-Lattice Relaxation in Liquids

As shown in the preceeding section, one can obtain rather direct information about the dynamic behaviour of molecules if the characteristic frequencies of the motion are comparable with frequency splittings in the NMR spectrum $\Delta\omega \cdot \tau \approx 1$. This is generally true for any kind of spectroscopy. The characteristic times for hindered rotation of molecules in liquids, however, are much shorter than the characteristic times of NMR ω^{-1}, not to mention $\Delta\omega^{-1}$. Therefore, NMR often is called a "slow" method. The reorientation of the molecules in a liquid, however, does not occur with a well-defined angular frequency and, if we would perform a Fourier analysis of this motion, we would find appreciable intensities at frequencies much lower than the "characteristic" ones. This provides the basis for obtaining even quantitative information about the reorientation by NMR, mainly through the standard relaxation mechanisms dipole-dipole and quadrupole coupling. For nuclei like ^{13}C and ^{15}N, anisotropic shielding and spin-rotation interaction also can become important for relaxation. About five years ago, the experimental material of ^{13}C relaxation in liquids still was rather small. Today the number of papers dealing with the measurement and interpretation of ^{13}C spin-lattice relaxation times has increased so drastically we will not even try to give a survey of the literature. Collecting references here, moreover, seems to be unnecessary since nuclear spin relaxation has been treated adequately in the "Specialist Periodical Reports" [*54*] by Boden, Tomlinson, and Zeidler, respectively. The chapters in this series [*54*] on Fourier Transform NMR contain numerous references on the various experimental techniques to measure T_1; therefore, we will not discuss experimental aspects either. Instead, we restrict ourselves to giving a few selected examples for the different relaxation mechanisms. We would like to show, in particular, what kind of information about the reorientation in a liquid can be gathered exploiting the different relaxation mechanisms. We are concerned mainly about the method and much less about the particular system being studied. Most of the examples, therefore, deal with rather small molecules. Based on studies of such simple systems, the main characteristics of ^{13}C and ^{15}N relaxation now are well understood. This provides the basis to use these isotopes for studying more complex systems.

By selecting the experimental conditions (e.g. nucleus, temperature, magnetic field strength), we now are able to select the mechanism which will give us the most reliable information about a particular problem. Only if the possibilities offered by NMR are exploited fully in this way, this "slow" method will be able to compete with other methods used to study the dynamic behaviour of liquids. This includes, in particular, the new developments in light and neutron scattering and picosecond spectroscopy, but also included are the analysis of IR and Raman band shapes and dielectric relaxation. A comparison of the different methods has been given by Steele in his recent review [*55*]. We also refer to the book "Organic Liquids" [*56*] based on the lectures of a recent EUCHEM conference on the subject. It gives a survey about the current state of the art for the different techniques, and moreover, contains an outstanding comparison of the different methods in the lecture by Bratos.

6.3.1. Separation of the Various Contributions to the Total Relaxation Rate

In Chapter 4, we have treated the various relaxation mechanisms separately. The total relaxation rate observed is given by the sum of the contributions of the different mechanisms (cf. Eq. (4.24) and its discussion). Often a single mechanism will predominate (e.g., the dipole-dipole coupling for $I = 1/2$ nuclei at low temperatures or the quadrupole coupling for $I > 1/2$ nuclei). If different mechanisms give comparable contributions, their separation is not trivial. It can be achieved, however, if T_1 is studied as a function of temperature and frequency and if, moreover, the nuclear Overhauser effect is determined. Rather than discussing the separation of the various contributions to the total relaxation rate generally, we will illustrate the various techniques by a few examples.

6.3.1.1. Nuclear Overhauser Effect, Separation for $\lambda = D$

Since for spin pairs like ^{13}C—^{1}H or ^{15}N—^{1}H etc. the scalar coupling is completely negligible compared with the dipolar interaction between the spins, the nuclear Overhauser effect, likewise, is only due to the dipolar interaction and, therefore, allows the separation of $1/T_1^D$ from all other contributions. We will not go into details here but refer the reader to the literature [*57–61*] and only give the result for a spin pair I, S (e.g., $I = {}^{13}C$, $S = {}^{1}H$.

The nuclear Overhauser enhancement, NOE (the change in the integrated NMR absorption intensity of a nuclear spin when the NMR absorption of another spin is saturated), is usually defined by:

$$NOE = 1 + \eta. \tag{6.4}$$

In our notation, we obtain for a spin pair [*61*]:

$$\eta = T_1 \hbar^2 \gamma_S^3 \gamma_I [-\frac{1}{3} g_2^D (\omega_S - \omega_I) + 2 g_2^D (\omega_I + \omega_S)], \tag{6.5}$$

which reduces for extreme narrowing (where g_2^D is independent of frequency) to:

$$\eta = \frac{1}{2} \frac{T_1}{T_1^D} \frac{\gamma_S}{\gamma_I}, \tag{6.6}$$

where T_1^D is given by T_1^{II} of Table 4.10. Through measurement of both T_1 and the nuclear Overhauser effect, one can determine the part of the relaxation rate due to the dipolar interaction alone:

$$\frac{1}{T_1^D} = \frac{1}{T_1} \frac{2\gamma_I}{\gamma_S} \eta. \tag{6.6a}$$

If we deal with dipolar relaxation only, the NOE is maximum; for ^{13}C, ^{1}H: $\eta = 1.988$. It should be noted that Eqs. (6.6) and (6.6a) hold for extreme narrowing and for a single spin pair only. If the I spin is coupled to more than one S spin, the maximum nuclear Overhauser effect will be unchanged although the relaxation time T_1 will depend on the number of S spins. This was illustrated experimentally byKuhlmann *et al.* [*58*] by studying the ^{13}C NMR of adamantane. For the integrated intensities of the CH_2 and the CH carbons, they found a ratio of 1.50 ± 0.08 reflecting the ratio of 6/4 of the corresponding carbon sites in the molecule, whereas the relaxation times were 11.4 ± 1 s and 20.5 ± 2 s, respectively. The ratio of the relaxation times, therefore, is $T_1^{CH}/T_1^{CH_2} = 1.80 \pm 0.25$ in excellent agreement with the ratio of 1.82 predicted by assuming only intramolecular dipole-dipole coupling but taking in account the next nearest hydrogen atoms [*58*].

For methyl carbons, on the other hand, despite the presence of three proton spins, the ^{13}C relaxation is not due solely to dipolar coupling. For CH_3OH, for example, Lyerla *et al.* [*58*] found $\eta = 0.73 \pm 0.07$ at room temperature, showing only 36% of the total relaxation rate is given by the dipolar contribution, whereas, 64% are due to other mechanisms (e.g., spin-rotation interaction).

6.3.1.2. Field Dependent Studies, Separation for $\lambda = CS, J$

For extreme narrowing, the only field and, therefore, frequency dependent contributions to the relaxation rate are relaxation through anisotropic shielding ($\lambda = CS$) and relaxation through scalar coupling ($\lambda = J$). An especially clear-cut example for the separation of $\lambda = CS$ is provided by the ^{13}C relaxation of liquid CS_2 as a function of frequency and temperature [*1*]. Here, only two relaxation mechanisms are operative; relaxation through anisotropic shielding and through spin-rotation interaction. Moreover, for CS_2 (being a linear molecule), both $\overleftrightarrow{\sigma}$ and c must be axially symmetric and $c_{\|} = 0$ [cf. Eq. (5.5)]. The experimental spin-lattice relaxation rates at frequencies of 14, 32, and 62 MHz, respectively, are plotted in Fig. 6.19a [*1*]. At low temper-

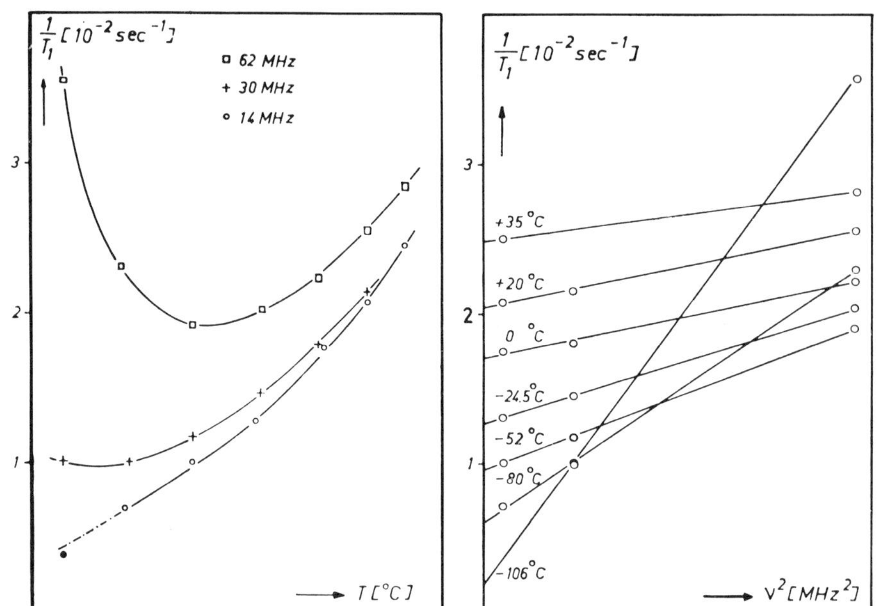

Fig. 6.19. Temperature and frequency dependence of ^{13}C relaxation rates in liquid CS_2 [*1*]. [From Spieß, H. W. *et al.*: J. Magn. Reson. 5, 101 (1971)]

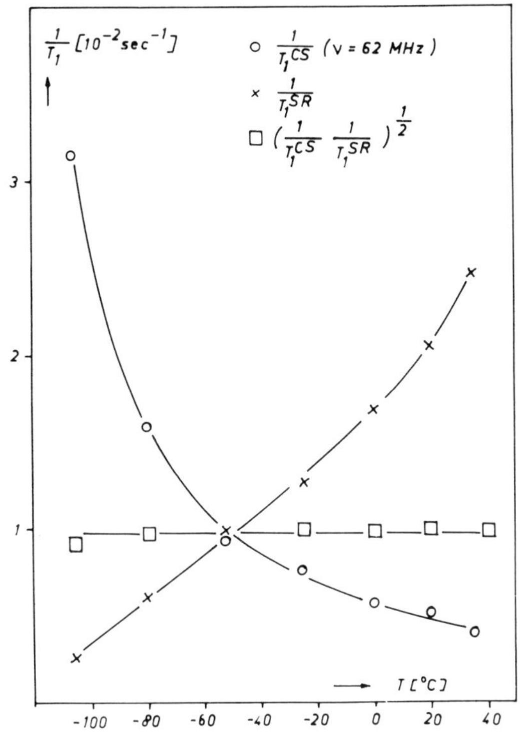

Fig. 6.20. ^{13}C relaxation rates due to anisotropic shielding at 62 MHz and through spin-rotation interaction in liquid CS_2 [*1*]. [From Spieß, H. W., *et al.*: J. Magn. Reson. 5, 101 (1971)]

atures, the relaxation rate increases drastically with increasing frequency, as expected for $\lambda = CS$, $1/T_1^{CS} \propto \omega^2$ since $\omega^2 \tau_2^2 \ll 1$ (cf. Table 4.4), whereas, $1/T_1^{SR}$ is independent of frequency (cf. Table 4.7). Accordingly, by plotting $1/T_1$ vs. ω^2, we obtain straight lines (Fig. 6.19b) and the two contributions can easily be separated. The result is given in Fig. 6.20. Both of the separated rates cover about the same range, whereas, the product of the two rates is indeed independent of temperature. This is to be expected if the reorientation of the CS_2 molecules in the liquid follows the diffusion model since from Tables 4.4 and 4.7, together with Eq. (4.106) we have for a linear molecule:

$$\frac{1}{T_1^{CS}} \cdot \frac{1}{T_1^{SR}} = \frac{4}{135} \frac{\omega^2 \theta^2}{\hbar^2} c^2 \Delta\sigma^2. \tag{6.7}$$

The product of the individual rates thus is independent of τ_2, τ_J, and the temperature. From the product of the experimental relaxation rates and the theoretical relation, Eq. (5.12), we were able to determine c and $\Delta\sigma$ for ^{13}C in CS_2 [*1*]. The value we found for $\Delta\sigma = 438 \pm 44$ ppm, shortly afterwards was confirmed quantitatively by solid state NMR [*2*] $\Delta\sigma = 425 \pm 15$ ppm. The relaxation study, therefore, did not only check the relation between τ_2 and τ_J, given by the diffusion model qualitatively, but confirmed Eq. (4.106) quantitatively (cf. also Fig. 4.10).

The *scalar* contribution to $1/T_1$ also can cause a field dependence of the total relaxation rate, even under extreme narrowing conditions (cf. Section 4.2.3.1) since

T_1^S and T_2^S will typically be much larger than τ_2. In fact, the scalar contribution to the relaxation rate for light nuclei can only be expected to be able to compete with other relaxation mechanisms if this rate is much closer to its maximum value than the other rates; $(\omega_I - \omega_S)^2 T_2^{S^2}$, therefore, should be comparable with unity. For ^{13}C, this relaxation mechanism has to be considered for the spin pair ^{13}C, ^{79}Br, in particular, since the Larmor frequencies for these two isotopes differ by only 0.35%. An experimental example is provided by the ^{13}C relaxation in CH_3Br, which was observed to be field dependent [*62*]: $T_1 = 8.5$ s at 15 MHz and $T_1 = 11.6$ s at 25 MHz. This frequency dependence of T_1, most probably, is due to scalar relaxation. In natural abundance, however, only approximately half of the bromine spins are ^{79}Br and the difference in Larmor frequency for ^{13}C and ^{81}Br, having about the same natural abundance as ^{79}Br, is more than 20 times as large (7.2%). Therefore, if scalar relaxation is significant, it will, due to the presence of the two bromine isotopes, also lead to nonexponential relaxation.

For heavy nuclei, scalar relaxation can give appreciable contributions. Therefore, for the ^{119}Sn relaxation in $ISn(CH_3)_3$ studied recently by Saluvere [*63*] as a function of temperature and frequency, both frequency dependences, due to relaxation through anisotropic shielding and due to scalar relaxation of ^{119}Sn and ^{127}I, are observed.

6.3.1.3. ^{13}C Relaxation in Acetone, $\lambda = D, CS, SR$

For the ^{13}C relaxation of the carbonyl carbon in acetone, three mechanisms are equally important: relaxation through dipolar and spin-rotation interaction and through anisotropic shielding at high magnetic fields. In fields up to 8.5 *T*, we find frequency dependent relaxation rates for both $(CH_3)_2CO$ and $(CD_3)_2CO$ (cf. Fig. 6.21). Extrapolation to zero frequency yields $1/T_1^{CS}$ and the frequency independent relaxation rate indicated by the dashed line which is in good agreement with earlier work of Olivson and Lippmaa [*64*]. Since $\Delta\sigma$ is known from solid state measurements (cf. Fig. 6.4) from $1/T_1^{CS}$, we readily obtain the correlation time for rotations about axes in the molecular plane (see below).

Assuming isotropic motion for the moment, we can calculate the contribution $1/T_1^D$ due to the *intra*molecular dipole-dipole interaction. The result is shown in Fig. 6.21. This intramolecular contribution dominates the frequency independent relaxation rate at low temperatures. This already shows that *inter*molecular contributions to the dipolar relaxation are not important here. Subtracting $1/T_1^D$ from the frequency independent relaxation rate yields the contribution of the spin-rotation interaction $1/T_1^{SR}$ which also gives a straight line in Fig. 6.21. At higher temperatures (above approximately 250 K), it dominates the total rate. Although in $(CH_3)_2CO$ all three mechanisms $\lambda = D, CS, SR$ give equally strong contributions in the temperature range studied, separation of the individual rates is straightforward and we immediately obtain the relevant result, that here the dipolar relaxation must be dominated by the intramolecular coupling. There is no evidence for a significant intermolecular contribution. This statement is possible only, because we have combined relaxation data in liquids with results from solid state NMR. In fact, only recently Goldammer *et al.* [*65*] presented evidence of a strong intermolecular contribution to the dipolar rate in acetone. The "discrepancy" stems solely from the less reliable determination of τ_2

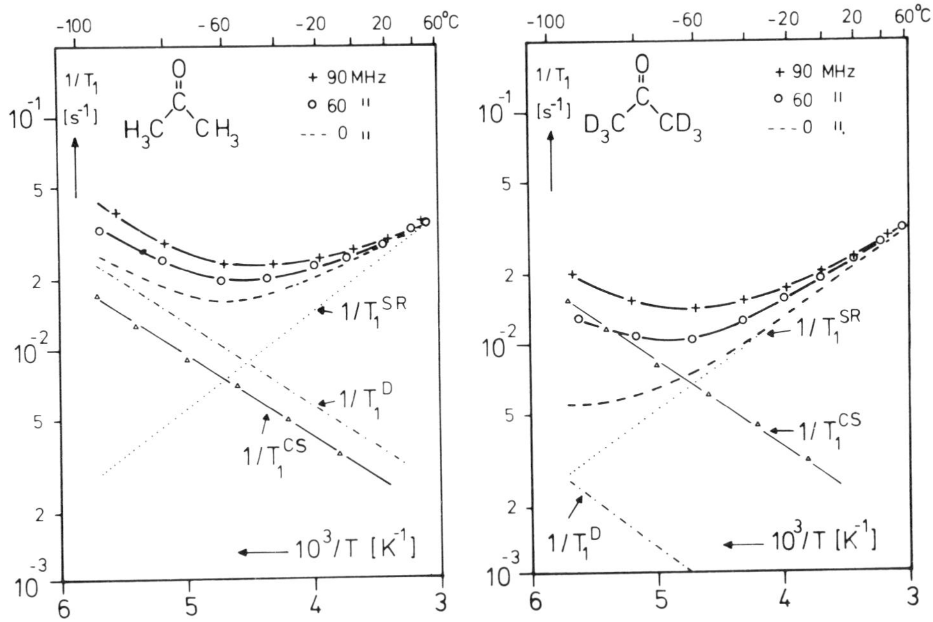

Fig. 6.21. ^{13}C relaxation rates of the carbonyl carbon in liquid acetone-h_6 and $-d_6$.

in [65]; the experimental data presented there and our data in Fig. 6.21 are in perfect agreement. For deuterated acetone, the situation is similar but the contribution of $1/T_1^D$ is much smaller (cf. Fig. 6.21b).

We would like to stress again that the separation of the different relaxation mechanisms is by no means trivial; yet it can be achieved. Typically for ^{13}C or ^{15}N in organic compounds, only the contribution of the dipole-dipole and the spin-rotation interaction has to be separated unless one works in high magnetic fields of superconducting magnets. This separation can be achieved either through temperature dependent studies or through determination of the nuclear Overhauser effect. A trivial statement does not seem to be out of place here: the more independent experimental material one can gather, the easier it is to avoid mistakes which otherwise are hard to exclude.

6.3.2. Relaxation through Dipole-Dipole and Quadrupole Coupling

Relaxation through spin-rotation interaction and anisotropic shielding offer possibilities to obtain especially interesting information about the reorientation of molecules in liquids and probably are overemphasized in this review. Besides these, one should never forget relaxation by dipole-dipole and quadrupole coupling as being, *by far, the most important* relaxation mechanisms. We, therefore, have labeled these two mechanisms here as standard relaxation mechanisms.

6.3.2.1. Intramolecular Contributions

The *quadrupole coupling* can be considered as the prototype of an intramolecular coupling where we have in mind ^{2}D, ^{14}N, ^{17}O in organic molecules in particular. Moreover, because of the strength of the quadrupole coupling [*66, 67*], it typically completely dominates the total relaxation rate. There is no need to separate it from other competing mechanisms. If the quadrupole coupling constant is known, we then readily obtain reliable values for the correlation time τ_2.

In liquids of low viscosity, the extreme narrowing condition is fulfilled ($\omega^2\tau_l^2 \ll 1$). Moreover, we only have to deal with irreducible tensors of second rank here. Then the Fourier transform of the autocorrelation function, Eq. (4.101), reduces to:

$$g_2^\lambda = \sum_{k=-2}^{2} c_{2k}^\lambda \tau_{2k}. \tag{6.8}$$

The reduced $\tilde{g}_2^\lambda = g_2^\lambda/\delta_\lambda^2$ then can be interpreted as effective correlation times τ_c^{eff}. For the analysis of the relaxation rates, the simple expressions of Tables 4.4 and 4.6 also can be used for anisotropic motion, provided $\eta_\lambda = 0$ by setting $\omega^2\tau_2^2 = 0$ and $\tau_2 = \tau_c^{\text{eff}}$. The correlation time τ_c^{eff} can be different for different couplings since the c_{2k}^λ can be different (cf. Table 4.9). Therefore, in our experimental examples, we will also use the familiar labels τ_c and τ_c^{eff}.

The analysis of the relaxation rates is especially straightforward if both the coupling tensor and the diffusion tensor have axial symmetry. Then τ_c^{eff} only depends on:

$$\rho = \frac{\tau_{c_\perp}}{\tau_{c_\parallel}} = \frac{D_\parallel}{D_\perp}, \tag{6.10}$$

and the angle β, between Z and the axis of symmetry of the diffusion tensor. From Table 4.9 we then deduce:

$$\tau_c^{\text{eff}} = \tau_{c_\perp}\left\{1 - \frac{3(\rho-1)}{5+\rho}\sin^2\beta\left[1 - \frac{3(\rho-1)}{2(2\rho+1)}\sin^2\beta\right]\right\}. \tag{6.11}$$

Quadrupolar relaxation rates of ^{2}D, ^{14}N, ^{17}O, and ^{35}Cl, therefore, offer the possibility to study anisotropic rotational motions in liquids since, for different nuclear sites in the molecule, β can be different (cf. the review articles [*68–70*]).

One of the pioneer works in this field is the study of *acetonitrile* – d_3 by Bopp [*71*] where for ^{14}N : $\beta = 0$, and for ^{2}D : $\beta = 109.5°$. The result is shown in Fig. 6.22, giving both the experimental data and the roational diffusion constants as obtained by Bopp [*71*]. Clearly, rotation about the threefold axis of the molecule is about an order of magnitude more rapid than rotation about axes perpendicular to this. Since acetonitrile is one of the liquids studied most extensively by various techniques, we will come back to this system later.

Completely analogous to quadrupolar relaxation, we can determine τ_c^{eff} from *dipolar* relaxation rates also. For ^{13}C or ^{15}N directly bound to hydrogens, the con-

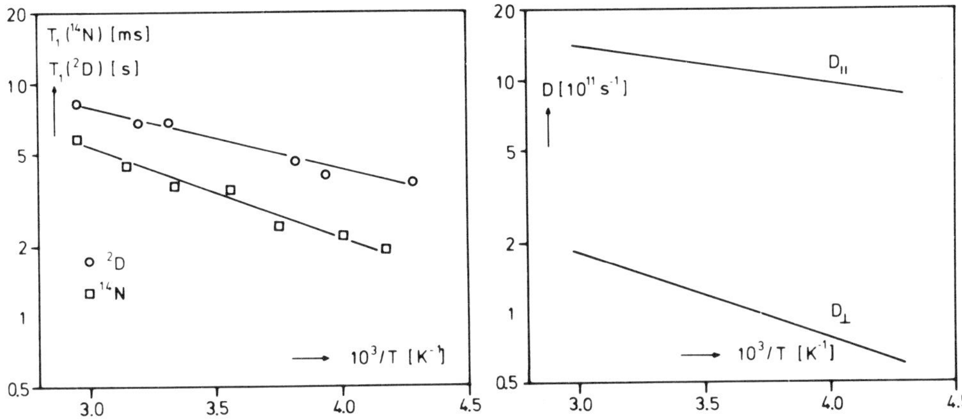

Fig. 6.22a and b. (a) ^{14}N and ^{2}D relaxation times in liquid CD_3CN (note different scales), and (b) rotational diffusion constants extracted from these data (Bopp [*71*]. [From Bopp, T. T.: J. Chem. Phys. 47, 3621 (1967)]

tributions of these protons predominate because $1/T_1^D \propto r^{-6}$ and the C–H or N–H distances are extremely short (approximately 1.1 Å only). Therefore, in these cases, the dipolar coupling also is purely intramolecular. Even dissolved oxygen that may be present and which, in proton relaxation due to its electronic paramagnetism, largely increases the relaxation rates observed only has a minor effect on ^{13}C relaxation rates of C–H carbons [*72*]. If one wants to analyze ^{13}C relaxation rates quantitatively, however, one, nevertheless, carefully has to remove oxygen from the sample. Slight differences, for example, of relaxation rates for different carbon positions in a molecule necessarily need not indicate anisotropic motion if the data are taken on a sample that has not been degassed [*73*].

We will give examples of studying anisotropic motions by exploiting relaxation through anisotropic shielding later. This mechanism *alone,* apart from the experimental difficulties, would not allow determination of anisotropic motion. Only *together* with correlation times (e.g., from ^{13}C–^{1}H dipolar coupling), do we get valuable information. Dipolar and quadrupolar relaxation, therefore, not only represent by far the most important relaxation mechanisms but, naturally, they provide the basis for obtaining information through the other mechanisms.

Let us discuss two more examples here:

i) Mobility along an Aliphatic Chain

By use of Fourier transform spectroscopy, the relaxation rates for the different carbon positions in a molecule can be determined simultaneously [*74*]. An elementary description is given in the book of Farrar and Becker [*75*]. For the ^{13}C nuclei along an aliphatic chain, the relaxation rate is dominated by the intramolecular dipole-dipole coupling and different relaxation times are direct evidence for different mobility of the CH_2 groups. As one of the pioneer examples, the T_1 values and effective correlation times for the different carbon positions in 1–*decanol* [*76*] are given in Fig. 6.23. The closer the carbon atoms are to the polar group of the molecule the

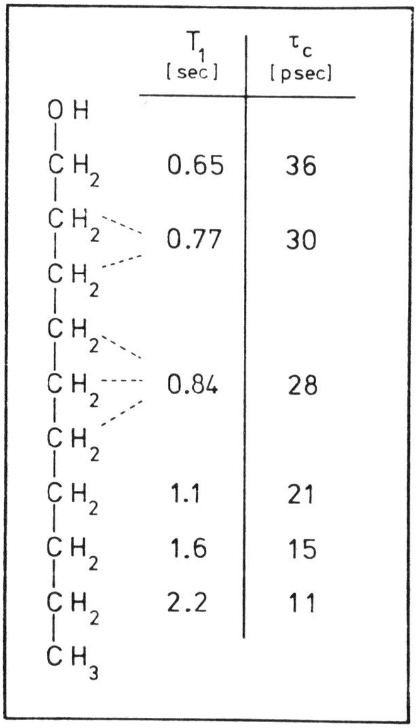

Fig. 6.23. ^{13}C relaxation times T_1 and rotational correlation times τ_C for the different CH_2 groups in 1-decanol at room temperature (Doddrell and Allerhand [*76*])

shorter is T_1 and, correspondingly, the longer is the correlation time τ_C (cf. Table 4.4). Obviously, the aliphatic end of such a chain is much more flexible than the polar one due to hydrogen bonding. A detailed analysis for such a molecule will be rather difficult; the qualitative result is unambiguous, however. A considerable number of related studies have since been performed (the references are available in [*54*]).

ii) Isotope Effect on Correlation Times

As noted above, one can obtain reliable values for reorientational correlation times τ_C from ^2D quadrupolar relaxation. Because of the low natural abundance of ^2D (0.015%), one then, typically, will use deuterated compounds. Therefore, one should investigate whether the values of τ_C found for the deuterated species can be used without modification for the normal protonated species. This question can be tackled by combining ^{13}C and ^2D relaxation studies. The dipolar coupling of the spin pairs $^{13}C-^2D$ and $^{13}C-^1H$, respectively can be described by axially symmetric second rank tensors ($\eta_D = 0$) (cf. Table 2.3). For the quadrupole coupling of ^2D in a C–D bond, likewise, $\eta_Q \approx 0$ [*77*]. Therefore, the correlation times obtained from ^{13}C dipolar relaxation and from ^2D quadrupolar relaxation of a $^{13}C-^2D$ group should be identical. The values of τ_C found in two independent ways can then be compared with τ_C obtained from dipolar relaxation of the corresponding $^{13}C-^1H$ group.

Experiments of this kind were performed on normal and perdeuterated *toluene* [*78*]. For the C–D groups of the phenyl ring, we found experimentally:

$$^{13C}T_1^D = 365\ ^{2D}T_1^Q, \tag{6.12}$$

and, with the known C–D distance and the experimental value of the quadrupole coupling constant [77], we calculate:

$$^{13C}T_1^D = 334\ ^{2D}T_1^Q. \tag{6.12a}$$

The numerical values in Eqs. (6.12) and (6.12a) differ somewhat from those given in [78] since the quadrupole coupling constant of ^{2}D in toluene has been determined in the meantime [77] and additional ^{13}C relaxation measurements at 90 MHz allowed a more precise determination of T_1^D. The good agreement of the numerical factors in Eqs. (6.12) and (6.12a) shows that we can determine effective correlation times for the deuterated phenyl ring C_6D_5 from either ^{13}C or ^{2}D relaxation (cf. Fig. 6.24). The correlation times of C_6H_5 determined from the ^{13}C data of normal toluene [78], however, are systematically shorter than those of C_6D_5 at the same temperature by approximately 20–25% over the entire temperature range from 180 K–330 K. This difference in τ_c corresponds to the difference in the mean moment of inertia of C_6H_5 and C_6D_5. A similar isotope effect has been found from analysis of Raman band shapes of C_6H_6 and C_6D_6 [79].

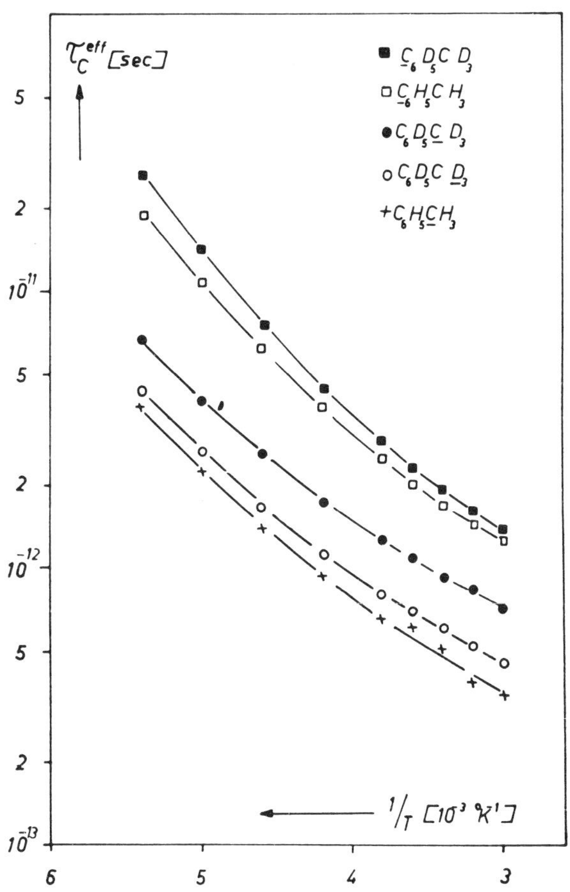

Fig. 6.24. Effective rotational correlation times for the phenyl ring and the methyl group in liquid loluene-h$_8$ and $-$d$_8$ from ^{13}C dipolar and ^{2}D quadrupolar relaxation [78]. [From Spieß, H. W., *et al.*: J. Magn. Reson. 9, 444, (1973)]

The differences in correlation time for the methyl groups CH_3 and CD_3, respectively, are even larger (cf. Fig. 6.24). Moreover, we obtain different values for τ_c^{eff} from ^{2}D and ^{13}C. This difference can be explained if one assumes the Z axis of the field gradient tensor at the deuteron site forms an angle of about 4° with C–D bond direction (for details cf. [*78*] and Section 6.3.4).

Contrary to toluene, we did not find such an isotope effect in acetone for rotations about axes in the molecular plane. This can be read directly from Fig. 6.21 since the relaxation rates due to anisotropic shielding are equal for acetone-h_6 and acetone–d_6. The detailed analysis of the data for $\lambda = CS$ and $\lambda = SR$ shows that for both species, the correlation times τ_c and τ_J do not fulfill Eq. (4.106) if we use the moment of inertia of the free molecule, the deviations being about 60%. This might, but need not be indicative of intermolecular interactions of the acetone molecules in the liquid not accounted for by the diffusion model.

These examples show special care must be taken if one wants to use correlation times determined in a deuterated species for the protonated species as well. There can be appreciable isotope effects on correlation times for one system while they can be negligible for another.

6.3.2.2. Intermolecular Dipole-Dipole Coupling

Considering intramolecular contributions of the dipole-dipole coupling is justified for C–H or N–H groups but not, however, for isolated ^{13}C or ^{15}N. For this reason, we have emphasized, in Section 6.3.1.3 that for the carbonyl carbon in acetone, the intramolecular coupling dominates. Contrary to this in *pyridine,* we found the ^{15}N–^{1}H coupling to be dominated almost completely by the intermolecular contribution [*80*]. The experimental relaxation rates are plotted in Fig. 6.25. At high frequencies (14 and 30 MHz for ^{15}N), the relaxation at low temperatures is strongly frequency dependent showing that under these conditions relaxation through anisotropic shielding dominates although we are dealing with a sample containing protons. The contribution of $\lambda = CS$ can be separated by extrapolation to $\nu = 0$ (cf. Fig. 6.19b) and this yields the dotted line in Fig. 6.25. This relaxation rate is dominated by spin-rotation interaction at high temperatures and by dipolar coupling at lower temperatures, respectively, as indicated by the opposite temperature dependence. The relaxation rate due to intramolecular coupling readily can be determined experimentally from ^{13}C relaxation. At temperatures below 260 K, that are of interest here, the ^{13}C relaxation is dominated by the intramolecular dipole-dipole coupling in the C–H group. Since the C–H and N–H internuclear vectors all lie in the molecular plane, the correlation times for C–H and N–H dipolar relaxation will be equal, even for anisotropic motion discussed in the following section. The intramolecular contribution to the ^{15}N dipolar relaxation rates are about an order of magnitude below the rates actually observed. Through combination of relaxation studies of different nuclei, therefore, we obtain the interesting result that, in liquid pyridine, there exists a strong intermolecular dipole-dipole coupling indicating interaction between the nitrogen atom and protons of adjacent pyridine molecules in the liquid. In view of the lone pair at the nitrogen atom, this seems to be plausible.

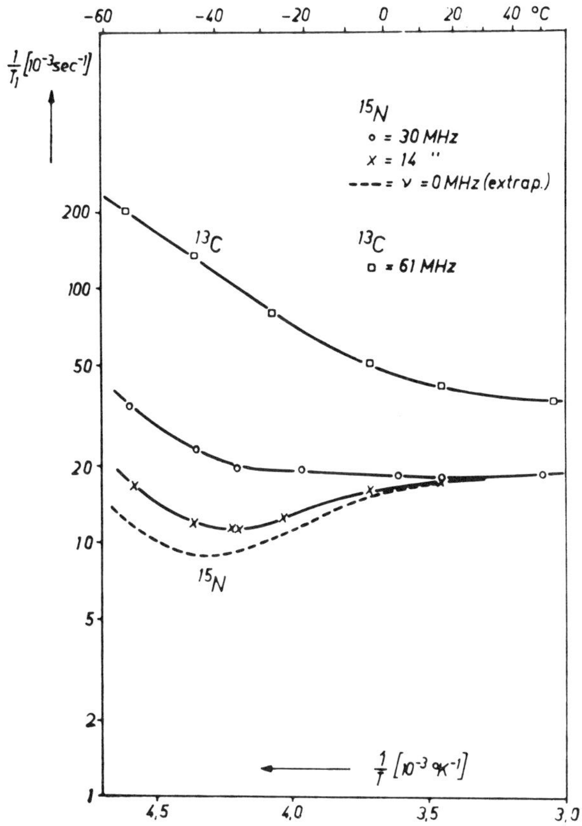

Fig. 6.25. [15]N and ^{13}C relaxation rates in liquid pyridine [*80*]. [From Schweitzer, D., Spieß, H. W.: J. Magn. Reson. 15, 529 (1974)]

Most recently, Hertz and coworkers [*81*] have demonstrated, in a beautiful work on *acetonitrile,* how, through measurements of $^{13}C-^{1}H$ intermolecular relaxation rates, one can obtain information about pair correlation functions. They determined the $^{13}C-^{1}H$ intermolecular rates from solutions of ^{13}C labeled $D_3C^{13}CN$ in H_3CCN and were able to show that there exists preferred ordering in the liquid, as suggested also from X-ray (and most recently from neutron scattering), and depolarized light scattering data [*82–84*] (see also below).

6.3.3. Anisotropic Motion of Planar Molecules

For planar molecules, one easily can imagine anisotropic motion in the liquid. For benzene, as an example, it is known from a combination of ^{13}C relaxation and depolarized light scattering [*85*] that, at room temperature, the rotation about the sixfold axis is about 3.5 times as fast as rotation about axes in the molecular plane, therefore, $\rho = 3.5$ [cf Eq. (6.11)]. Even higher values of ρ were deduced from combining Raman band shape analysis with ^{2}D relaxation [*79*]. Measurement of the anisotropy of the motion, using NMR alone, is straightforward if, for one interaction

λ_1, the unique axis is perpendicular to the molecular plane, whereas, for another interaction λ_2, it is in the plane. Then the correlation times obtained from the relaxation rates of these two mechanisms are largely different (cf. Fig. 4.8). For an axially symmetric diffusion tensor then, from λ_1, we get $\tau_\perp$ directly and, from λ_2, we get a combination of $\tau_\parallel$ and $\tau_\perp$ [cf. Table 4.9 and Eq. (6.11)].

For the standard relaxation mechanisms $\lambda = D$ and $\lambda = Q$, the unique directions for planar molecules all lie in the molecular plane. Then one always determines a combination of $\tau_\parallel$ and $\tau_\perp$ and a separation is difficult. It nevertheless, can be achieved for both dipolar and quadrupolar relaxation. In the case of the dipolar relaxation, one has to determine the relaxation rates for the individual lines in a multiplet and make use of cross correlation terms (cf. Refs. [*43–46*] of Chapter 4). For *trisubstituted benzenes,* Vold *et al.* [86] were able to determine all three principal values of the diffusion tensor from proton relaxation as shown in Fig. 6.26.

In the case of the quadrupolar coupling, the existence of an asymmetry parameter $\eta_Q \neq 0$ allows a different combination of $\tau_\parallel$ and $\tau_\perp$ as for $\eta_Q = 0$ (cf. Table 4.9). This has been used by Kintzinger and Lehn [*87*] to study anisotropic motion of heterocycles by combining ^{14}N and ^{2}D relaxation.

For the $\overleftrightarrow{\sigma}$ tensor, on the other hand, the unique axis in planar molecules typically is perpendicular to the molecular plane as discussed in Section 6.1.2. Therefore, by determining effective correlation times from relaxation through anisotropic shielding, on the one hand, and from dipolar or quadrupolar relaxation, on the other hand, we can utilize fully the maximum difference of correlation times possible [cf. Eq. (6.11)] since, for $\lambda = CS$, we have $\beta = 0$, and, for $\lambda = D, Q,$ we have $\beta = \pi/2$. In this way, it is

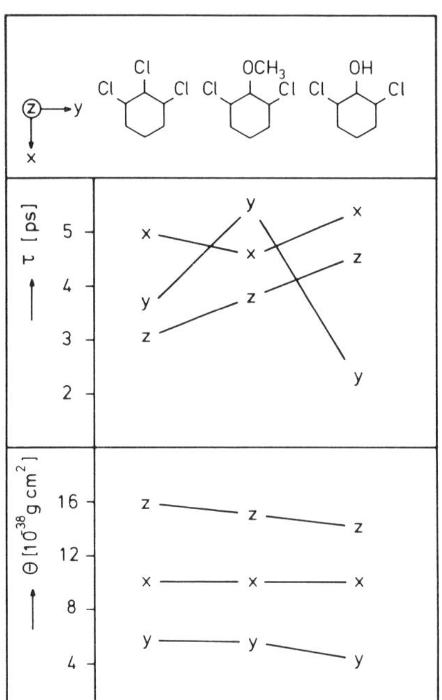

Fig. 6.26. Rotational correlation times of trisubstituted benzenes at room temperature, extracted from proton ralaxation (Vold *et al.* [*86*]). Also shown are the moments of inertia calculated from the molecular geometry

possible to study anisotropic motions over a considerable temperature range with a single experimental set-up and using the same sample throughout. As mentioned above, anisotropic motion also can be studied by combining nuclear relaxation with other techniques (e.g., light scattering [*84, 85*] or Raman band shape analysis [*79*]) which also allows the determination of $\tau_\perp$. These techniques are rather elegant. One should never forget, however, that anisotropic motion often reflects itself in relatively small differences of the correlation times only. As different methods often involve different sources of systematic errors for the absolute values of τ, such a combination calls for special care. Whereas, for intramolecular couplings in NMR, the relaxation rate depends only on the *auto*correlation function of the molecules, the correlation time determined by depolarized Rayleigh scattering, τ_{LS}, also is sensitive to *long range* order in the liquid, if present. For an axially symmetric diffusion tensor, Keyes and Kivelson [*88*] showed, that τ_{LS} and $\tau_\perp$ (the correlation time for reorientation of a single molecule) are proportional to each other:

$$\tau_{LS} = g^{(2)}\,\tau_\perp, \tag{6.13}$$

where $g^{(2)}$ is a generalized Kirkwood structure coefficient [*84, 88*]. The difference of τ_{LS} and $\tau_\perp$ has been demonstrated recently by Versmold [*84*] who determined τ_{LS} by depolarized Rayleigh scattering and $\tau_\perp$ from ^{14}N quadrupole relaxation of mixtures of *acetonitrile* and CCl_4 (cf. Fig. 6.27). Only for infinite dilution, τ_{LS} and $\tau_\perp$ are identical, whereas, for higher concentrations of CH_3CN, the pair correlation of the acetonitrile molecules renders τ_{LS} considerably longer than $\tau_\perp$.

Let us now discuss two examples where relaxation through anisotropic shielding has been used to study anisotropic motion.

i) In *pyridine*, we determined $\tau_\perp$ from relaxation through anisotropic shielding of ^{15}N [*80*], using $\Delta\sigma$ as determined in the solid (cf. Fig. 6.9). From ^{13}C relaxation, we obtained τ_c^{eff} for the dipolar relaxation ^{13}C—^1H. These two correlation times were

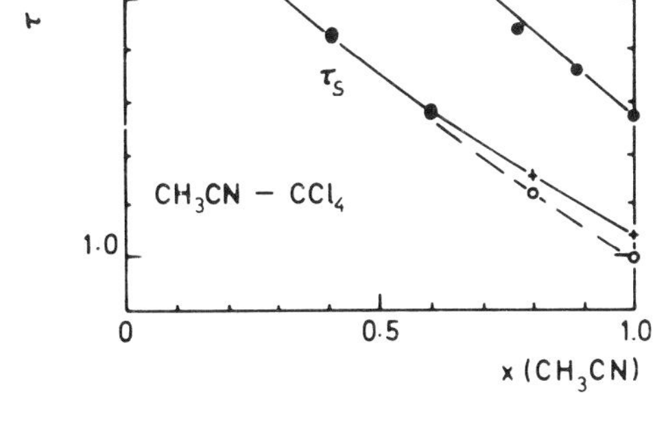

Fig. 6.27. Correlation times in mixtures of CH_3CN and CCl_4 at room temperature (Versmold [*84*]). • τ_{LS} as obtained from depolarized Rayleigh scattering, + $\tau_S = \tau_\perp$ as obtained from ^{14}N quadrupolar relaxation, ○ viscosity (Courtesy of H. Versmold)

found to differ by no more then 20% in the temperature range of 223–260 K and, from this, one calculates $\rho = \tau_\perp/\tau_\parallel = 1.4$. The anisotropy of the rotational motion in pyridine, therefore, is considerably smaller than it is in benzene. This result, at first sight, seems to be in contradiction to an earlier study of molecular motion in liquid pyridine [87] where, at 253 K, $\rho = 3$ was deduced from ^{14}N and ^{2}D relaxation data. This result had been obtained using the ^{2}D quadrupole coupling constant of benzene since the value for pyridine had not yet been determined. Using the value for e^2qQ/h, since obtained experimentally in pyridine [89], this discrepancy is completely removed since then at 253 K, τ_c^{eff} for ^{14}N and ^{2}D from the relaxation data of Ref. [87] are found to be equal. This further supports our statement that, in pyridine contrary to benzene, the reorientation is more or less isotropic.

This shows that the anisotropy of the rotational motion of a molecule in a liquid is not determined solely by different principal moments of inertia, as suggested by the extension of the J-diffusion model to symmetric tops (cf. Section 4.2.4.3). According to Eq. (4.111), the anisotropy of the motion should be equal in benzene and pyridine since the ratio of the moments of inertia $\theta_\parallel/\theta_\perp$ is the same in both cases. Similarly, for the trisubstituted benzenes, the large variation of τ_y does not correspond at all to a similar variation in θ_{yy} (cf. Fig. 6.26).

ii) Contrary to pyridine in *toluene,* we found the reorientation of the phenyl ring to be highly anisotropic, as evidenced by analysis of the ^{13}C relaxation rates shown in Fig. 6.28 for the temperature range 183–250 K. In normal toluene, the dipolar coupling ^{13}C–^{1}H is so large that relaxation through anisotropic shielding cannot compete. In deuterated toluene, the dipolar coupling is reduced sufficiently so a significant frequency dependence is observed between 14, 61, and 90 MHz. Since we only measured an average over the different ^{13}C positions of the ring, we

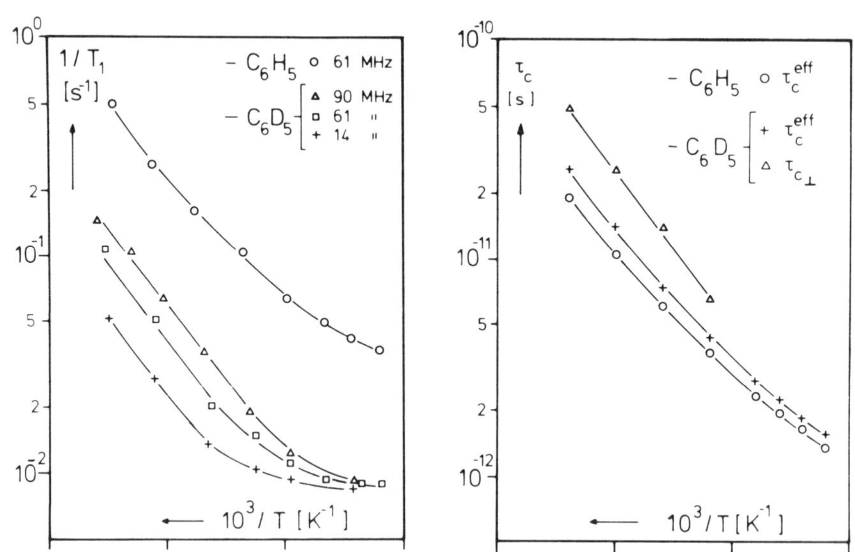

Fig. 6.28. ^{13}C relaxation rates and correlation times for the phenyl ring in toluene

can determine only the difference between $\tau_{\parallel}$ for rotations about the normal to the phenyl ring and $\tau_{\perp}$ (the mean value of the other two correlation times), although toluene is an asymmetric top. The correlation time $\tau_{\perp}$, obtained from $1/T_1^{CS}$, is about twice as long as τ_c^{eff} for $^{13}C{-}^{2}D$ dipolar coupling (cf. Fig. 6.28b). This gives $\rho = \tau_{\perp}/\tau_{\parallel} \approx 3.5$ constant in the temperature range 183–250 K. This result is in accordance with the recent study of toluene through ^{13}C relaxation and depolarized Rayleigh scattering [*85*] at room temperature, showing that all three principal rotational diffusion constants differ significantly. Similarly, from ^{2}D relaxation of selectively deuterated toluene at lower temperatures, the two in plane diffusion constants were found to be different [*90*].

Therefore, by use of superconducting magnets allowing measurement of ^{13}C relaxation rates up to 90 MHz, relaxation through anisotropic shielding can be exploited to study anisotropic rotational motion in liquids. In order to get complete information, the lines for different carbon positions of the aromatic rings have to be resolved in the spectrum; therefore, deuteron decoupling is needed. Our example of acetone shows, however, similar studies can be performed for carbonyl groups in compounds containing remote protons. Only standard proton decoupling, therefore, is needed in this case. Likewise for ^{15}N, due to the large shielding anisotropies, this relaxation mechanism is even more important and can be detected in normal protonated samples.

6.3.4. Internal Rotation of Methyl Groups

For molecules containing methyl groups, the internal rotation of the methyl groups about its C_3 axis can be considerably faster than the rotation of the remainder of the molecule. This is the case in liquid *toluene* as we will see below. In order to study the *rapid* internal motion, the standard relaxation mechanisms $\lambda = D, Q$ are not well suited, as shown by calculating $\tau_c{}^{\text{eff}}$ according to Eq. (6.11), the result is plotted in Fig. 6.29. For a methyl group, the angle β between Z and C_3 is the tetrahedral angle for ^{2}D relaxation or ^{13}C intramolecular dipole-dipole relaxation, whereas, for intramolecular proton relaxation $\beta = \pi/2$. Clearly, the diffusion model strictly will not be adequate to describe such a rapid rotation (see below). The general considerations discussed here, however, can be expected to be largely independent of the particular model used.

According to Fig. 6.29 the values calculated for $\rho = 50$ and $\rho \to \infty$ are so close it seems to be almost impossible to get ρ from a ratio $\tau_c^{\text{eff}}/\tau_{c_\perp}$, determined experimentally, since even minor errors in τ_c^{eff} or $\tau_{c_\perp}$ can change the result for ρ drastically. Here, one should not think only of the experimental uncertainty of the T_1 measurement but also of the inherent uncertainties of the determination of τ_c from relaxation data (e.g., the coupling parameters *in the liquid* are never known with such extreme accuracy that would be needed). For $^{13}C{-}^{1}H$ dipole-dipole or ^{2}D quadrupole coupling, where β is close to the tetrahedral angle, the result, moreover, for ρ also is very sensitive against the angle β (cf. Fig. 6.29). This is the reason, for the methyl group in toluene, we found different effective correlation times from ^{13}C dipolar and ^{2}D quadrupolar relaxation (cf. discussion of Fig. 6.24).

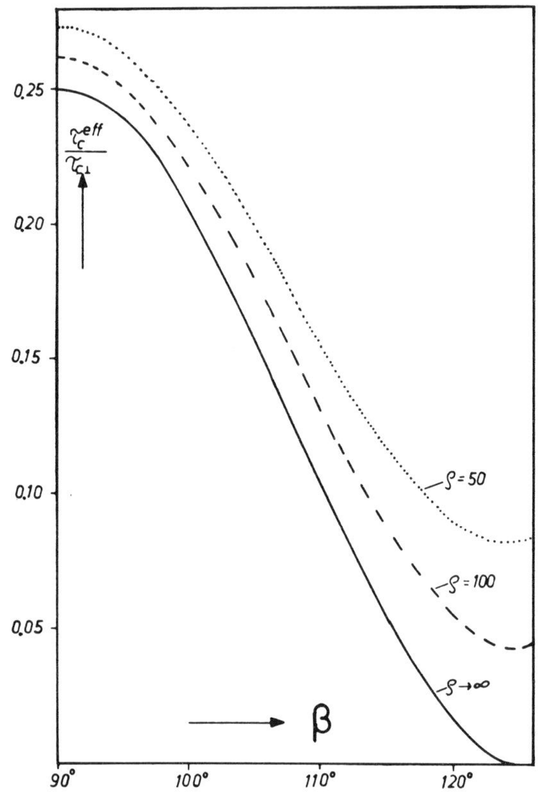

Fig. 6.29. Ratio of effective correlation times τ_c^{eff} and $\tau_{c_\perp}$ for rapid internal rotation as calculated from Eq. (6.11). [From Spieß, H. W., *et al.*: J. Magn. Reson. 9, 444, (1973)]

The main reason for the difficulties mentioned here is the fact, for rapid internal rotation ($\rho \gg 1$), the correlation time τ_c^{eff} does not approach zero but approached a limiting value, determined by $\tau_{c_\perp}$. Therefore, τ_c^{eff} contains only little information about the exact value of ρ. Moreover, the relaxation rate decreases for $\lambda = D, CS, J$ if the speed of the rotation increases. If, by use of Eq. (6.11), one obtains values of $\rho > 10$, one should be especially careful. It then will be advisable to rely on the significant result $\tau_\parallel \ll \tau_\perp$ rather than try to extract unreliable numbers for ρ.

The spin-rotational relaxation rate, on the other hand, provides a means to study such fast rotations quantitatively. In Fig. 6.30 we have plotted already separated ^{13}C relaxation rates $1/T_1^D$ and $1/T_1^{SR}$ for CH_3 and CD_3 in toluene [*78*]. In order to analyze our data according to Table 4.7, we need the spin-rotation constant $c_\parallel$ which we assume to be equal to the spin-rotation constant c of ^{13}C in methane. By use of Eq. (5.12) together with ^{13}CO as reference compound, we obtain $c = -108.7$ kHz for CH_4 and $c = -54.3$ kHz for CD_4. With these values the data were analyzed to yield τ_J and its temperature dependence [*78*]. Later the spin-rotation constant in methane was determined by molecular beam resonance [*91*] ($c = -100.1$ kHz). The deviation is only 8%, again showing the usefulness of Eq. (5.12). The values found for τ_J in this way can be interpreted within the J-diffusion model as mean time between two collisions of the methyl group. If we assume

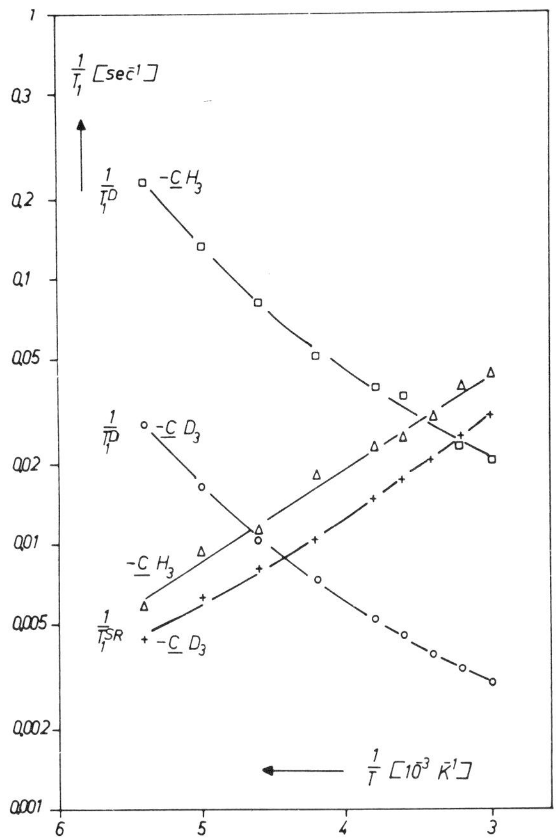

Fig. 6.30. [13]C dipolar and spin-rotational relaxation rates for the methyl group CH_3 and CD_3 in liquid toluene [78]. [From Spieß, H. W., *et al.*: J. Magn. Reson. 9, 444, (1973)]

that the methyl group rotates with the mean rotational frequency of the free rotor at the same temperature, we can get a very direct picture of the rotation of the methyl group by calculating the mean angle $\overline{\vartheta}$. The methyl group rotates between two subsequent collisions: $\overline{\vartheta} = \tau_J \cdot \sqrt{kT/\theta}$. The result is depicted in Fig. 6.3 for CH_3 and CD_3.

For comparison, Fig. 6.31 also contains $\overline{\vartheta}$ values for CS_2 obtained from the data presented in Fig. 6.19. In Section 6.3.1.2, we had mentioned already that our CS_2 data were consistent with the diffusion model. Accordingly, here we find $\overline{\vartheta}$ to be below 10° in the entire temperature range between melting point and boiling point. We would like to point out this in no way proves the validity of the diffusion model as such. Through our independent determination of τ_c and τ_J, however, we can check, whether, within a certain model, its assumptions–here: small rotation angles $\overline{\vartheta}$ between collisions–are actually fulfilled for a certain system.

Contrary to CS_2, the methyl group of toluene on the average already rotates by 25° between two subsequent collisions at the melting point and this angle increases with increasing temperature so $\overline{\vartheta}$ = 130° at 335 K. This shows that here the diffusion model cannot be expected to give an adequate description of the internal motion. Most remarkable is the fact that the mean rotation angle $\overline{\vartheta}$ is found to be equal for

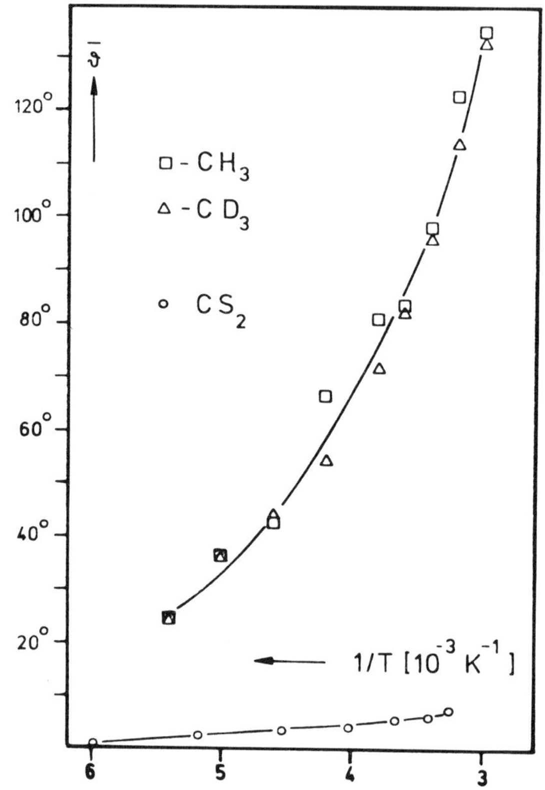

Fig. 6.31. Mean rotational angle between two subsequent collisions ($\bar{\vartheta}$) extracted from spin-rotational relaxation rates for the methyl group in toluene and for CS_2

CH_3 and CD_3 at the same temperature although relaxation rates, correlation times, and mean angular velocities of the free rotor are different for the two species.

Although the mean rotational angle $\bar{\vartheta}$ gives a very direct picture of the internal motion of the methyl group, this should not remain the only quantity we extract from our ^{13}C data, in particular, since $\bar{\vartheta}$ can only be given within a certain model for the internal rotation. Unfortunately, however, analysis of the spin-rotational relaxation rate only gives us $\tau_J^{(\text{internal})}$. In order to be able to compare the rotation of the methyl group with that of the phenyl ring, we need the correlation time $\tau_c^{(\text{internal})}$ which we can calculate from $\tau_J^{(\text{internal})}$ by use of the extended diffusion model for internal rotation (cf. Section 4.2.4.3 and Fig. 4.9). The values for $\tau_c^{(\text{internal})}$ obtained in this way [*78*] are plotted in Fig. 6.32, together with τ_c, for the phenyl ring obtained from ^{13}C dipolar relaxation. In the temperature range studies, $\tau_c^{(\text{internal})}$ and $\tau_J^{(\text{internal})}$ are not much different ($\tau_J^* \approx 1$), contrary to the assumptions of the classic diffusion model where one assumes $\tau_J \ll \tau_c$. The numerical values of $\tau_c^{(\text{internal})}$, of course, depend on the specific model used to calculate it. It can be expected, however, that in the range $\tau_J^* \approx 1$, any realistic model will give similar results. We, therefore, can now compare the correlation times τ_c for the phenyl ring and $\tau_c^{(\text{internal})}$. Both are determined independently. At the melting point, we find $\rho = \tau_c/\tau_c^{(\text{internal})} \approx 200$ and this ratio decreases

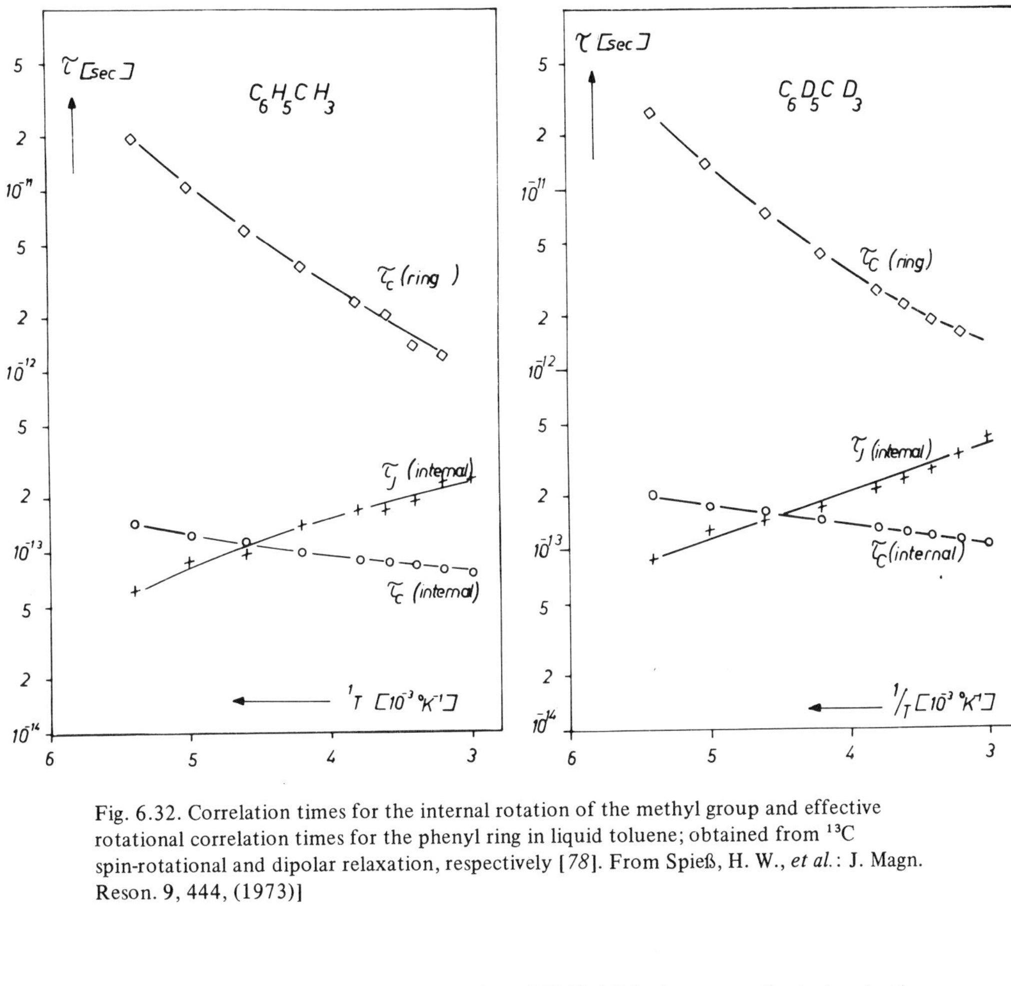

Fig. 6.32. Correlation times for the internal rotation of the methyl group and effective rotational correlation times for the phenyl ring in liquid toluene; obtained from ^{13}C spin-rotational and dipolar relaxation, respectively [*78*]. From Spieß, H. W., *et al.*: J. Magn. Reson. 9, 444, (1973)]

with increasing temperature so that $\rho = 13$ at 350 K. This decrease of ρ is due to the fact that the activation energy for the reorientation of the phenyl ring is much higher than that for the almost freely rotating methyl group.

6.4 Conclusion

The recent developments in NMR (Fourier spectroscopy, in particular, increase of resolution and sensitivity in *solids*, and application of high magnetic field not only offer new possibilities to obtain static parameters from NMR spectra, but they also offer new possibilities to obtain dynamic information, in particular about rotational motions of molecules. For such nuclei as, e.g., ^{13}C, ^{15}N, ^{19}F, ^{31}P, two relaxation mechanisms, known for a long time, now reach practical importance; relaxation by spin-rotation interaction and, with use of high magnetic fields, relaxation through anisotropic shielding.

Whereas spin-rotational relaxation, in general, is of importance in liquids or gases only, anisotropic shielding also provides a means to study slow rotational motions in *solids* since characteristic line shapes are observed. The shielding anisotropies for the $I = 1/2$ nuclei (e.g., ^{1}H, ^{13}C, ^{15}N, ^{19}F, ^{31}P) cover a wide range (from only a few ppm to about 1000 ppm) and, in addition, we nowadays can choose the magnetic field strength as large as 8.5 T. Therefore, line shape analysis can be exploited to obtain information about both the type and the frequency of the motion in the range from approximately 1–500 kHz. Similar line shapes are observed for ^{2}D (due to quadrupolar coupling) whereby the range of accessible frequencies is extended up to about 2 MHz.

In *liquids* of low viscosity, NMR, as a slow method, only allows determination of time integrals over correlation functions. Nevertheless, we can get valuable information about rotational motions since, by choosing the experimental conditions (e.g., isotope being studied, magnetic field strength, temperature, spin decoupling), we can select the relaxation mechanism which will give us the most reliable information about a particular problem.

Dipole-dipole coupling can be due to both *intra-* and *inter*molecular interactions. Contrary to proton relaxation where, typically, both contributions are equally important, the ^{13}C or ^{15}N relaxation rates for C–H or N–H groups are almost entirely dominated by the intramolecular dipole-dipole coupling to the directly bound proton. By analysing these data, therefore, we readily obtain reliable values of the correlation time of reorientation τ_c. Intermolecular contributions comparable with intramolecular ones seem to be rather exceptional. If encountered, they may be indicative of special intermolecular interactions; an interesting field in itself.

Relaxation through *anisotropic shielding* can be significant for ^{13}C or ^{15}N in molecules with π bonds only and if, in addition, high magnetic fields of 6–8 T are employed. By combination with solid state measurements, we again obtain τ_c. This mechanism is especially valuable if one wants to study anisotropic rotational motions since the unique direction of the shielding tensor often differs from those of the other couplings (in particular, in planar molecules).

For nuclei with spin $I > 1/2$ (e.g., ^{2}D, ^{14}N, ^{17}O) the relaxation rate is dominated by the *quadrupolar* coupling. This mechanism has been used extensively in nuclear relaxation studies for some time. Again, by combination with solid state or gas phase measurements, we obtain reliable values of τ_c.

For organic molecules in both cases (anisotropic shielding and quadrupole coupling), we are dealing with intramolecular interactions. Therefore, these mechanisms represent useful tools to study rotational motions rather than to elucidate special intermolecular interactions in the liquid. If such interactions exist, they will often change the coupling parameters from the values found in the solid or in the gas phase, making the analysis of the relaxation rates difficult.

The *spin-rotation* interaction gives appreciable contributions for rapid motion of the total molecule or for internal rotations (as in methyl groups). Because of the close relationship between magnetic shielding and spin-rotation interaction, this mechanism is much more important for ^{13}C and ^{15}N than it is for protons since, for these nuclei, not only the shielding effects are larger but also the spin-rotation interaction. This mechanism is not only suited to study fast motions but, from spin-rotational relaxation rates, one, in general, can obtain τ_J (the correlation time of angular momentum).

We emphasize again that τ_J is obtained as time integral over the correlation function of angular momentum. Naturally, therefore, we do not obtain information about the detailed form of this correlation function. Nevertheless, τ_J represents unique information since, by the independent measurements of τ_c and τ_J, we can, for example, test the validity of models developed to describe the rotation of molecules in a liquid. It certainly is noteworthy that the "slow" nuclear relaxation seems to represent the most reliable method to determine τ_J to date. From a process that often takes more than 30 s (the ^{13}C or ^{15}N relaxation time) and which, therefore, can be followed easily using a regular watch, one gets a measure of a correlation time which typically is less than 0.1 ps.

7. Appendix

A. Dipole-Dipole Coupling

In this appendix, we want to show that the dipolar coupling between two spins I and S is described by an *irreducible second rank tensor.* We start from:

$$\hbar\,\mathcal{H}_D = \frac{\vec{\mu}_I\vec{\mu}_S}{r_{IS}^3} - 3\,\frac{(\vec{\mu}_I\mathbf{r}_{IS})\cdot(\vec{\mu}_S\mathbf{r}_{IS})}{r_{IS}^5}\,. \tag{A.1}$$

In irreducible tensor notation for the **I**, **S**, and $\mathbf{r} = \mathbf{r}_{IS}$ operators $\mathcal{H}_D$ reads:

$$\mathcal{H}_D = \frac{\gamma_I\gamma_S\hbar}{r^5}\Big\{(-I_{+1}S_{-1} + I_0S_0 - I_{-1}S_{+1})\,r^2$$
$$-3\,(-I_{-1}r_{+1} + I_0r_0 - I_{+1}r_{-1})\,(-S_{-1}r_{+1} + S_0r_0 - S_{+1}r_{-1})\Big\}. \tag{A.2}$$

After performing the multiplications noting that the **r** operators commute among themselves and with the spin operators, we obtain:

$$\mathcal{H}_D = \frac{\gamma_I\gamma_S\hbar}{r^5}\Big\{(-I_{+1}S_{-1} + I_0S_0 - I_{-1}S_{+1})\,(-2\,r_{-1}r_{+1} + r_0^2)$$
$$-3\,[I_{-1}S_{-1}r_{+1}^2 + I_{+1}S_{+1}r_{-1}^2$$
$$-(I_{-1}S_0 + I_0S_{-1})\,r_{+1}r_0 - (I_{+1}S_0 + I_0S_{+1})\,r_{-1}r_0$$
$$+(I_{-1}S_{+1} + I_{+1}S_{-1})\,r_{+1}r_{-1} + I_0S_0r_0^2]\Big\}. \tag{A.3}$$

The spin operators now are sorted out to form the tensor operators T_{2m} (cf. Eq. (2.13)) and from the **r** operators we likewise form the tensor operators R_{2m}:

$$R_{20} = \frac{2}{\sqrt{6}} (r_{+1} r_{-1} + r_0^2)$$

$$R_{2\pm1} = \frac{2}{\sqrt{2}} r_0 r_{\pm1}; \qquad R_{2\pm2} = r_{\pm1}^2. \tag{A.4}$$

Some of the terms of Eq. (A.3) already are of the form $T_{2m}R_{2-m}$:

$$\begin{aligned}\mathcal{H}_D = \frac{\gamma_I\gamma_S\hbar}{r^5} & [-3(T_{22}R_{2-2} + T_{2-2}R_{22}) \\ & + 3(T_{21}R_{2-1} + T_{2-1}R_{21}) \\ & - (I_{+1}S_{-1} + I_{-1}S_{+1}) r_{+1}r_{-1} - 2 I_0 S_0 r_0^2 \\ & - 2 I_0 S_0 r_{+1} r_{-1} - (I_{+1}S_{-1} + I_{-1}S_{+1}) r_0^2]\end{aligned} \tag{A.5}$$

$$\begin{aligned} = -\frac{3\gamma_I\gamma_S\hbar}{r^5} & [T_{22}R_{2-2} + T_{2-2}R_{22} - (T_{21}R_{2-1} + T_{2-1}R_{21}) \\ & + \frac{1}{6} \underbrace{(I_{+1}S_{-1} + I_{-1}S_{+1} + 2 I_0 S_0)}_{\sim T_{20}} \underbrace{(2 r_{+1} r_{-1} + r_0^2)}_{\sim R_{20}}].\end{aligned} \tag{A.6}$$

therefore $\mathcal{H}_D$ is obtained in the form

$$\mathcal{H}_D = -\frac{3\gamma_I\gamma_S\hbar}{r^5} \sum_{m=-2}^{2} (-1)^m T_{2m} R_{2-m}, \tag{A.7}$$

there are no terms with $l \neq 2$.

B. Relation Between $\rho_{ab}^{(l)}$ and ρ_{lm}

In order to get the relation between the cartesian components $\rho_{ab}^{(l)}$ and the irreducible components ρ_{lm}, we consider the Hamiltonian in cartesian representation in the principal axes system

$$\mathcal{H} = \mathbf{V} \overleftrightarrow{\rho}\, \mathbf{U}. \tag{B.1}$$

U and **V** are vector operators, $\overleftrightarrow{\rho}$ is given by the expression of Eq. (2.30)

$$l = 0: \mathcal{H}^{(0)} = R(V_x U_x + V_y U_y + V_z U_z)$$

$$l = 1: \mathcal{H}^{(1)} = V_x(\rho_{xy} U_y + \rho_{xz} U_z) \tag{B.2}$$

$$+ V_y(-\rho_{xy} U_x + \rho_{yz} U_z)$$

$$+ V_z(-\rho_{xz} U_x - \rho_{yz} U_y)$$

$$= (V_x U_y - V_y U_x)\rho_{xy}$$

$$+ (V_x U_z - V_z U_x)\rho_{xz}$$

$$+ (V_y U_z - V_z U_y)\rho_{yz} \tag{B.3}$$

$$l = 2: \mathcal{H}^{(2)} = \delta[-V_x U_x \cdot \frac{1}{2}(1 + \eta) - V_y U_y \cdot \frac{1}{2}(1 - \eta) + V_z U_z]$$

$$= \frac{1}{2}\delta\,(-V_x U_x - V_y U_y - V_z U_z + 3\, V_z U_z) \tag{B.4}$$

$$+ \frac{1}{2}\delta\eta\,(- V_x U_x + V_y U_y).$$

In order to be able to compare these expressions with $\rho_{lm} T_{l-m}$, we have to express T_{lm}, likewise by cartesian components. These are obtained readily from Eqs. (2.11–2.13) by inserting the expressions for U_m and V_m given in Eq. (2.8). The result is

$$T_{00} = V_x U_x + V_y U_y + V_z U_z = \mathbf{V} \cdot \mathbf{U} \tag{B.5}$$

$$\begin{aligned} T_{10} &= \frac{i}{\sqrt{2}}\,(V_x U_y - V_y U_x) \\ T_{1\pm1} &= \frac{1}{2}[-([V_x \pm i V_y)\, U_z + V_z(U_x \pm i U_y)] \end{aligned} \tag{B.6}$$

$$\begin{aligned} T_{20} &= \frac{1}{\sqrt{6}}\,(- V_x U_x - V_y U_y - V_z U_z + 3\, V_z U_z) \\ T_{2\pm1} &= \frac{1}{2}[\mp V_z (U_x \pm i U_y) \mp (V_x \pm i V_y)\, U_z] \\ T_{2\pm2} &= [V_x U_x - V_y U_y \pm i(V_x U_y + V_y U_x)]. \end{aligned} \tag{B.7}$$

Now, by comparing Eqs. (B.5)–(B.7) with Eqs. (B.2)–(B.4), respectively, we derive

$$\begin{aligned} \rho_{00} &= R, & \rho_{20} &= \sqrt{\frac{3}{2}}\,\delta, \\ \rho_{1\pm1} &= (\rho_{xz} \pm i\,\rho_{yz}), & \rho_{2\pm1} &= 0, \\ \rho_{10} &= -i\sqrt{2}\,\rho_{xy}, & \rho_{2\pm2} &= -\frac{1}{2}\,\delta\eta. \end{aligned} \tag{B.8}$$

For an arbitrary axes system we likewise obtain [1]

$$T_{00} = \frac{1}{3}(T_{xx} + T_{yy} + T_{zz}), \qquad T_{20} = \frac{1}{\sqrt{6}}\,[3\,T_{zz} - (T_{xx} + T_{yy} + T_{zz})],$$

$$T_{10} = -\frac{i}{\sqrt{2}}\,(T_{xy} - T_{yx}), \qquad T_{2\pm1} = \mp\frac{1}{2}\,[T_{xz} + T_{zx} \pm i\,(T_{yz} + T_{zy})],$$

$$T_{1\pm1} = -\frac{1}{2}\,[T_{zx} - T_{xz} \pm i(T_{zy} - T_{yz})],\; T_{2\pm2} = \frac{1}{2}\,[T_{xx} - T_{yy} \pm i(T_{xy} + T_{yx})]. \quad (B.9)$$

C. Wigner Rotation Matrices

General expressions for the matrix elements $\mathfrak{D}^{(l)}_{m'm}(\alpha, \beta, \gamma)$ are given in every textbook on group theory. The definition of the eulerian angles is not uniform, however, and, consequently, the Wigner rotation matrix elements also differ. These angles can correspond to rotations of *functions* or of *coordinate* systems where the rotation axes can either be *fixed* in space or can be axes of a *rotated* coordinate system. The different conventions are compared, for example, by Tinkham ([2], p. 111 f.) and by Brink and Satchler ([3], p. 21). For different conventions, the Wigner matrices have different meanings, therefore, we reproduce the matrices used here explicitly. We mainly need the rotation matrices in order to obtain the operators R_{lm} from operators $\rho_{lm'}$, defined in the principal axes system according to

$$R_{lm} = \sum_{m'=-l}^{l} \rho_{lm'} \mathfrak{D}^{(l)}_{m'm}\,(\alpha, \beta, \gamma). \quad (C.1)$$

Let us label the coordinate system in which we want to know the R_{lm} by x, y, z and the principal axes system by X, Y, Z. Then the eulerian angles depend only on the orientation of the x, y, z system in the principal axes system. The eulerian angles β and α simply are the polar angles of the z axis in the principal axes system X, Y, Z (cf. Fig. C.1). The eulerian angle γ is the angle between the node-line K and the y axis taken in the x-y plane. Note that the angles correspond to positive rotations. This definition of the eulerian angles is the same as that of Rose [4], Tinkham [2],

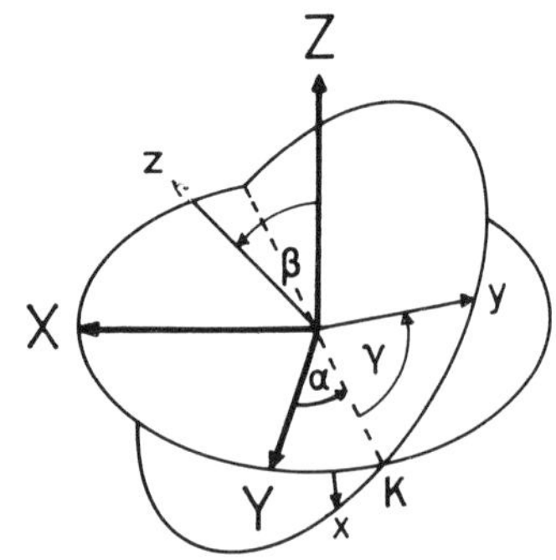

Fig. C.1. Definition of eulerian angles α, β, γ

and Brink and Satchler [3] where α, β, γ are introduced as rotation angles through which the original system X, Y, Z has to be rotated, in the positive sense, first about the original Z axis (through α), about the new Y' axis parallel K (through β), and about the final Z'' axis (through γ).

With this definition of the eulerian angles, the general expression for the matrix elements of $\mathfrak{D}^{(l)}_{m'm}(\alpha, \beta, \gamma)$ reads [2–4]

$$\mathfrak{D}^{(l)}_{m'm}(\alpha,\beta,\gamma) = e^{-im'\alpha} e^{-im\gamma} \sum_{\kappa} \frac{(-1)^{\kappa} \sqrt{(l+m)!(l-m)!(l+m')!(l-m')!}}{\kappa!(l+m-\kappa)!(l-m'-\kappa)!(\kappa+m'-m)!} \cdot \left(\cos\frac{\beta}{2}\right)^{2l-2\kappa-m'+m} \left(-\sin\frac{\beta}{2}\right)^{2\kappa+m'-m}. \quad (C.2)$$

The summation over κ runs over all values of κ for which the denominator is finite. Since we are dealing with integers l and m only, κ also must be an integer. Accordingly, the argument n of $n!$ must be a positive integer $n \geqslant 0$ in order that $n!$ be finite. For $l = 0$, we have $\mathfrak{D}^{(0)}_{00} = 1$. The rotation matrices $\mathfrak{D}^{(1)}$ and $\mathfrak{D}^{(2)}$ are given for convenience in Tables C.1 and C.2. In a slightly different form containing $\beta/2$, these matrices have been given, for example, by Brink and Satchler [3]. Finally, the relation to the spherical harmonics $Y_{lm}(\beta, \alpha)$ should be noted:

$$\mathfrak{D}^{(l)}_{m0}(\alpha, \beta, 0) = \sqrt{\frac{4\pi}{2l+1}}\, Y^{*}_{lm}(\beta, \alpha). \quad (C.2a)$$

Table C.1. Wigner rotation matrix $\mathfrak{D}^{(1)}$ (α, β, γ)

m' \ m	1	0	−1
1	$\frac{1+\cos\beta}{2}\, e^{-i(\alpha+\gamma)}$	$\frac{-1}{\sqrt{2}} \sin\beta\, e^{-i\alpha}$	$\frac{1-\cos\beta}{2}\, e^{i(-\alpha+\gamma)}$
0	$\frac{1}{\sqrt{2}} \sin\beta\, e^{-i\gamma}$	$\cos\beta$	$\frac{-1}{\sqrt{2}} \sin\beta\, e^{i\gamma}$
−1	$\frac{1-\cos\beta}{2}\, e^{i(\alpha-\gamma)}$	$\frac{1}{\sqrt{2}} \sin\beta\, e^{i\alpha}$	$\frac{1+\cos\beta}{2}\, e^{i(\alpha+\gamma)}$

The rotation matrices given here have been derived for the fundamental case that α, β, γ describe positive rotations of *functions* about *space fixed* axes of the X, Y, Z system where the rotation about γ is performed first. The angles α, β γ can be obtained from the *inverse* rotation of the coordinate system x, y, z through $-\alpha, -\beta, -\gamma$, rotation through $-\gamma$ performed first, about *rotated* axes (cf. [2] p. 112, and Eq. (2.3)). This gives the eulerian angles of Fig. C.1.

Table C.2. Wigner rotation matrix $\mathfrak{D}^{(2)}(\alpha, \beta, \gamma)$

m' \ m	2	1	0	−1	−2
2	$\left(\frac{1+\cos\beta}{2}\right)^2 e^{-2i(\alpha+\gamma)}$	$-\frac{1+\cos\beta}{2}\sin\beta\; e^{-i(2\alpha+\gamma)}$	$\sqrt{\frac{3}{8}}\sin^2\beta\; e^{-i2\alpha}$	$-\frac{1-\cos\beta}{2}\sin\beta\; e^{i(-2\alpha+\gamma)}$	$\left(\frac{1-\cos\beta}{2}\right)^2 e^{2i(-\alpha+\gamma)}$
1	$\frac{1+\cos\beta}{2}\sin\beta\; e^{-i(\alpha+2\gamma)}$	$\left[\cos^2\beta-\frac{1-\cos\beta}{2}\right] e^{-i(\alpha+\gamma)}$	$-\sqrt{\frac{3}{8}}\sin 2\beta\; e^{-i\alpha}$	$\left[\frac{1+\cos\beta}{2}-\cos^2\beta\right] e^{i(-\alpha+\gamma)}$	$-\frac{1-\cos\beta}{2}\sin\beta\; e^{i(-\alpha+2\gamma)}$
0	$\sqrt{\frac{3}{8}}\sin^2\beta\; e^{-i2\gamma}$	$\sqrt{\frac{3}{8}}\sin 2\beta\; e^{-i\gamma}$	$\frac{3\cos^2\beta-1}{2}$	$-\sqrt{\frac{3}{8}}\sin 2\beta\; e^{i\gamma}$	$\sqrt{\frac{3}{8}}\sin^2\beta\; e^{i2\gamma}$
−1	$\frac{1-\cos\beta}{2}\sin\beta\; e^{i(\alpha-2\gamma)}$	$\left[\frac{1+\cos\beta}{2}-\cos^2\beta\right] e^{i(\alpha-\gamma)}$	$\sqrt{\frac{3}{8}}\sin 2\beta\; e^{i\alpha}$	$\left[\cos^2\beta-\frac{1-\cos\beta}{2}\right] e^{i(\alpha+\gamma)}$	$-\frac{1+\cos\beta}{2}\sin\beta\; e^{i(\alpha+2\gamma)}$
−2	$\left(\frac{1-\cos\beta}{2}\right)^2 e^{2i(\alpha-\gamma)}$	$\frac{1-\cos\beta}{2}\sin\beta\; e^{i(2\alpha-\gamma)}$	$\sqrt{\frac{3}{8}}\sin^2\beta\; e^{i2\alpha}$	$\frac{1+\cos\beta}{2}\sin\beta\; e^{i(2\alpha+\gamma)}$	$\left(\frac{1+\cos\beta}{2}\right)^2 e^{2i(\alpha+\gamma)}$

Frequently, we will need the coupling tensors in cartesian representation in a molecule fixed axes system different from the principal axes system of $\rho_{AB}^{(2)}$. It is convenient to calculate the matrix elements $\rho_{ab}^{(1)}$ and $\rho_{ab}^{(2)}$ directly by a real orthogonal transformation with the eulerian angles α, β, γ, still defined by Fig. C.1:

$$\vec{\vec{\rho}}_{ab} = \mathbb{R}\,\vec{\vec{\rho}}_{AB}\,\mathbb{R}^{-1}, \tag{C.3}$$

with

$$\mathbb{R} = \begin{pmatrix} \cos\alpha\cos\beta\cos\gamma - \sin\alpha\sin\gamma & \sin\alpha\cos\beta\cos\gamma + \cos\alpha\sin\gamma & -\sin\beta\cos\gamma \\ -\cos\alpha\cos\beta\sin\gamma - \sin\alpha\cos\gamma & -\sin\alpha\cos\beta\sin\gamma + \cos\alpha\cos\gamma & \sin\beta\sin\gamma \\ \cos\alpha\sin\beta & \sin\alpha\sin\beta & \cos\beta \end{pmatrix} \tag{C.4}$$

For the traceless symmetric part, we obtain:

$$\vec{\vec{\rho}}_{ab} = \mathbb{R}\delta \begin{pmatrix} -\frac{1}{2}(1+\eta) & 0 & 0 \\ 0 & -\frac{1}{2}(1-\eta) & 0 \\ 0 & 0 & 1 \end{pmatrix} \mathbb{R}^{-1} \equiv \begin{pmatrix} \rho_{xx}^{(2)} & \rho_{xy}^{(2)} & \rho_{xz}^{(2)} \\ \rho_{xy}^{(2)} & \rho_{yy}^{(2)} & \rho_{yz}^{(2)} \\ \rho_{xz}^{(2)} & \rho_{yz}^{(2)} & \rho_{zz}^{(2)} \end{pmatrix} \tag{C.5}$$

with

$$\rho_{xx}^{(2)} = \frac{\delta}{2}\Big\{-1 + 3\sin^2\beta\cos^2\gamma - \eta[\cos 2\alpha\,(\cos^2\beta\cos^2\gamma - \sin^2\gamma) - \sin 2\alpha\sin 2\gamma\cos\beta]\Big\}$$

$$\rho_{yy}^{(2)} = \frac{\delta}{2}\Big\{-1 + 3\sin^2\beta\sin^2\gamma - \eta[\cos 2\alpha\,(\cos^2\beta\sin^2\gamma - \cos^2\gamma) + \sin 2\alpha\sin 2\gamma\cos\beta]\Big\} \tag{C.5'}$$

$$\rho_{zz}^{(2)} = \frac{\delta}{2}\Big\{3\cos^2\beta - 1 - \eta\sin^2\beta\cos 2\alpha\Big\}$$

$$\rho_{xy}^{(2)} = -\frac{\delta}{4}\Big\{3\sin^2\beta\sin 2\gamma - \eta[\cos 2\alpha\sin 2\gamma\,(1 + \cos^2\beta) + 2\sin 2\alpha\cos 2\gamma\cos\beta]\Big\}$$

$$\rho_{xz}^{(2)} = -\frac{\delta}{4}\Big\{3\sin 2\beta\cos\gamma + \eta[\cos 2\alpha\sin 2\beta\cos\gamma - 2\sin 2\alpha\sin\beta\sin\gamma]\Big\}$$

$$\rho_{yz}^{(2)} = \frac{\delta}{4}\Big\{3\sin 2\beta\sin\gamma + \eta[\cos 2\alpha\sin 2\beta\sin\gamma + 2\sin 2\alpha\sin\beta\cos\gamma]\Big\},$$

and for the antisymmetric part, likewise;

$$\overset{\rightrightarrows}{\rho}{}_{ab}^{(1)} = \mathbb{R}\begin{pmatrix} 0 & \rho_{XY} & \rho_{XZ} \\ -\rho_{XY} & 0 & \rho_{YZ} \\ -\rho_{XZ} & -\rho_{YZ} & 0 \end{pmatrix}\mathbb{R}^{-1} \equiv \begin{pmatrix} 0 & \rho_{xy}^{(1)} & \rho_{xz}^{(1)} \\ -\rho_{xy}^{(1)} & 0 & \rho_{yz}^{(1)} \\ -\rho_{xz}^{(1)} & -\rho_{yz}^{(1)} & 0 \end{pmatrix}, \tag{C.6}$$

with

$$\rho_{xy}^{(1)} = \rho_{XY}\cos\beta - \rho_{XZ}\sin\alpha\sin\beta + \rho_{YZ}\cos\alpha\sin\beta$$

$$\rho_{xz}^{(1)} = -\rho_{XY}\sin\beta\sin\gamma + \rho_{XZ}\,(\cos\alpha\cos\gamma - \sin\alpha\cos\beta\sin\gamma) + \rho_{YZ}\,(\sin\alpha\cos\gamma + \cos\alpha\cos\beta\sin\gamma)$$

$$\rho_{yz}^{(1)} = -\rho_{XY}\sin\beta\cos\gamma - \rho_{XZ}\,(\cos\alpha\sin\gamma + \sin\alpha\cos\beta\cos\gamma) + \rho_{YZ}\,(-\sin\alpha\sin\gamma + \cos\alpha\cos\beta\cos\gamma). \tag{C.6'}$$

D. Factors of Proportionality a_{lF}

In our derivation of the relaxation rates for the dipolar and the J-coupling between like spins I^i and I^j, we encounter factors of proportionality [cf. Eq. (4.56)]

$$a_{1F} = \frac{1}{4\sqrt{2}}\,[F(F+1) - 2I(I+1)]$$

$$a_{2F} = \frac{1}{2\sqrt{10}}\,[3\,I(I+1) - \tfrac{1}{2}F(F+1)], \tag{D.1}$$

where F is an integer $0 \leqslant F \leqslant 2\,I$. In the high temperature approximation, each factor a_{lF} is weighted by $(2\,F+1)\,F\,(F+1)$. The factor $(2\,F+1)$ gives the number of substates and the factor $F\,(F+1)$ gives the strength of the coupling. For the weighted average, we obtain:

$$\overline{a_1} = \frac{1}{4\sqrt{2}}\left[\frac{\sum\limits_{F=0}^{2I}(2\,F+1)\,F^2(F+1)^2}{I(I+1)\sum\limits_{F=0}^{2I}(2F+1)F(F+1)} - 2\right] I\,(I+1)$$

$$\overline{a_2} = \frac{1}{2\sqrt{10}}\left[3 - \frac{\sum\limits_{F=0}^{2I}(2\,F+1)\,F^2(F+1)^2}{2I(I+1)\sum\limits_{F=0}^{2I}(2F+1)F(F+1)}\right] I\,(I+1) \tag{D.2}$$

By repeated complete induction, one proves:

$$\frac{\sum\limits_{F=0}^{2I}(2\,F+1)\,F^2(F+1)^2}{2I(I+1)\sum\limits_{F=0}^{2I}(2F+1)F(F+1)} = \frac{4}{3}, \tag{D.3}$$

and this yields

$$\overline{a_1} = \frac{1}{4\sqrt{2}}\,[\tfrac{8}{3} - 2]\,I\,(I+1) = \frac{1}{6\sqrt{2}}\,I\,(I+1)$$

$$\overline{a_2} = \frac{1}{2\sqrt{10}}\,[3 - \tfrac{4}{3}]\,I\,(I+1) = \frac{\sqrt{10}}{12}\,I\,(I+1). \tag{D.4}$$

References

Chapter 1

1. Bloembergen, N., Purcell, E. M., Pound, R. V.: Phys. Rev. **73**, 679 (1948)
2. Anderson, J. P., Weiss, P. R.: J. Phys. Soc. Japan **9**, 316 (1954)
3. Abragam, A.: The principles of nuclear magnetism. London: Oxford University Press 1961
4. Waugh, J. S. Huber, L. M., Haeberlen, U.: Phys. Rev. Letters **20**, 180 (1968)
5. Pines, A., Gibby, M. G., Waugh, J. S.: J. Chem. Phys. **56**, 1776 (1972)
6. Spiess, H. W., Grosescu, R., Haeberlen, U.: Chem. Phys. **6**, 226 (1974).
7. Haeberlen, U.: Advances in magnetic resonance, Suppl. 1. New York: Academic Press 1976
8. Mehring, M.: NMR basic principles and progress, Vol. 11. Berlin-Heidelberg-New York: Springer 1976
9. Ernst, R. R., Anderson, W. A.: Rev. Sci. Instr. **37**, 93 (1966)
10. Kivelson, D., Ogan, K.: Advances in magnetic resonance, Vol. 7, p. 71. New York: Academic Press 1974

Chapter 2

1. Abragam, A., Price, M. H. L.: Proc. Roy. Soc. London A **205**, 135 (1951)
2. Abragam, A.: The principles of nuclear magnetism. London: Oxford University Press 1961
3. Haeberlen, U.: Advances in magnetic resonance, Suppl. 1. New York: Academic Press 1976
4. Mehring, M.: NMR basic principles and progress, Vol. 11. Berlin-Heidelberg-New York: Springer 1976.
5. Rose, M. E.: Elementary theory of angular momentum. New York: J. Wiley 1957.
6. Edmonds, A. R.: Angular momentum in quantum mechanics. Princeton: Princeton University Press 1957
7. Tinkham, M.: Group theory and quantum mechanics. New York: McGraw-Hill 1964
8. Brink, D. M., Satchler, G. R.: Angular momentum. London: Oxford University Press 1962
9. Heine, V.: Group theory. New York: Pergamon Press 1960
10. Sillescu, H.: Kernmagnetische Resonanz. Berlin-Heidelberg-New York: Springer 1966
11. Weil, J. A., Buch, T., Clapp, J. E.: Advances in magnetic resonance, Vol. 6, p. 183. New York: Academic Press 1973
12. Buckingham, A. D., Love, I.: J. Mag. Res. **2**, 338 (1970)
13. Buckingham, A. D., Malm, S. M.: Mol. Phys. **22**, 1127 (1971)
14. Van Vleck, J. H.: The theory of electric and magnetic susceptibility. London: Oxford University Press 1932

Chapter 3

1. Emsley, J. W., Feeney, J., Sutcliff, L. H., Editors: Progress in nuclear magnetic resonance spectroscopy. Oxford: Pergamon Press 1966–1976
2. Abragam, A.: The principles of nuclear magnetism. London: Oxford University Press 1976
3. Cohen, M. H., Reif, F.: Solid state physics, Vol. 5, p. 321. New York: Academic Press 1957
4. Weil, J. A., Buch, T., Clapp, J. E.: Advances in magnetic resonance, Vol. 6, p. 183. New York: Academic Press 1973
5. Haeberlen, U.: Advances in magnetic resonance, Suppl. 1. New York: Academic Press 1976
6. Mehring, M.: NMR basic principles and progress, Vol. 11. Berlin-Heidelberg-New York: Springer 1976
7. Bloembergen, N., Rowland, T. J.: Acta Met. **1**, 731 (1953)
8. Bloembergen, N., Rowland, T. J.: Phys. Rev. **55**, 1679 (1955)

9. Hartmann, H., Fleissner, M., Sillescu, H.: Theor. Chim. Acta **3**, 347 (1965)
10. Pake, G. E.: J. Chem. Phys. **16**, 327 (1948)
11. Hentschel, R., Schlitter, J., Sillescu, H., Spiess, H. W.: J. Chem. Phys. **6**, 56 (1978)
12. Kanert, O., Mehring, M.: NMR basic principles and progress, Vol. 3, p. 1. Berlin-Heidelberg-New York: Springer 1971
13. Das, T. P., Hahn, E. L.: Solid state physics, Suppl. 1. New York: Academic Press 1958
14. Lucken, E. A. C.: Nuclear quadrupole coupling constants. New York: Academic Press 1969
15. Spiess, H. W., Haas, H., Hartmann, H.: J. Chem. Phys. **50**, 3057 (1969)
16. Spiess, H. W., Sheline, R. K.: J. Chem. Phys. **54**, 1099 (1971)
17. Creel, R. B., von Meerwall, E., Griffin, C. F., Barnes, R. G.: J. Chem. Phys. **58**, 4930 (1973)
18. von Meerwall, E., Creel, R. B., Griffin, C. F., Segel, S. L.: J. Chem. Phys. **59**, 5350 (1973)
19. Van der Hart, D. L., Gutowsky, H. S., Farrar, T. C.: J. Am. Chem. Soc. **89**, 5056 (1967)
20. Sheldrick, G. M.: Chem. Comm. **1967**, 751
21. Stoll, M. E., Vaughan, R. W., Saillant, R. B., Cole, T.: J. Chem. Phys. **61**, 2896 (1974)
22. Spiess, H. W., Haeberlen, U., Zimmermann, H.: J. Mag. Res. **25**, 55 (1977)

Chapter 4

1. Andrew, E. R., Eades, R. G.: Proc. Roy. Soc. London A **218**, 537 (1953)
2. Abragam, A.: The principles of nuclear magnetism. London: Oxford University Press 1961
3. Anderson, J. P., Weiss, P. R.: J. Phys. Soc. Japan **9**, 316 (1954)
4. Spiess, H. W.: Chem. Phys. **6**, 217 (1974)
5. Mehring, M.: NMR basic principles and progress, Vol. 11. Berlin-Heidelberg-New York: Springer 1976
6. Gordon, R. G., McGinnis, R. P.: J. Chem. Phys. **49**, 2455 (1968)
7. Wilkinson, J. H.: The algebraic eigenvalue problem, Chapter 8. London: Oxford University Press 1965.
8. Margenau, H., Murphy, G. M.: The mathematics of physics and chemistry. New York: D. Van Nostrand Comp. 1943
9. Sillescu, H.: J. Chem. Phys. **54**, 2110 (1971)
10. Sillescu, H.: Ber. Bunsenges. Phys. Chem. **75**, 283 (1971)
11. Freed, J. H., Bruno, G. V., Polaszek, C. F.: J. Phys. Chem. **75**, 3385 (1971)
12. Hwang, J. S., Mason, R. P., Hwang, L. P., Freed, J. H.: J. Phys. Chem. **79**, 489 (1975)
13. Debye, P.: Polare Molekeln. Leipzig: S. Hirzel 1929
14. Stearn, A. E., Eyring, H.: J. Chem. Phys. **5**, 113 (1937)
15. Alexander, S., Baram, A., Luz, Z.: Mol. Phys. **27**, 441 (1974)
16. Baram, A., Luz, Z., Alexander, S.: J. Chem. Phys. **64**, 4321 (1976)
17. Chiba, T.: J. Chem. Phys. **39**, 947 (1963)
18. Bloembergen, N., Purcell, E. M., Pound, R. V.: Phys. Rev. **73**, 679 (1948)
19. Kubo, R., Tomita, K.: J. Phys. Soc. Japan **9**, 888 (1954)
20. Woessner, D. E.: J. Chem. Phys. **36**, 1 (1962)
21. Woessner, D. E.: J. Chem. Phys. **37**, 647 (1962)
22. Woessner, D. E., Snowden, B. S., Meyer, G. H.: J. Chem. Phys. **50**, 719 (1969)
23. Huntress, W. T.: Advances in magnetic resonance, Vol. 4, p. 1. New York: Academic Press 1970
24. Versmold, H.: Z. Naturforsch. **25a**, 367 (1970)
25. Sillescu, H.: Adv. Mol. Rel. Proc. **3**, 91 (1972)
26. Haeberlen, U., Waugh, J. S.: Phys. Rev. **185**, 420 (1969)
27. Blicharski, J. S.: Z. Naturforsch. **27a**, 1355, 2456 (1972)
28. Lynden-Bell, R. M.: Mol. Phys. **29**, 301 (1975)
29. Blicharski, J. S.: Acta Phys. Polon. **36**, 211 (1969)
30. Werbelow, L. G., Marshall, A. G.: J. Mag. Res. **11**, 299 (1973), Mol. Phys. **28**, 113 (1974)

31. Deutch, J. M.: Electron spin relaxation in liquids (Muus, L. T., Atkins, P. W., eds. New York: Plenum Press 1972
32. Freeman, R., Wittekoek, S.: J. Mag. Res. **1**, 238 (1969)
33. Redfield, A. G.: IBM Journal **1**, 19 (1957)
34. Slichter, C. P.: Principles of magnetic resonance. New York: Harper & Row, 1963
35. Edmonds, A. R.: Angular momentum in quantum mechanics. Princeton: Princeton University Press 1957
36. Heine, V.: Group theory. New York: Pergamon Press 1960
37. Solomon, I.: Phys. Rev. **99**, 559 (1955)
Solomon, I., Bloembergen, N.: J. Chem. Phys. **25**, 261 (1956)
38. Macker, E. L., McLean, C.: J. Chem. Phys. **42**, 4254 (1965)
39. Hoffman, R. A., Forsén, S.: J. Chem. Phys. **45**, 2049 (1966)
40. Alger, T. D., Collings, S. W., Grant, D. M.: J. Chem. Phys. **54**, 2820 (1971)
Alger, T. D., Freeman, R., Grant, D. M.: J. Chem. Phys. **57**, 2168 (1972)
41. Buchner, W., Emmerich, B.: J. Mag. Res. **4**, 90 (1971)
42. Buchner, W.: J. Mag. Res. **12**, 79, 83 (1973); **17**, 229 (1975)
43. Werbelow, L. G., Grant, D. M.: J. Chem. Phys. **63**, 544 (1975)
44. Bovée, W. M. M. J.: Mol. Phys. **29**, 1673 (1975)
45. Vold, R. L., Vold, R. R., Canet, D.: J. Chem. Phys. **66**, 1202 (1977)
46. Werbelow, L. G., Grant, D. M.: Advances in magnetic resonance, Vol. 9, p. 190. New York: Academic Press 1977
47. Noggle, J. H., Schirmer, R. E.: The nuclear Overhauser effect. New York: Academic Press 1971
48. Hausser, K. H., Stehlik, D.: Advances in magnetic resonance, Vol. 3, p. 79. New York: Academic Press 1968
49. McClung, R. E. D.: J. Chem. Phys. **51**, 3842 (1969)
McClung, R. E. D., Versmold, H.: J. Chem. Phys. **57**, 2596 (1972)
50. Hubbard, P. S.: Phys. Rev. **131**, 1155 (1963)
51. Bender, H. J., Zeidler, M. D.: Ber. Bunsenges. Phys. Chem. **75**, 236 (1971)
52. Schweitzer, D., Spiess, H. W.: J. Mag. Res. **15**, 529 (1974)
53. Wang, C. H.: J. Mag. Res. **9**, 75 (1973)
Wang, C. H.: Mol. Phys. **28**, 801 (1974)
54. Hertz, H. G.: Progress in nuclear magnetic resonance spectroscopy, Vol. 3, p. 159. Oxford: Pergamon Press 1967
55. Hertz, H. G.: Ber. Bunsenges. Phys. Chem. **80**, 950 (1976)
56. Hertz, H. G.; Tutsch, R.: Ber. Bunsenges. Phys. Chem. **80**, 1268 (1976)
Bender, H. J., Hertz, H. G.: Ber. Bunsenges. Phys. Chem. **81**, 468 (1977)
57. Zeidler, M. D.: Mol. Phys. **30**, 1441 (1975)
58. Willenberg, B., Sillescu, H.: Makromol. Chem. **178**, 2401 (1977)
59. Favro, L. D.: Phys. Rev. 119, 53 (1960)
60. Perrin, F.: J. Phys. Radium **5**, 497 (1934),
Perrin, F.: J. Phys. Radium **7**, 1 (1965)
61. Zwanzig, R.: Ann. Rev. Phys. Chem. **16**, 67 (1965)
62. Kivelson, D.: Mol. Phys. **28**, 321 (1974),
Kivelson, D.: J. Chem. Phys. **63**, 5034 (1975)
63. Gordon, R. G.: J. Chem. Phys. **44**, 1830 (1966)
64. Spiess, H. W., Schweitzer, D., Haeberlen, U.: J. Mag. Res. **9**, 444 (1973)
65. Fixman, M., Rider, K.: J. Chem. Phys. **51**, 2425 (1969)
66. Farrar, T. C., Maryott, A. A., Malmberg, M. S.: Ann. Rev. Phys. Chem. **23**, 193 (1972)
67. Abramowitz, M., Stegun, I. A. (eds.): Handbook of mathematical functions. Washington DC: GPO 1964
68. Spiess, H. W., Schweitzer, D., Haeberlen, U., Hausser, K. H.: J. Mag. Res. **5**, 101 (1971)
69. Maryott, A. A., Farrar, T. C., Malmberg, M. S.: J. Chem. Phys. **54**, 64 (1971)
70. DeZwaan, J., Finney, R. J., Jonas, J.: J. Chem. Phys. **60**, 3223 (1974)
71. McClung, R. E. D.: J. Chem. Phys. **57**, 5478 (1972)

72. Bull, T. E.: J. Chem. Phys. **65**, 4802 (1976)
73. Steele, W. A.: Adv. Chem. Phys. **34**, 1 (1976)
74. Hwang, L. P., Freed, J. H.: J. Chem. Phys. **63**, 118 (1975)
75. Lindenberg, K., Cukier, R. I.: J. Chem. Phys. **62**, 3271 (1975)
76. Alder, B. J., Wainwright, T. E.: J. Chem. Phys. **33**, 1439 (1960)
77. Rahman, A., Stillinger, F. H.: J. Chem. Phys. **55**, 3336 (1971)

Chapter 5

1. Noack, F.: NMR basic principles and progress, Vol. 3, p. 83. Berlin-Heidelberg-New York: Springer 1971
2. Ramsey, N. F.: Molecular beams. London: Oxford University Press 1956
3. Townes, C. H., Schawlow, A. L.: Microwave spectroscopy. New York: McGraw-Hill 1955
4. Gordy, W., Cook, R. L.: Microwave molecular spectra, in technique of organic chemistry, Vol. IV, Part II. New York: Interscience 1970
5. Flygare, W. H.: Chem. Rev. **74**, 653 (1974)
6. Spiess, H. W., Garrett, B. B., Sheline, R. K., Rabideau, S. W.: J. Chem. Phys. **51**, 1201 (1969)
7. Diehl, P., Khetrapal, C. L.: NMR basic principles and progress, Vol. 1, p. 1. Berlin-Heidelberg-New York: Springer 1969
8. Khetrapal, C. L., Kunwar, A. C., Tracey, A. S., Diehl, P.: NMR basic principles and progress, Vol. 9, p. 1. Berlin-Heidelberg-New York: Springer 1975
9. Appleman, B. R., Dailey, B. P.: Advances in magnetic resonance, Vol. 7, p. 231. New York: Academic Press 1974
10. Yannoni, C. S.: J. Am. Chem. Soc. **91**, 4611 (1969)
11. Montana, A. J., Dailey, B. P.: Mol. Phys. **30**, 1521 (1975)
12. Montana, A. J., Dailey, B. P.: J. Mag. Res. **21**, 25 (1976)
    Montana, A. J., Dailey, B. P.: J. Mag. Res. **22**, 117 (1976)
13. Gerritsen, J., MacLean, C.: J. Mag. Res. **5**, 44 (1971)
14. Schumann, C., Dreeskamp, H., Hildenbrand, K.: J. Mag. Res. **18**, 97 (1975)
15. Goldanskii, V. I., Herber, R. H.: Chemical applications of Mössbauer spectroscopy. New York: Academic Press 1968
16. Greenwood, N. N., Gibb, T. C.: Mössbauer spectroscopy. London: Chapman & Hall 1971
17. Gruvermann, J. I., Editor: Mössbauer effect methodology. New York: Plenum Press 1965–1974
18. Haas, H., Recknagel, E., Spellmeyer, B.: Proc. 18th Congress Ampere, p. 259, Nottingham 1974
19. Shirley, D. A., Haas, H.: Ann. Rev. Phys. Chem. **23**, 385 (1972)
20. Stöckmann, H. J., Ackermann, H., Dubbers, D., Grupp, M., Heitjans, P.: Z. Physik **269**, 47 (1974)
21. Ackermann, H., Dubbers, D., Grupp, M., Heitjans, P., Stöckmann, H. J., Wanczek, K. P., Lecker, R., Leckebusch, R.: Z. Naturforsch. **31a**, 1298 (1976)
22. Haeberlen, U.: Advances in magnetic resonance, Suppl. 1. New York: Academic Press 1976
23. Mehring, M.: NMR basic principles and progress, Vol. 11. Berlin-Heidelberg-New York: Springer 1976
24. Andrew, E. R.: Progress in nuclear magnetic resonance spectroscopy, Vol. 8. Oxford: Pergamon Press 1971
25. Slucher, R. E., Hahn, E. L.: Phys. Rev. **166**, 332 (1968)
    Koo, J.: Ph. D. Thesis, University of California, Berkeley (unpublished)
26. Blinc, R.: Advances in nuclear quadrupole resonance, Vol. 2. London: Heyden 1975
27. Edmonds, D. T.: Physics Reports **29C**, 4 (1977)
28. Ramsey, N. F.: Phys. Rev. **A1**, 1320 (1970)
29. Spiess, H. W., Grosescu, R., Haeberlen U.: Chem. Phys. **6**, 226 (1974)
30. Kempf, J., Spiess, H. W., Haeberlen, U., Zimmermann, H.: Chem. Phys. Letters **17**, 39 (1972)

31. Hartmann, S. R., Hahn, E. L.: Phys. Rev. **128**, 2042 (1962)
32. Pines, A., Gibby, M. G., Waugh, J. S.: J. Chem. Phys. **56**, 1776 (1972)
33. Pines, A., Gibby, M. G., Waugh, J. S.: J. Chem. Phys. **59**, 569 (1973)
34. Bleich, H. E., Redfield, A. G.: J. Chem. Phys. **55**, 5405 (1971)
35. Yannoni, C. S., Bleich, H. E.: J. Chem. Phys. **55**, 5406 (1971)
36. Grannell, P. K., Mansfield, P., Whittaker, M. A.: Phys. Rev. **B8**, 4149 (1973)
37. Pines, A., Ruben, D. J., Vega, S., Mehring, M.: Phys. Rev. Letters **36**, 110 (1976)
38. Schaefer, J., Stejskal, E. O., Buchdahl, R.: Macromolecules **8**, 291 (1975)
Schaefer, J., Stejskal, E. O., Buchdahl, R.: Macromolecules **10**, 384 (1977)
39. Lippmaa, E., Alla, M., Tuherm, T.: Proc. 19th Congress Ampere, p. 113, Heidelberg 1976
40. Waugh, J. S., Huber, L. M., Haeberlen, U.: Phys. Rev. Letters **20**, 180 (1968)
41. Haeberlen, U., Waugh, J. S.: Phys. Rev. **175**, 453 (1968)
42. Haeberlen, U., Waugh, J. S.: Phys. Rev. **185**, 420 (1969)
43. Spiess, H. W., Zimmermann, H., Haeberlen, U.: Chem. Phys. **12**, 123 (1976)
44. Haeberlen, U., Sagnowski, S. F., Aravamudhan, S., Post, H.: J. Mag. Res. **25**, 307 (1977)
45. Van Hecke, P., Spiess, H. W., Haeberlen, U.: J. Mag. Res. **22**, 103 (1976)
46. Pake, G. E.: J. Chem. Phys. **16**, 327 (1948)
47. Reeves, L. W.: Progress in nuclear magnetic resonance spectroscopy, Vol. 4, p. 193. Oxford: Pergamon Press 1969
48. Van Hecke, P., Spiess, H. W., Haeberlen, U., Haussühl, S.: J. Mag. Res. **22**, 93 (1976)
49. Halstead, T. K., Spiess, H. W., Haeberlen, U.: Mol. Phys. **31**, 1569 (1976)
50. Griffin R. G., Pines, A., Waugh, J. S.: J. Chem. Phys. **63**, 3676 (1975)
51. Stoll, M. E., Vaughan, R. W., Saillant, R. B., Cole, T.: J. Chem. Phys. **61**, 2896 (1974)
52. Müller, L., Kumar, A., Baumann, T., Ernst, R. R.: Phys. Rev. Letters **32**, 1402 (1974)
53. Waugh, J. W.: Proc. Natl. Acad. Sci. USA **73**, 1394 (1976)
54. Hester, R. K., Ackerman, J. L., Cross, V. R., Waugh, J. S.: Phys. Rev. Letters **34**, 993 (1975)
Hester, R. K., Ackerman, J. L., Cross, V. R., Waugh, J. S.: J. Chem. Phys. **63**, 3606 (1975)
55. Hester, R. K., Ackerman, J. L., Neff, B. L., Waugh, J. S.: Phys. Rev. Letters **36**, 1081 (1976)
56. Stoll, M. E., Vega, A. L., Vaughan, R. W.: J. Chem. Phys. **65**, 4093 (1976)
57. Aue, W. P., Karhan, J., Ernst, R. R.: J. Chem. Phys. **64**, 4226 (1976)
58. Abragam, A.: The principles of nuclear magnetism. London: Oxford University Press 1961
59. Lucken, E. A. C.: Nuclear quadrupole coupling constants. New York: Academic Press 1969
60. Moccia, R., Zandomeneghi, M.: Advances in nuclear quadrupole resonance spectroscopy, Vol. 2, p. 135. London: Heyden 1975
61. Murrell, J. N.: Progress in nuclear magnetic resonance spectroscopy, Vol. 6. Oxford: Pergamon Press 1970
62. Strong, A. B., Ikenberry, D., Grant, D. M.: J. Mag. Res. **9**, 145 (1973)
63. Van Vleck, J. H.: The theory of electric and magnetic susceptibility. London: Oxford University Press 1932.
64. Lipscomb, W. N., Advances in magnetic resonance, Vol. 2, p. 137. New York: Academic Press 1966
65. Pople, J. A., McIver, J. W., Ostlund, N. S.: J. Chem. Phys. **49**, 2960 (1968)
66. Ditchfield, R.: Chem. Phys. Letters **15**, 203 (1972)
Ditchfield, R.: J. Chem. Phys. **65**, 3123 (1976)
67. Flygare, W. H.: J. Chem. Phys. **41**, 793 (1964)
68. Flygare, W. H., Goodisman, J.: J. Chem. Phys. **49**, 3122 (1968)
69. Gierke, T. D., Flygare, W. H.: J. Am. Chem. Soc. **94**, 7277 (1972)
70. Schweitzer, D., Spiess, H. W.: J. Mag. Res. **15**, 529 (1974)
71. Ozier, I., Crapo, L. M., Ramsey, N. F.: J. Chem. Phys. **49**, 2314 (1968)
72. Huo, W. N.: J. Chem. Phys. **43**, 624 (1965)
73. Bender, H. J., Zeidler, M. D.: Ber. Bunsenges. Phys. Chem. **75**, 236 (1971)
74. Buckingham, A. D., Love, I.: J. Mag. Res. **2**, 338 (1970)
75. Pople, J. A., Santry, D. P.: Mol. Phys. **8**, 1 (1964)

Chapter 6

1. Spiess, H. W., Schweitzer, D., Haeberlen, U., Hausser, K. H.: J. Mag. Res. **5**, 101 (1971)
2. Pines, A., Rhim, W. K., Waugh, J. S.: J. Chem. Phys. **54**, 5438 (1971)
3. Mehring, M.: NMR basic principles and progress, Vol. 11. Berlin-Heidelberg-New York: Springer 1976
4. Haeberlen, U.: Advances in magnetic resonance, Suppl. 1. New York: Academic Press 1976
5. Haeberlen, U., Kohlschütter, U., Kempf, J., Spiess, H. W., Zimmermann, H.: Chem. Phys. **3**, 248 (1974)
   Sagnowski, S. F., Haeberlen, U., Aravamudhan, S.: J. Mag. Res. (in press) (1977)
6. Spiess, H. W., Zimmermann, H., Haeberlen, U.: J. Mag. Res. **25**, 55 (1977)
7. Spiess, H. W., Zimmermann, H., Haeberlen, U.: Chem. Phys. **12**, 123 (1976)
8. Post, H., Haeberlen, U.: Private communication
9. Schreiber, L. B., Vaughan, R. W.: Chem. Phys. Letters **28**, 586 (1974)
10. Sagnowski, S., Aravamudhan, S., Haeberlen, U.: J. Chem. Phys. **66**, 4697 (1977)
11. Achlama, A. M., Kohlschütter, U., Haeberlen, U.: Chem. Phys. **7**, 287 (1975)
12. Pines, A., Ruben, D. J., Vega, S., Mehring, M.: Phys. Rev. Letters **36**, 110 (1976)
13. Van Hecke, P., Spiess, H. W., Haeberlen, U.: J. Mag. Res. **22**, 103 (1976)
14. Gierke, T. D., Flygare, W. H.: J. Am. Chem. Soc.: **94**, 7277 (1972)
15. Malli, G., Fraga, S.: Theor. Chim. Acta **5**, 284 (1966)
16. Ditchfield, R.: Chem. Phys. **2**, 400 (1973)
17. Pople, J. A., Schneider, W. G., Bernstein, H. J.: High Resolution NMR. New York: McGraw-Hill 1959
18. Osborn, J. A.: Phys. Rev. **67**, 351 (1945)
19. Haeberlen, U.: Private communication
20. Waugh, J. S., Gibby, M. G., Pines, A., Kaplan, S.: Proc. 17th Congress Ampere, p. 11. Turku 1972
21. Kempf, J., Spiess, H. W., Haeberlen, U., Zimmermann, H.: Chem. Phys. **4**, 269 (1974)
22. Lauterbur, P. C.: Phys. Rev. Letters **1**, 343 (1958)
23. Ozier, I., Crapo, L. M., Ramsey, N. F.: J. Chem. Phys. **49**, 2314 (1968)
24. Spiess, H. W., Grosescu, R., Haeberlen, U.: Chem. Phys. **6**, 226 (1974)
25. Pausak, S., Pines, A., Waugh, J. S.: J. Chem. Phys. **59**, 591 (1973)
26. Kempf, J. Spiess, H. W., Haeberlen, U., Zimmermann, H.: Chem. Phys. Letters **17**, 39 (1972)
27. Chang, J. J., Griffin, R. G., Pines, A.: J. Chem. Phys. **60**, 2561 (1974)
   Chang, J. J., Griffin, R. G., Pines, A.: J. Chem. Phys. **62**, 4923 (1975)
28. Griffin, R. G., Pines, A., Pausak, S., Waugh, J. S.: J. Chem. Phys. **63**, 1267 (1975)
   Griffin, R. G., Ruben, D. J.: J. Chem. Phys. **63**, 1272 (1975)
29. Baker, M. R., Anderson, C. H., Ramsey, N. F.: Phys. Rev. **133A**, 1533 (1964)
30. Bhattacharya, P. K., Dailey, B. P.: J. Chem. Phys. **59**, 5820 (1973)
31. Kaplan, S., Pines, A., Griffin, R. G., Waugh, J. S.: Chem. Phys. Letters **25**, 78 (1974)
32. Gibby, M. G., Griffin, R. G., Pines, A., Waugh, J. S.: Chem. Phys. Letters **17**, 80 (1972)
33. Schweitzer, D., Spiess, H. W.: J. Mag. Res. **16**, 243 (1974)
34. Schweitzer, D., Spiess, H. W.: J. Mag. Res. **15**, 529 (1974)
35. Ando, I., Nishioka, A., Kondo, M.: Chem. Phys. Letters **25**, 212 (1974)
   Ando, I., Nishioka, A.: Bull. Chem. Soc. Japan **48**, 841 (1975)
36. Flygare, W. H., Goodisman, J.: J. Chem. Phys. **49**, 3122 (1968)
37. Appleman, B. R., Dailey, B. P.: Advances in magnetic resonance, Vol. 7, p. 231. New York: Academic Press 1974
38. Spiess, H. W.: Chem. Phys. **6**, 217 (1974)
39. Ailion, D. C.: Advances in magnetic resonance, Vol. 5, p. 177. New York: Academic Press 1974
40. Lauer, O., Stehlik, D., Hausser, K. H.: J. Mag. Res. **6**, 524 (1972)
41. Pines, A. *et al.*: J. Chem. Phys. (to be published)
42. McKnett, C. L., Dybowski, C. R., Vaughan, R. W.: J. Chem. Phys. **63**, 4578 (1975)

43. Ruben, D. J., Wemmer, D. E., Pines, A.: Proc. 19th Congress Ampere, p. 518. Heidelberg 1976
44. Baram, A., Luz, Z., Alexander, S.: J. Chem. Phys. **64**, 4321 (1976)
45. Hensen, K., Riede, W. O., Sillescu, H., v. Wittgenstein, A.: J. Chem. Phys. **61**, 4365 (1974)
46. Freed, J.: Spin labeling, ed. L. J. Berliner, Chapter 3. New York: Academic Press 1976
47. Cotton, F. A., Danti, A., Waugh, J. S., Fessenden, R. W.: J. Chem. Phys. **29**, 1427 (1958)
48. Spiess, H. W., Mahnke, H.: Ber. Bunsenges. Phys. Chem. **76**, 990 (1972)
49. Burdett, J. K., Grzybowski, J. M., Poliakoff, M., Turner, J. J.: J. Am. Chem. Soc. **98**, 5728 (1976)
50. Berglund, B., Tegenfeld, J.: J. Mag. Res. (in press) (1977)
51. NMR basic principles and progress, Vol. 4. Berlin-Heidelberg-New York: Springer 1971
52. Zachmann, H. G.: Kunststoff Handbuch, Vol. I. p. 169. München: Carl Hanser 1975
53. Hentschel, D., Diploma thesis, Mainz 1977
Hentschel, D., Sillescu, H., Spiess, H. W.: To be published
54. Harris, R. K. (ed.): Nuclear magnetic resonance. A specialist periodical report, Vols. 1–5. London: The Chemical Society, Burlington House 1972–1976
55. Steele, W. A.: Adv. Chem. Phys. **34**, 1 (1976)
56. Lippert, E., Bratos, S., Buckingham, A. D. (eds.): Organic liquids. New York: John Wiley & Sons 1978
57. Noggle, J. H., Schirmer, R. E.: The nuclear Overhauser effect. New York: Academic Press 1971
58. Kuhlmann, K. F., Grant, D. M., Harris, R. K.: J. Chem. Phys. **52**, 3439 (1970), Lyerla, J. R., Grant, D. M., Harris, R. K.: J. Phys. Chem. **75**, 585 (1971)
59. Campbell, I. D., Freeman, R.: J. Chem. Phys. **58**, 2666 (1973)
60. Freeman, R., Hill, H. D. W., Tomlinson, G. L., Hall, L. D.: J. Chem. Phys. **61**, 4466 (1974)
61. Doddrell, D., Glushki, V., Allerhand, A.: J. Chem. Phys. **56**, 3683 (1972)
62. Lyerla, J. R., Grant, D. M., Bertrand, R. D.: J. Phys. Chem. **75**, 3967 (1971)
63. Saluvere, T.: Private communication
64. Olivson, A., Lippmaa, E.: Chem. Phys. Letters **11**, 241 (1971)
65. V. Goldammer, E., Lüdemann, H. D., Müller, A.: J. Chem. Phys. **60**, 4590 (1974)
66. Edmonds, D. T.: Physic Reports **29C**, 4 (1977)
67. Lucken, E. A. C.: Nuclear quadrupole coupling constants. New York: Academic Press 1969
68. Huntress, W. T.: Advances in magnetic resonance, Vol. 4, p. 1. New York: Academic Press 1970
69. Sillescu, H.: Adv. Mol. Relax. Proc. 3, 91 (1972)
70. Jonas, J.: Advances in magnetic resonance, Vol. 6, p. 73. New York: Academic Press 1973
71. Bopp, T. T.: J. Chem. Phys. **47**, 3621 (1967)
72. Jäckle, H., Haeberlen, U., Schweitzer, D.: J. Mag. Res. **4**, 198 (1971)
73. Levy, G. C., Cargioli, J. D., Anet, F. A. L.: J. Am. Chem. Soc. **95**, 1527 (1973)
Levy, G. C., Holak, T., Steigel, A.: J. Am. Chem. Soc. **98**, 495 (1976)
74. Vold, R. L., Waugh, J. S., Klein, M. P., Phelps, D. E.: J. Chem. Phys. **48**, 3831 (1968)
75. Farrar, T. C., Becker, E. D.: Pulse and Fourier transform NMR. New York: Academic Press 1971
76. Doddrell, D., Allerhand, A.: J. Am. Chem. Soc. **93**, 1558 (1971)
77. Barnes, R. G., Bloom, J. W.: J. Chem. Phys. **57**, 3082 (1972)
78. Spiess, H. W., Schweitzer, D., Haeberlen, U.: J. Mag. Res. **9**, 444 (1973)
79. Gillen, K. T., Griffiths, J. S.: Chem. Phys. Letters **17**, 359 (1972)
80. Schweitzer, D., Spiess, H. W.: J. Mag. Res. **15**, 529 (1974)
81. Kratochwill, A., Hertz, H. G.: Ber. Bunsenges. Phys. Chem. (to be published)
82. Kratochwill, A., Weidner, J. U., Zimmermann, H.: Ber. Bunsenges. Phys. Chem. 77, 408 (1973)
83. Bertagnolli, H., Zeidler, M. D.: Mol. Phys. (to be published)
84. Versmold, H.: Ber. Bunsenges. Phys. Chem. (to be published)
85. Bauer, D. R., Alms, G. R., Brauman, J. I., Pecora, R.: J. Chem. Phys. **61**, 2255 (1974)
86. Vold, R. L., Vold, R. R., Canet, D.: J. Chem. Phys. **66**, 1202 (1977)

87. Kintzinger, J. P., Lehn, J. M.: Mol. Phys. **22**, 273 (1971)
Kintzinger, J. P., Lehn, J. M.: Mol. Phys. **27**, 491 (1974)
88. Keyes, T., Kivelson, D.: J. Chem. Phys. **56**, 1057 (1972)
Keyes, T.: Mol. Phys. **23**, 737 (1972)
89. Rinné, M., Depireux, J.: Advances in nuclear quadrupole resonance, Vol. 1, p. 357. London: Heyden 1974
90. Rinné, M., Depireux, J.: Proc. 19th Congress Ampere, p. 633. Heidelberg 1976
91. Ozier, I., Vitkevitch, Ramsey, N. F.: Proc. 27th Symposium on Molecular Structure. Ohio 1972

Chapter 7

1. Cook, R. L., De Lucia, F. C.: Amer. J. Phys. **39**, 1433 (1971)
2. Tinkham, M.: Group theory and quantum mechanics. New York: McGraw Hill 1964
3. Brink, D. M., Satchler, G. R.: Angular momentum. London: Oxford University Press 1962
4. Rose, M. E.: Elementary theory of angular momentum. New York: John Wiley 1957

NMR
Basic Principles and Progress
Grundlagen und Fortschritte

Editors: P. Diehl, E. Fluck, R. Kosfeld

Volume 10

Van der Waals Forces and Shielding Effects

1975. 13 figures, 46 tables. II, 118 pages
ISBN 3-540-07340-X
Contents: Historical Development (up to 1961). Continuum Models. Pair Interaction Models. Other Experimental Proton Data on σ_W. The Physical Nature of the Field F^2 and of the Associated Excitation Energy. The Site Factor. The Repulsion Effect. The Effect of Higher Order Dispersion Terms. The Parameters B. σ_W in Dense Media. The Temperature Dependence of σ_W. Factor Analysis. ^{19}F σ_W Studies. σ_W of Nuclei other than ^1H and ^{19}F. Alternative Referencing Systems. On the Required Molecular Parameters and Physical Constants.

Volume 11
M. Mehring

High Resolution NMR Spectroscopy in Solids

1976. 104 figures. XI, 246 pages
ISBN 3-540-07704-9
Contents: Introduction. Nuclear Spin Interactions in Solids. Multiple-Pulse NMR Experiments. Double Resonance Experiments. Magnetic Shielding Tensor. Spin-Lattice Relaxation in Line Narrowing Experiments. Appendix.

Volume 12
B. Lindman, S. Forsén

Chlorine, Bromine and Iodine NMR Physico-Chemical and Biological Applications

1976. 74 figures, 29 tables. XIV, 368 pages
ISBN 3-540-07725-1
Contents: Introductory Aspects. Relaxation in Molecules or Ions with Covalently Bonded Halogens. Shielding Effects in Covalent Halogen Compounds. Scalar Spin Couplings. Relaxation of Chloride, Bromide and Iodide Ions. Shielding of Halide Ions. Quadrupole Splittings in Liquid Crystals. Halide Ions in Biological Systems. Studies of the Perchlorate Ion.

Volume 13

Introductory Essays

Editor: M. M. Pintar
1976. 48 figures. XI, 154 pages
ISBN 3-540-07754-5
Contents: A Guide to Relaxation Theory. Thermodynamics of Spin Systems in Solids. Coherent Averaging and Double Resonance in Solids. Macroscopic Dipole Coherence Phenomena. Nuclear Spins and Non Resonant Electromagnetic Phenomena. Nuclear Spin Relaxation in Molecular Hydrogen. Longitudinal Nuclear Spin Relaxation Time Measurements in Molecular Gases. Spin-Lattice Relaxation in Nematic Liquid Crystals. Via the Modulation of the Intramolecular Dipolar Interactions by Order Fluctuations. Studies of Molecular Tunnelling. Effect of Molecular Tunnelling on NMR Absorption and Relaxation in Solids. How to Build a Fourier Transform NMR Spectrometer for Biochemical Applications.

Volume 14
H. Nöth, B. Wrackmeyer

Nuclear Magnetic Resonance Spectroscopy of Boron Compounds

1978. 1 figure, 96 tables. XII, 461 pages
ISBN 3-540-08456-8
Contents: Introduction and Scope. – Nuclear Magnetic Properties of Boron. – ^{11}B NMR or Two-Coordinate Boron. – ^{11}B Chemical Shifts of Three Coordinate Boron. – ^{11}B NMR of Transition Metal Boron Compounds. – ^{11}B NMR of Diborane and Derivatives. – ^{11}B NMR of Tetracoordinate Boron. – Spin-Spin Coupling Constants ^nJ (^{11}BX). – Tables of ^{11}B NMR Data. – References. – Author Index.

Springer-Verlag
Berlin
Heidelberg
New York